INTEGRAL AND FINITE DIFFERENCE INEQUALITIES AND APPLICATIONS

NORTH-HOLLAND MATHEMATICS STUDIES 205

(Continuation of the Notas de Matemática)

Editor: Jan van Mill
Faculteit der Exacte Wetenschappen
Amsterdam, The Netherlands

ELSEVIER

Amsterdam – Boston – Heidelberg – London – New York – Oxford
Paris – San Diego – San Francisco – Singapore – Sydney – Tokyo

INTEGRAL AND FINITE DIFFERENCE INEQUALITIES AND APPLICATIONS

B.G. PACHPATTE
57 Shri Niketan Colony
Near Abhinay Talkies
Aurangabad 431 001
Maharashtra
India

ELSEVIER

Amsterdam – Boston – Heidelberg – London – New York – Oxford
Paris – San Diego – San Francisco – Singapore – Sydney – Tokyo

Elsevier
Radarweg 29, PO Box 211, 1000 AE Amsterdam, The Netherlands
The Boulevard, Langford Lane, Kidlington, Oxford OX5 1GB, UK

First edition 2006

Library of Congress Cataloging-in-Publication Data
A catalog record for this book is available from the Library of Congress

British Library Cataloguing in Publication Data
A catalogue record for this book is available from the British Library

ISBN-13: 978-0-444-52762-2
ISSN (Series): 0304-0208

Transferred to Digital Printing 2008

To the memory of my mother

Preface

Inequalities have proven to be one of the most important and far-reaching tools for the development of many branches of mathematics. There are many types of inequalities of importance. Integral and finite difference inequalities with explicit estimates are powerful mathematical appartus which aid the study of the qualitative behavior of solutions of various types of differential, integral and finite difference equations. Because of its usefulness and importance, such inequalities have attracted much attention and a great number of papers, surveys and monographs have appeared in the literature. The extensive surveys of such inequalities which are adequate in many applications may be found in the monographs [34] and [42] up to the years of their publications.

Inequalities with explicit estimates are particularly fascinating and have numerous applications. The variety of nonlinear problems is evergrowing, and new methods have to be found to study them. By the desire to widen the scope of such inequalities, recently many papers have appeared which deal with the large number of inequalities applicable in situations in which the earlier inequalities do not apply directly. I believe that these inequalities will strongly influence further research into the topic for a long time to come.

The present monograph is an attempt to present some of the more recent developments related to integral and finite difference inequalities with explicit estimates. The literature in this field is extensive and as yet scattered in the original papers in the journals. The rapid development of this area and the variety of applications force us to be quite selective. We only concentrate on recent advances not covered in the earlier monographs [34] and [42] by the author. Our choices reflect our interests and what we know, as well as those results we consider potentially applicable in a wider range of applications. We do not claim to include all the recent results about such inequalities, but at least to cover those results that have a considerable variety of applications.

This monograph will be of interest to mathematicians whose work involves differential, integral and finite difference equations and numerical analysis. For researchers working in these areas, it will be a valuable source of reference and inspiration. All the material included is presented in an elementary way and the book can be used as a text for advanced graduate cources. It will also be of interest to researchers in mathematical analysis, statistics, computer science and other areas of applied science and engineering.

It is my pleasure to acknowledge the fine cooperation and assistance provided by Jan van Mill, Arjen Sevenster, (Mrs.) Andy Deelen and the editorial and production staff of Elsevier Science. Finally, I wish to express my greatful appreciation to my family members for their understanding, patience and constant encourgement during the writing of the book.

B.G. Pachpatte

Contents

Introduction

It is a well known truth that the inequalities have always been of great importance for the development of many branches of mathematics. Indeed, this importance seems to have increased considerably during the last century and the theory of inequalities nowadays may be regarded as an independent branch of mathematics. This field is dynamic and experiencing an explosive growth in both theory and applications. A particular feature that makes the study of this interesting topic so fascinating arises from the numerous fields of applications. As a response to the needs of diverse applications, a large variety of inequalities have been proposed and studied in the literature, see [1-85] and the references given therein. This theory did not just add new objects of study, but also brought with it some new insights and new techniques which are instrumental in solving many important problems.

The integral inequalities of various types have been widely studied in most subjects involving mathematical analysis. They are particulary useful for approximation theory and numerical analysis in which estimates of approximation errors are involved. In recent years, the application of integral inequalities has greatly expanded and they are now used not only in mathematics but also in the areas of physics, technology and biological sciences. The theory of differential and integral inequalities has gained increasing significance in the last century as is apparent from the large number of publications on the subject. With the growing range of applications, the theory of integral inequalities enjoy a rapid increase of interest and widespread recognition as an important area of mathematical analysis.

Many nonlinear dynamical systems are too complicated to be effectively analized. In many situations, we are interested in knowing qualitative properties of solutions without explicit knowledge of the solution process. Having knowledge of the existence of solutions of the system, the integral inequalities with explicit estimates serve as an important tool in their analysis. In fact, the integral inequalities with explicit estimates and fixed point theorems are powerful tools in nonlinear analysis. The theory of integral inequalities with explicit estimates has emerged as an interesting and fascinating topic of applicable analysis with a wide range of applications. One can hardly imagine the development of the theory of differential and integral equations without such inequalities. As the literature is extensive and spans more than a century, it will be helpful to summarize some fundamental known inequalities.

An early significant result in this area and certainly a keystone in the development of the theory of differential equations can be stated as follows:

If u is a continuous function defined on $[a, a+h]$ and

$$0 \leq u(t) \leq \int_a^t (c + du(s))\, ds,$$

for $t \in [a, a+h]$ where c, d are nonnegative constants, then for the function $u(t)$ one has the estimate

$$u(t) \leq c \exp(dh),$$

for t in the same interval.

The above inequality was discovered by Gronwall [16] in 1919 while investigating the dependence of a system of differential equations with respect to a parameter and now known in general as Gronwall's inequality. However, it seems that the idea of such an inequality was grounded in the work of Peano [80] in 1885-86. Gronwall might not have thought that this discovery would be an object for such great interest in the future. Gronwall's inequality, like the fundamental inequalities as, the arithmetic mean and geometric mean inequality, the Hölder's (in particular, Cauchy-Schwarz) inequality and the Minkowoski inequality caught the fancy of a number of research workers and a large number of papers which deal with various generalizations, extensions and numerious variants have appeared in the literature, see [1-9,11,12,14,15,17,19,20-28,33-79,84,85] and the references cited therein.

In 1956, Bihari [8] gave a nonlinear generalization of Gronwall's inequality, of fundamental importance in the study of nonlinear problems and is known as Bihari's inequality. Another important development that also started almost simultaneously, when Wendroff has given some important extensions of Gronwall's inequality in two independent variables, see [4, p. 154]. The main result due to Wendroff can be stated as follows.

Let $u(x, y), c(x, y)$ be nonnegative continuous functions defined for $x, y \in R_+$. If

$$u(x, y) \leq a(x) + b(y) + \int_0^x \int_0^y c(s, t)\, u(s, t)\, dt ds,$$

for $x, y \in R_+$, where $a(x), b(y)$ are positive continuous functions for $x, y \in R_+$ having derivatives such that $a'(x) \geq 0, b'(y) \geq 0$ for $x, y \in R_+$, then

$$u(x, y) \leq E(x, y) \exp\left(\int_0^x \int_0^y c(s, t)\, dt ds\right),$$

for $x, y \in R_+$,where

$$E(x,y) = \frac{[a(x) + b(0)][a(0) + b(y)]}{[a(0) + b(0)]},$$

for $x, y \in R_+$.

The above inequality has its orgin in the field of partial differential equations and provides a very useful and inspiring integral inequality of fundamental importance. Indeed, the well known book 'Inequalities' by Beckenbach and Bellman [4] is certainly to be credited for bringing to the notice a fundamental unpublished work of Wendroff. Since the publication of the book [4] in 1961, a great interest in such kinds of inequalities has certainly contributed to the development of the theory of certain partial differential and integral equations, see [3,34] and the references given therein.

The well known Gronwall's inequality and its nonlinear version due to Bihari [8] are not directly applicable to studing integral equations with weakly singular kernels. In the theory of such problems, Henry [17] proposes a method to estimate solutions of linear integral inequality with weakly singular kernel. In 1997, Medveď [24] proposed a new approach for obtaining explicit estimates on the inequalities of the form

$$u(t) \leq a(t) + \int_0^t (t-s)^{\beta-1} f(s)\, w(u(s))\, ds,$$

and its variants and generalizations, where $0 < \beta < 1$. The case $\beta = 1$, a, f, u continuous, nonnegative, w linear is covered by the Gronwall's inequality and the case $\beta = 1$, w continuous, nonnegative, nonlinear is covered by the Bihari result [8]. The resulting estimates obtained in [17,24-28] play the same role in the theory of parabolic partial differential equations; see [25,27,28].

In the study of qualitative behavior of solutions of certain nonlinear differential and integral equations some specific types of inequalities are needed in various situations. To name a few, the following inequality which provides an explicit bound on unknown function has played a very important role in the study of various classes of differential and integral equations; see [33,34].

If u, f are nonnegative continuous functions on R_+, $c \geq 0$ is a constant, and

$$u^2(t) \leq c + 2\int_0^t f(s)\, u(s)\, ds,$$

for $t \in R_+$,then

$$u(t) \leq \sqrt{c} + \int_0^t f(s)\,ds,$$

for $t \in R_+$.

The striking feature of this inequality is that it is applicable in situations for which the well known Gronwall and Bihari inequalities do not apply directly. For a detailed account on such inequalities and some applications, see [34]. The explicit bounds on the integral inequalities of the form

$$u(t) \leq c + \int_\alpha^t a(s)\,u(s)\,ds + \int_\alpha^\beta b(s)\,u(s)\,ds,$$

for $t \in [\alpha, \beta]$, under some suitable conditions on the functions involved, are also equally important in the study of certain classes of differential and integral equations. It appears that Gamidov [15] first initiated the study of obtaining explicit upper bounds on such inequalities while studying the boundary value problems for higher order differential equations.

The theory of retarded differential equations is the object of many works for more than a century. There are many ideas and techniques that have been outlined to study such equations, see [7,13,18,19] and the references cited therein. Inspired by the important role played by the integral inequalities with explicit estimates in the theory of differential and integral equations, some researchers have obtained analogues of such inequalities, which can be used as tools in the study of retarded differential and integral equations, see [21,22,43,47,58,61,64,69,77] and [3, pp. 142-145]. There is no doubt that the retarded integral inequalities with explicit estimates will continue to play an important role in the study of various types of retarded differential and integral equations.

During the past few decades some researchers have shown interest in developing the theory of the advanced type of differential equations. If we compare some fundamental aspects on the advanced type of equations with retarded type including ordinary differential equations, it seems, however, to be difficult to apply the fixed point theorems to the advanced types. If the uniqueness of the solutions is not guaranteed, it is convenient to consider the maximal and minimal solutions. As for the advanced types, however, the same methods as in the theory of retarded types may not be possible. In the study of retarded types of differential and integral equations, some retarded integral inequalities with explicit estimates play an important role. It seems, however, not to be easy to obtain such inequalities for advanced types. See [82]. We would like to mention here that another interesting but challenging problem associated with the study of differential equations in which the derivatives depend not only on constant values of unknown function from the past, but also on those from the future. The main advantage of such equations is that it enables the formulation of initial value problems that can be extended to the past as well as to the future, that is for all real time t. Numerous models related to such equations remain to be studied for which the above noted basic problems remain open.

Many physical problems, arising in a wide variety of applications are governed by both ordinary and partial finite difference equations. The theory of finite difference equations, the methods used in their solutions and their wide applications has drawn much attention in recent years.Through the widespread use of computers in recent years and renewed interest in numerical techniques, it seems that the theory of difference equations will quite likely be a fruitful source for future research. We hope that the tools developed in this theory may shed some light in the development of various fields of applied sciences as well.

As can be anticipated, since the integral inequalities with explicit estimates are so important in the study of properties of solutions of differential and integral equations, their finite difference (or discrete) analogues should also be useful in the study of properties of solutions of finite difference equations. The finite difference version of the well known Gronwall inequality seems to have appeared first in the work of Mikeladze [29] in 1935. It is well recognized that the discrete version of Gronwall's inequality provides a very useful and important tool in proving convergence of the discrete variable methods. In view of wider applications, finite difference inequalities with explicit estimates have been generalized, extended and used considerably in the development of the theory of finite difference equations. A large number of related results can be found in the references [1,3,5,12,42].

The lasting influence of integral and finite difference inequalities with explicit estimates, in the development of the theory of differential, integral and finite difference equations is enormous. Since about 1980, the subject has undergone explosive growth and attracted many researchers by its usefulness and basic character. Indeed, a particular feature that makes such inequalities so fascinating arises from the numerous fields of applications. The variety of nonlinear problems is evergrowing, and new methods have to be found for each of them. During nearly one hundred year history, the subject has been reflected in a great number of books and papers dedicated to such inequalities and applications. See [1,3,12,14,23,34,42] and the references given therein. The theory of such inequalities is basic and important and will no doubt continue to serve as an indispensable tool in future investidations.

In 1998 and 2002, the author wrote the monographs [34] and [42], which are devoted to the integral and finite difference inequalities with explicit estimates. Dictated by the need of various types of inequalities while studying many systems arising from diverse applications, such inequalities have received considerable attention during the past few years and a number of papers have appeared in the literature. This monograph is an outgrowth of the author's recent work, among many others in this area, tracing back to his earlier books mentioned above. As the literature is extensive, our focus in this monograph is mainly the results which have quite recently appeared and which are adequate in new applications in the development of the theory of differential, integral and finite difference equations. In fact it brings readers to the forefront of current research in this prosperous field and complement the results in monographs [34] and [42] in various ways. The selection of the material is largely influenced by my interests and the content consists predominantly of my own work.

This monograph is written with a view to provide basic tools for researchers working in mathematical analysis and applications, and those concentrating on differential, integral and finite difference equations. Of course, many generalizations, extensions, variants and applications of the results presented here are also possible. Naturally, these considerations will make the analysis more complicated, and leave it to the reader to fill in where needed. The book is self-contained and thus should be useful for those who are interested in learning or applying the inequalities with explicit estimates in their studies. In addition, it can be used as a text for advanced graduate cources and will serve as a reference in the field of system theory. I hope that, it will convince the reader that the integral, and finite difference inequalities with explicit estimates constitute a very useful tool in the study of various types of differential, integral and finite difference equations and will be a valuable source for a long time to come.

The present monograph consists of five chapters and references. Chapters 1 and 2 present a large number of basic linear and nonlinear integral inequalities involving functions of one and two independent variables, which in turn can be used as powerful tools in the study of various classes of differential and integral equations. Chapter 3 contains many new linear and nonlinear retarded integral inequalities involving functions of one and two independent variables which are useful in the study of various types of retarded differential and integral equations. Chapters 4 and 5 deals with the new linear and nonlinear finite difference inequalities involving functions of one and two independent variables, which find important applications in the study of different types of finite difference equations. Each chapter contains a section on basic applications of some of the inequalities therein. Regarding the list of references, I would like to mention that a large number of references on the topics discussed here are provided in the books [34] and [42] by the present author; see also [1,3,4,12,14,23,32,84] and the references given there. Without any intention of being complete, here only those references from the recent journal literature which are used in the text are given.

Throughout, we shall use the following notations and definitions.

Let R denotes the set of real numbers and $R_+ = [0, \infty)$, $Z = \{0, \pm 1, \pm 2, ...\}$, $N = \{1, 2, ...\}$, $N_0 = \{0, 1, 2, ...\}$, $N_{\alpha,\beta} = \{\alpha, \alpha+1, ..., \alpha+n = \beta\}$ for $n \in N, \alpha \in N_0, \beta \in N$ such that $\alpha \leq \beta$. The derivative of a function $u(t)$ for $t \in R$ is denoted by $u'(t)$ or $\frac{d}{dt}u(t)$. The partial derivatives of a function $z(x, y)$ for $x, y \in R$ with respect to x, y and xy are denoted by $D_1 z(x, y)$ or $z_x(x, y)$ or $\frac{\partial}{\partial x} z(x, y)$, $D_2 z(x, y)$ or $z_y(x, y)$ or $\frac{\partial}{\partial y} z(x, y)$ and $D_1 D_2 z(x, y) = D_2 D_1 z(x, y)$ or $z_{xy}(x, y)$ or $\frac{\partial^2}{\partial y \partial x} z(x, y)$. For the functions $w(m)$, $z(m, n)$ for $m, n \in Z$, we define the operators $\Delta, \Delta_1, \Delta_2$ by $\Delta w(m) = w(m+1) - w(m)$, $\Delta_1 z(m, n) = z(m+1, n) - z(m, n)$, $\Delta_2 z(m, n) = z(m, n+1) - z(m, n)$ respectively and $\Delta_2 \Delta_1 z(m, n) = \Delta_2 (\Delta_1 z(m, n))$. Let $C(A, B), C^1(A, B), D(A, B)$ denote the class of continuous functions, the class of continuous and differentiable functions, the class of functions from the set A to the set B respectively. We use the usual conventions that the empty sums and products are taken to be 0 and 1 respectively. Furthermore, throughout the work, we shall assume that all the integrals, sums and products involved exist on the respective domains of their definitions and are finite, and hence converge, so we shall omit such types of conditions. The notations, definitions, and symbols used in the work are standard and are explained, if necessary,at appropriate places.

Chapter 1

Integral inequalities in one variable

1.1 Introduction

During the past few decades abundance of applications is stimulating a rapid development of the theory of differential and integral equations. A variety of new methods and tools are developed by different investigators to study various types of differential and integral equations. The method of integral inequalities with explicit estimates is a very powerful tool in studying various properties of solutions of differential and integral equations. Motivated by the desire to apply such inequalities to numerious applications, in the past few years, a number of new inequalities have been investigated in [24-28,44,45,50-55]. In this chapter we present some fundamental integral inequalities recently established in the literature, which can be used as handy tools in the analysis of certain classes of differential and integral equations. Some immediate applications are also given.

1.2 Basic nonlinear integral inequalities

The explicit bounds given by the well known Gronwall-Bellman [16,6] inequality and its nonlinear generalization due to Bihari [8] (see also, LaSalle [20]) are used to a considerable extent in the study of differential and integral equations. In this section we present some useful generalizations and variants of the above mentioned inequalities.

We shall start with the following generalization of Bihar's inequality (see [34, p. 107]).

Theorem 1.2.1. Let $u(t), a(t), a'(t) \in C(R_+, R_+)$, $k(t,\sigma)$, $\frac{\partial}{\partial t}k(t,\sigma) \in C(D, R_+)$ where $D = \{(t,\sigma) \in R_+^2 : 0 \le \sigma \le t < \infty\}$. Let $g \in C(R_+, R_+)$ be a nondecreasing function, $g(u) > 0$ on $(0,\infty)$. If

$$u(t) \le a(t) + \int_0^t k(t,\sigma) g(u(\sigma))\, d\sigma, \tag{1.2.1}$$

for $t \in R_+$, then for $0 \le t \le t_1;\ t, t_1 \in R_+$,

$$u(t) \le G^{-1}\left[G(a(t)) + \int_0^t A(s)\, ds\right], \tag{1.2.2}$$

where

$$A(t) = k(t,t) + \int_0^t \frac{\partial}{\partial t} k(t,\sigma) d\sigma, \tag{1.2.3}$$

$$G(r) = \int_{r_0}^r \frac{ds}{G(s)}, r > 0, \tag{1.2.4}$$

$r_0 > 0$ is arbitrary and G^{-1} is the inverse of G and $t_1 \in R_+$ is chosen so that

$$G(a(t)) + \int_0^t A(s)\, ds \in Dom\left(G^{-1}\right),$$

for all $t \in R_+$ lying in the interval $0 \le t \le t_1$.

Proof. We note that, since $a'(t) \ge 0$, the function $a(t)$ is monotonically increasing. Let $a(t) > 0$ for $t \in R_+$ and define a function $z(t)$ by the right hand side of (1.2.1). Then $z(0) = a(0)$, $u(t) \le z(t)$, $z(t)$ is positive and by hypotheses, it is nondecreasing and

$$z'(t) = a'(t) + k(t,t)\, g(u(t)) + \int_0^t \frac{\partial}{\partial t} k(t,\sigma) g(u(\sigma))\, d\sigma$$

$$\le a'(t) + A(t)\, g(z(t)). \tag{1.2.5}$$

From (1.2.4), (1.2.5) the fact that $a(t) \le z(t)$ and the nondecreasing character of g we have

$$\frac{d}{dt} G(z(t)) = \frac{z'(t)}{g(z(t))} \le \frac{a'(t) + A(t)\, g(z(t))}{g(z(t))}$$

$$\leq \frac{a'(t)}{g(a(t))} + A(t)$$
$$= \frac{d}{dt} G(a(t)) + A(t). \tag{1.2.6}$$

By setting $t = s$ in (1.2.6) and integrating it from 0 to t , $t \in R_+$ and using the fact that $z(0) = a(0)$ we have

$$G(z(t)) \leq G(a(t)) + \int_0^t A(s)\,ds. \tag{1.2.7}$$

From (1.2.7) and the hypotheses on G we observe that

$$z(t) \leq G^{-1}\left[G(a(t)) + \int_0^t A(s)\,ds\right]. \tag{1.2.8}$$

Using (1.2.8) in $u(t) \leq z(t)$ we get the required inequality in (1.2.2).

If $a(t)$ is nonnegative, we carry out the above procedure with $a(t) + \varepsilon$ instead of $a(t)$,where $\varepsilon > 0$ is an arbitrary small constant, and subsequently pass to the limit $\varepsilon \to 0$ to obtain (1.2.2). The subinterval $0 \leq t \leq t_1$ is obvious.

Remark 1.2.1. We note that the inequality established in Theorem 1.2.1 is a slight variant of the inequality given by Pachpatte in [68]. In the special case when $a(t) = c$ (a nonnegative constant), $k(t,\sigma) = f(\sigma)$ and hence $\frac{\partial}{\partial t}k(t,\sigma) = 0$, the inequality in Theorem 1.2.1 reduces to the Bihari's inequality, see [8]. If we take $g(u) = u$ in Theorem 1.1, then the bound obtained in (1.2.2) reduces to

$$u(t) \leq a(t)\exp\left(\int_0^t A(s)\,ds\right),$$

for $t \in R_{+.}$ In this case Theorem 1.2.1 is a generalization of the well known Gronwall-Bellman inequality, see [16,6].

The inequalities in the following theorem are established by Pachpatte in [55].

Theorem 1.2.2. Let $u(t), k(t,\sigma), \frac{\partial}{\partial t}k(t,\sigma)$ be as in Theorem 1.2.1 and $c \geq 0$ is a constant.

(a_1) If

$$u^2(t) \leq c + \int_0^t k(t,\sigma)u(\sigma)\,d\sigma, \tag{1.2.9}$$

for $t \in R_+$, then

$$u(t) \leq \sqrt{c} + \frac{1}{2}\int_0^t A(s)\,ds, \tag{1.2.10}$$

for $t \in R_+$, where $A(t)$ is given by (1.2.3).

(a_2) Let $g(u)$ be as in Theorem 1.2.1. If

$$u^2(t) \leq c + \int_0^t k(t,\sigma)u(\sigma)\,g(u(\sigma))\,d\sigma, \tag{1.2.11}$$

for $t \in R_+$, then for $0 \leq t \leq t_2$; $t, t_2 \in R_+$,

$$u(t) \leq G^{-1}\left[G\left(\sqrt{c}\right) + \frac{1}{2}\int_0^t A(s)\,ds\right], \tag{1.2.12}$$

where G, G^{-1}, A are as in Theorem 1.2.1, and $t_2 \in R_+$ is chosen so that

$$G\left(\sqrt{c}\right) + \frac{1}{2}\int_0^t A(s)\,ds \in Dom\left(G^{-1}\right),$$

for all $t \in R_+$ lying in the interval $0 \leq t \leq t_2$.

Proof. (a_1) Let $c > 0$ and define a function $z(t)$ by the right hand side of (1.2.9). Then $z(0) = c$, $u(t) \leq \sqrt{z(t)}$, $z(t)$ is positive and nondecreasing for $t \in R_+$ and

$$z'(t) = k(t,t)\,u(t) + \int_0^t \frac{\partial}{\partial t}k(t,\sigma)u(\sigma)\,d\sigma$$

$$\leq k(t,t)\sqrt{z(t)} + \int_0^t \frac{\partial}{\partial t}k(t,\sigma)\sqrt{z(\sigma)}d\sigma$$

$$\leq A(t)\sqrt{z(t)}, \tag{1.2.13}$$

which implies

$$\sqrt{z(t)} \leq \sqrt{c} + \frac{1}{2}\int_0^t A(s)\,ds. \tag{1.2.14}$$

Using (1.2.14) in $u(t) \leq \sqrt{z(t)}$, we get the desired inequality in (1.2.10. The proof of the case when $c \geq 0$ can be completed as mentioned in the proof of Theorem 1.2.1.

(a_2) Let $c > 0$ and define a function $w(t)$ by the right hand side of (1.2.11). Then $w(0) = c$, $u(t) \leq \sqrt{w(t)}$, $w(t)$ is positive, nondecreasing for $t \in R_+$ and as in the proof of (1.2.13) we get

$$w'(t) \leq A(t)\sqrt{w(t)}g\left(\sqrt{w(t)}\right), \tag{1.2.15}$$

which implies

$$\sqrt{w(t)} \leq \sqrt{c} + \int_0^t A(s)\, g\left(\sqrt{w(t)}\right) ds. \tag{1.2.16}$$

Now an application of Bihari's inequality given in Theorem 1.3.1 in [34] yields

$$\sqrt{w(t)} \leq G^{-1}\left[G\left(\sqrt{c}\right) + \frac{1}{2}\int_0^t A(s)\, ds\right]. \tag{1.2.17}$$

Using (1.2.17) in $u(t) \leq \sqrt{w(t)}$, we get the required inequality in (1.2.12). The proof of the case when $c \geq 0$ follows as mentioned in the proof of Theorem 1.2.1. The subinterval $0 \leq t \leq t_2$ is obvious.

Remark 1.2.2. If we take $k(t,\sigma) = f(\sigma)$ and hence $\frac{\partial}{\partial t}k(t,\sigma) = 0$ in Theorem 1.2.2, then the bounds obtained in (1.2.10), (1.2.12) reduces to

$$u(t) \leq \sqrt{c} + \frac{1}{2}\int_0^t f(\sigma) d\sigma,$$

$$u(t) \leq G^{-1}\left[G\left(\sqrt{c}\right) + \frac{1}{2}\int_0^t f(\sigma)\, d\sigma\right],$$

respectively. We note that, by following the proof of Theorem 1.2.1 one can very easily obtain the bounds on inequalities (1.2.9), (1.2.11) when the constant c is replaced by the function $a(t)$, where $a(t)$ is as in Theorem 1.2.1.

In [24] Medveď defined a special class of nonlinear functions and developed a method to estimate solutions for nonlinear integral inequalities with singular kernels and the nonlinearity of that class. The class of functions defined in [24] is as follows.

Let $q > 0$ be a real number and $0 < T \leq \infty$. We say that a function $w : R_+ \to R$ satisfies a condition (q), if

$$e^{-qt}\left[w(u)\right]^q \leq R(t)\, w\left(e^{-qt}u^q\right), \tag{q}$$

for all $u \in R_+, t \in [0,T)$, where $R(t)$ is continuous, nonnegative function.

Remark 1.2.3. If $w(u) = u^m, m > 0$, then

$$e^{-qt}[w(u)]^q = e^{(m-1)qt} w\left(e^{-qt}u^q\right),$$

for any $q > 1$, i.e., the condition q is satisfied with $R(t) = e^{(m-1)qt}$. For $w(u) = u + au^m$, where $0 \le a \le 1, m \ge 1$ the function w satisfies the condition (q) with $q > 1$ and $R(t) = 2^{q-1}e^{qmt}$, see [24].

The following theorems are proved in Medveď [24].

Theorem 1.2.3. Let $0 < T \le \infty$, $u(t), b(t), a(t), a'(t) \in C([0,T), R_+)$; $w \in C(R_+, R)$ be a nondecreasing function, $w(0) = 0, w(u) > 0$ on $(0,T)$ and

$$u(t) \le a(t) + \int_0^t (t-s)^{\beta-1} b(s) w(u(s)) ds, \tag{1.2.18}$$

for $t \in [0,T)$ where $\beta > 0$ is a constant. Then the following assertions hold:

(i) Suppose $\beta > \frac{1}{2}$ and w satisfies the condition (q) with $q = 2$. Then

$$u(t) \le e^t \left\{\Omega^{-1}\left[\Omega\left(2a(t)^2\right) + g_1(t)\right]\right\}^{\frac{1}{2}}, \tag{1.2.19}$$

for $t \in [0, T_1]$, where

$$g_1(t) = \frac{\Gamma(2\beta - 1)}{4^{\beta-1}} \int_0^t R(s) b(s)^2 ds,$$

Γ is the gamma function, $\Omega(v) = \int\limits_{v_0}^{v} \frac{ds}{w(s)}, v_0 > 0, \Omega^{-1}$ is the inverse of Ω, and $t_1 \in R_+$ is such that $\Omega\left(2a(t)^2\right) + g_1(t) \in Dom\left(\Omega^{-1}\right)$ for all $t \in [0, T_1]$.

(ii) Let $\beta \in \left(0, \frac{1}{2}\right]$ and w satisfies the condition (q) with $q = z + 2$, where $z = \frac{1-\beta}{\beta}$ i.e., $\beta = \frac{1}{z+1}$. Let Ω, Ω^{-1} be as in part (i). Then

$$u(t) \le e^t \left\{\Omega^{-1}\left[\Omega\left(2^{q-1}a(t)^q\right) + g_2(t)\right]\right\}^{\frac{1}{q}}, \tag{1.2.20}$$

for $t \in [0, T_1]$, where

$$g_2(t) = 2^{q-1} K_z^q \int_0^t R(s) b(s)^q ds,$$

$$K_z = \left[\frac{\Gamma(1 - \alpha p)}{p^{1-\alpha p}}\right]^{\frac{1}{p}}, \alpha = \frac{z}{z+1}, p = \frac{z+2}{z+1}, \tag{1.2.21}$$

and $T_1 \in R_+$ is such that $\Omega\left(2^{q-1}a(t)^q\right) + g_2(t) \in Dom\left(\Omega^{-1}\right)$ for all $t \in [0, T_1]$.

Proof. First we shall prove the assertion (i). Using the Cauchy-Schwarz inequality we obtain from (1.2.18)

$$u(t) \leq a(t) + \int_0^t (t-s)^{\beta-1} e^s b(s) e^{-s} w(u(s))\, ds$$

$$\leq a(t) + \left[\int_0^t (t-s)^{2\beta-2} e^{2s} ds\right]^{\frac{1}{2}} \left[\int_0^t b(s)^2 e^{-2s} w(u(s))^2 ds\right]^{\frac{1}{2}}. \quad (1.2.22)$$

For the first integral in (1.2.22) we have the estimate

$$= \int_0^t (t-s)^{2\beta-2} e^{2s} ds = \int_0^t \tau^{2\beta-2} e^{2(t-\tau)} d\tau$$

$$= e^{2t} \int_0^t \tau^{2\beta-2} e^{-2\tau} d\tau = \frac{2e^{2t}}{4^\beta} \int_0^{2t} \sigma^{2\beta-2} e^{-\sigma} d\sigma$$

$$< \frac{2e^{2t}}{4^\beta} \Gamma(2\beta-1).$$

Therefore we obtain from (1.2.22)

$$u(t) \leq a(t) + \left[\frac{2e^{2t}}{4^\beta}\Gamma(2\beta-1)\right]^{\frac{1}{2}} \left[\int_0^t b(s)^2 e^{-2s} w(u(s))^2 ds\right]^{\frac{1}{2}}.$$

Using the well known consequence of the Jensen inequality:

$$\left(\sum_{i=1}^n a_i\right)^r \leq n^{r-1} \sum_{i=1}^n a_i^r, \quad (1.2.23)$$

(where $a_i \geq 0, r > 0$ are real numbers, see [30,65]), with $n = 2, r = 2$ we obtain

$$u^2(t) \leq 2a(t)^2 + \frac{e^{2t}\Gamma(2\beta-1)}{4^{\beta-1}} \int_0^t b(s)^2 e^{-2s} w(u(s))^2 ds, \quad (1.2.24)$$

and applying the condition (q) with $q = 2$ we have

$$v(t) \leq \alpha(t) + K \int_0^t b(s)^2 R(s) w(v(s))\, ds,$$

where

$$v(t) = \left(e^{-t}u(t)\right)^2, \alpha(t) = 2a(t)^2, K = \frac{\Gamma(2\beta-1)}{4^{\beta-1}}. \quad (1.2.25)$$

Now proceeding as in the proof of Theorem 1.2.1 we obtain

$$v(t) \le \Omega^{-1}\left[\Omega(\alpha(t)) + g_1(t)\right]. \tag{1.2.26}$$

From (1.2.25) and (1.2.26) we get (1.2.19).

Next, we prove the assertion (ii). Obviously, $\beta - 1 = -\alpha = -\frac{z}{z+1}$. Let p, q be as in the statement of theorem. Then $\frac{1}{p} + \frac{1}{q} = 1$ and using the Hölder's integral inequality we obtain from (1.2.18)

$$u(t) \le a(t) + \int_0^t (t-s)^{\beta-1} b(s) w(u(s))\, ds$$

$$= a(t) + \int_0^t (t-s)^{-\alpha} e^{s} b(s) e^{-s} w(u(s))\, ds$$

$$\le a(t) + \left[\int_0^t (t-s)^{-\alpha p} e^{ps} ds\right]^{\frac{1}{p}} \left[\int_0^t b(s)^q e^{-qs} w(u(s))^q\, ds\right]^{\frac{1}{q}}. \tag{1.2.27}$$

For the first integral in (1.2.27) we have the estimate

$$\int_0^t (t-s)^{-\alpha p} e^{ps} ds = e^{pt} \int_0^t \tau^{-\alpha p} e^{-p\tau} d\tau$$

$$= \frac{e^{pt}}{p^{1-\alpha p}} \int_0^{pt} \sigma^{-\alpha p} e^{-\sigma} d\sigma < \frac{e^{pt}}{p^{1-\alpha p}} \Gamma(1-\alpha p).$$

Obviously, $1 - \alpha p = \frac{1}{(z+1)^2} > 0$ and so $\Gamma(1-\alpha p) \in R$. Thus (1.2.27) and the condition (q) yield

$$u(t) \le a(t) + e^t K_z \left[\int_0^t b(s)^q R(s) w\left(e^{-qs} u(s)^q\right) ds\right]^{\frac{1}{q}}, \tag{1.2.28}$$

where K_z is defined by (1.2.21). Now using the inequality (1.2.23) with $n = 2, r = q$ we obtain

$$u(t)^q \le 2^{q-1} a(t)^q + 2^{q-1} e^{qt} K_z^q \int_0^t b(s)^q R(s) w\left(e^{-qs} u(s)^q\right) ds, \tag{1.2.29}$$

and this yields

$$v(t) \le \phi(t) + 2^{q-1} K_z^q \int_0^t b(s)^q R(s) w(v(s))\, ds, \tag{1.2.30}$$

where

$$v(t) = \left(e^{-t}u(t)\right)^q, \phi(t) = 2^{q-1}a(t)^q. \tag{1.2.31}$$

Now by proceeding as in the proof of Theorem 1.2.1 we obtain

$$v(t) \leq \Omega^{-1}\left[\Omega(\phi(t)) + g_2(t)\right]. \tag{1.2.32}$$

The required inequality in (1.2.20) follows from (1.2.31) and (1.2.32).

As a consequence of Theorem 1.2.3 we have

Theorem 1.2.4. Let $0 < T \leq \infty$, $u(t), b(t), a(t), a'(t)$ be as in Theorem 1.2.3 and

$$u(t) \leq a(t) + \int_0^t (t-s)^{\beta-1} b(s) u(s) ds, \tag{1.2.33}$$

where $\beta > 0$. Then the following assertions hold:

(i) If $\beta > \frac{1}{2}$, then

$$u(t) \leq \sqrt{2}\, a(t) \exp\left[\frac{2\Gamma(2\beta-1)}{4^\beta}\int_0^t b(s)^2 ds + t\right], \tag{1.2.34}$$

for $t \in [0, T)$.

(ii) If $\beta = \frac{1}{z+1}$ for some $z \geq 1$, then

$$u(t) \leq \left(2^{q-1}\right)^{\frac{1}{q}} a(t) \exp\left[\frac{2^{q-1}}{q} K_z^q \int_0^t b(s)^q ds + t\right], \tag{1.2.35}$$

for $t \in [0, T)$, where K_z is defined by (1.2.11), $q = z + 2$.

Theorem 1.2.5. Let $0 < T \leq \infty$, $u(t), b(t), a(t), a'(t)$ and $w(u)$ be as in Theorem 1.2.3 and

$$u^2(t) \leq a(t) + \int_0^t (t-s)^{\beta-1} b(s) w(u(s)) ds, \tag{1.2.36}$$

where $\beta > 0$ is a constant. Then the following assertions hold:

(i) Suppose $\beta > \frac{1}{2}$ and w satisfies the condition (q) with $q = 2$. Then

$$u(t) \leq e^t \left\{ \Lambda^{-1} \left[\Lambda\left(2a(t)^2\right) + K \int_0^t b(s)^2 R(s)\, ds \right] \right\}^{\frac{1}{4}}, \tag{1.2.37}$$

for $t \in [0, T_1]$, where

$$K = \frac{\Gamma(2\beta - 1)}{4^{\beta-1}}, \Lambda(v) = \int_{v_0}^{v} \frac{d\sigma}{w(\sqrt{\sigma})}, v_0 > 0, \tag{1.2.38}$$

$T_1 \in R_+$ is such that $\Lambda\left(2a(t)^2\right) + K \int_0^t b(s)^2 R(s)\, ds \in Dom\left(\Lambda^{-1}\right)$ for all $t \in [0, T_1]$, Γ is the gamma function.

(ii) Let $\beta \in \left(0, \frac{1}{2}\right]$ and w satisfies the condition (q) with $q = z + 2$, where $z = \frac{1-\beta}{\beta}$ i.e., $\beta = \frac{1}{z+1}$. Let Λ, Λ^{-1} be as in part (i). Then

$$u(t) \leq e^t \left\{ \Lambda^{-1} \left[\Lambda\left(2^{q-1} a(t)^q\right) + 2^{q-1} K_z^q \int_0^t b(s)^q R(s)\, ds \right] \right\}^{\frac{1}{2q}}, \tag{1.2.39}$$

for $t \in [0, T_1]$, where

$$K_z = \left[\frac{\Gamma(1 - \beta p)}{p^{1-\beta p}} \right]^{\frac{1}{p}}, \beta = \frac{1}{z+1}, p = \frac{z+2}{z+1}, \tag{1.2.40}$$

$T_1 \in R_+$ is such that $\Lambda\left(2^{q-1} a(t)^q\right) + 2^{q-1} K_z^q \int_0^t b(s)^q R(s)\, ds \in Dom\left(\Lambda^{-1}\right)$ for all $t \in [0, T_1]$.

Proof. First we prove the assertion (i). Following the proof of Theorem 1.2.3 one can show that

$$v^2(t) \leq \alpha(t) + K \int_0^t b(s)^2 R(s)\, w(v(s))\, ds, \tag{1.2.41}$$

where

$$v(t) = \left(e^{-t} u(t)\right)^2, \alpha(t) = 2a(t)^2, K = \frac{\Gamma(2\beta - 1)}{4^{\beta-1}}. \tag{1.2.42}$$

Define by $e(t)$ the right hand side of (1.2.41). Then $v(t) \leq \sqrt{e(t)}$. Now by following the proof of Theorem 1.2.1 with suitable modifications we obtain

$$e(t) \leq \Lambda^{-1} \left[\Lambda(\alpha(t)) + K \int_0^t b(s)^2 R(s)\, ds \right],$$

and thus we have

$$v(t) \leq \sqrt{e(t)} \leq \left\{\Lambda^{-1}\left[\Lambda(\alpha(t)) + K\int_0^t b(s)^2 R(s)\,ds\right]\right\}^{\frac{1}{2}}. \tag{1.2.43}$$

From (1.2.42) and (1.2.43) we get (1.2.37).

Now we shall prove the assertion (ii). Following the proof of assertion (ii) of Theorem 1.2.3 one can show that

$$v^2(t) \leq \phi(t) + 2^{q-1}K_z^q \int_0^t b(s)^q R(s)\, w(v(s))\,ds, \tag{1.2.44}$$

where

$$v(t) = \left(e^{-t}u(t)\right)^q, \phi(t) = 2^{q-1}a(t)^q, \tag{1.2.45}$$

and K_z is given as in (1.2.40). Following the procedure from the proof of assertion (i) we obtain

$$v(t) \leq \left\{\Lambda^{-1}\left[\Lambda(\phi(t)) + 2^{q-1}K_z^q\int_0^t b(s)^q R(s)\,ds\right]\right\}^{\frac{1}{2}}. \tag{1.2.46}$$

From (1.2.45) and (1.2.46) we obtain (1.2.39).

Remark 1.2.4. We note that in the book [17] Henry obtained by an iterative argument an estimate on the inequality of the form (1.2.33). The analysis used in the proof of Theorems 1.2.3 and 1.2.5 is based on the method developed by Medveď in [24]. For the application to global existence of solutions and a stability theorem for a class of parabolic partial differential equations, see [25,28].

1.3 More nonlinear integral inequalities

This section deals with some more nonlinear integral inequalities established by Pachpatte in [35,45] which claims their orgins in the inequalities given by Ou-Iang [33] and Dafermos [10], see also [34].

In [35] Pachpatte proved the inequalities in the following two theorems.

Theorem 1.3.1. Let $u(t), a(t), b(t), g(t), h(t) \in C(R_+, R_+)$ and $p > 1$ be a real constant.

(a_1) If

$$u^p(t) \le a(t) + b(t) \int_0^t [g(s) u^p(s) + h(s) u(s)]\, ds, \tag{1.3.1}$$

for $t \in R_+$, then

$$u(t) \le \left\{ a(t) + b(t) \int_0^t \left[g(s) a(s) + h(s) \left(\frac{p-1}{p} + \frac{a(s)}{p} \right) \right] \right.$$

$$\left. \times \exp\left(\int_s^t b(\sigma) \left(g(\sigma) + \frac{h(\sigma)}{p} \right) d\sigma \right) ds \right\}^{\frac{1}{p}}, \tag{1.3.2}$$

for $t \in R_+$.

(a_2) Let $c(t)$ be a real-valued positive continuous and nondecreasing function defined on R_+. If

$$u^p(t) \le c^p(t) + b(t) \int_0^t [g(s) u^p(s) + h(s) u(s)]\, ds, \tag{1.3.3}$$

for $t \in R_+$, then

$$u(t) \le c(t) \left\{ 1 + b(t) \int_0^t \left[g(s) + h(s) c^{1-p}(s) \right] \right.$$

$$\left. \times \exp\left(\int_s^t b(\sigma) \left(g(\sigma) + \frac{h(\sigma)}{p} c^{1-p}(\sigma) \right) d\sigma \right) ds \right\}^{\frac{1}{p}}, \tag{1.3.4}$$

for $t \in R_+$.

(a_3) Let $k(t,s)$ and its partial derivative $\frac{\partial}{\partial t} k(t,s)$ be real-valued nonnegative continuous functions for $0 \le s \le t < \infty$. If

$$u^p(t) \le a(t) + b(t) \int_0^t k(t,s) [g(s) u^p(s) + h(s) u(s)] ds, \tag{1.3.5}$$

for $t \in R_+$, then

$$u(t) \le \left\{ a(t) + b(t) \int_0^t B(\sigma) \exp\left(\int_\sigma^t A(\tau) d\tau \right) d\sigma \right\}^{\frac{1}{p}}, \tag{1.3.6}$$

for $t \in R_+$, where

$$A(t) = k(t,t)\,b(t)\left(g(t) + \frac{h(t)}{p}\right)$$

$$+ \int_0^t \frac{\partial}{\partial t} k(t,s)\,b(s)\left(g(s) + \frac{h(s)}{p}\right) ds, \tag{1.3.7}$$

$$B(t) = k(t,t)\left(g(t)\,a(t) + h(t)\left(\frac{p-1}{p} + \frac{a(t)}{p}\right)\right)$$

$$+ \int_0^t \frac{\partial}{\partial t} k(t,s)\left(g(s)\,a(s) + h(s)\left(\frac{p-1}{p} + \frac{a(s)}{p}\right)\right) ds, \tag{1.3.8}$$

for $t \in R_+$.

Proof. (a_1) Define a function $z(t)$ by

$$z(t) = \int_0^t \left[g(s)\,u^p(s) + h(s)\,u(s)\,ds\right]. \tag{1.3.9}$$

Then $z(0) = 0$ and (1.3.1) can be written as

$$u^p(t) \le a(t) + b(t)\,z(t). \tag{1.3.10}$$

From (1.3.10) and using the elementary inequality, see [30, p. 30]

$$x^{\frac{1}{p}} y^{\frac{1}{q}} \le \frac{x}{p} + \frac{y}{q}, \tag{1.3.11}$$

where $x \ge 0, y \ge 0$ and $\frac{1}{p} + \frac{1}{q} = 1$, we observe that

$$u(t) \le \left(a(t) + b(t)\,z(t)\right)^{\frac{1}{p}} (1)^{1/\left(p/p-1\right)}$$

$$\le \frac{p-1}{p} + \frac{a(t)}{p} + \frac{b(t)}{p} z(t). \tag{1.3.12}$$

Differentiating (1.3.9) and using (1.3.10) and (1.3.12) we get

$$z'(t) \le b(t)\left(g(t) + \frac{h(t)}{p}\right) z(t)$$

$$+ \left[g(t)\,a(t) + h(t)\left(\frac{p-1}{p} + \frac{a(t)}{q}\right)\right]. \tag{1.3.13}$$

The inequality (1.3.13) implies the estimate

$$z(t) \leq \int_0^t \left[g(s)\, a(s) + h(s) \left(\frac{p-1}{p} + \frac{a(s)}{p} \right) \right]$$

$$\times \exp\left(\int_s^t b(\sigma) \left(g(\sigma) + \frac{h(\sigma)}{p} \right) d\sigma \right) ds. \tag{1.3.14}$$

The required inequality (1.3.2) follows from (1.3.14) and (1.3.10).

(a_2) Since $c(t)$ is a positive, continuous and nondecreasing function for $t \in R_+$, from (1.3.3) we observe that

$$\left(\frac{u(t)}{c(t)} \right)^p \leq 1 + b(t) \int_0^t \left[g(s) \left(\frac{u(s)}{c(s)} \right)^p + h(s)\, c^{1-p}(s) \left(\frac{u(s)}{c(s)} \right) \right] ds.$$

Now an application of the inequality given in (a_1) yields the desired result in (1.3.4).

(a_3) Define a function $z(t)$ by

$$z(t) = \int_0^t k(t,s) \left[g(s)\, u^p(s) + h(s)\, u(s) \right] ds. \tag{1.3.15}$$

Then as in the proof of part (a_1), from (1.3.15) we see that the inequalities (1.3.10) and (1.3.12) hold. Differentiating (1.3.15) and using (1.3.10), (1.3.12) and the fact that $z(t)$ is monotonic nondecreasing in t we get

$$z'(t) = k(t,t) \left[g(t)\, u^p(t) + h(t)\, u(t) \right]$$

$$+ \int_0^t \frac{\partial}{\partial t} k(t,s) \left[g(s)\, u^p(s) + h(s)\, u(s) \right] ds$$

$$\leq k(t,t) \left[g(t) \left(a(t) + b(t)\, z(t) \right) + h(t) \left(\frac{p-1}{p} + \frac{a(t)}{p} + \frac{b(t)}{p} z(t) \right) \right]$$

$$+ \int_0^t \frac{\partial}{\partial t} k(t,s) \left[g(s) \left(a(s) + b(s)\, z(s) \right) \right.$$

$$\left. + h(s) \left(\frac{p-1}{p} + \frac{a(s)}{p} + \frac{b(s)}{p} z(s) \right) \right] ds$$

$$\leq \left[k(t,t)\, b(t) \left(g(t) + \frac{h(t)}{p} \right) + \int_0^t \frac{\partial}{\partial t} k(t,s)\, b(s) \left(g(s) + \frac{h(s)}{p} \right) \right] z(t)$$

$$+k(t,t) \left(g(t)\, a(t) + h(t) \left(\frac{p-1}{p} + \frac{a(t)}{p} \right) \right)$$

$$+ \int_0^t \frac{\partial}{\partial t} k(t,s) \left(g(s)\, a(s) + h(s) \left(\frac{p-1}{p} + \frac{a(s)}{p} \right) \right) ds$$

$$= A(t)\, z(t) + B(t). \tag{1.3.16}$$

The inequality (1.3.16) implies the estimate

$$z(t) \leq \int_0^t B(\sigma) \exp \left(\int_\sigma^t A(\tau) d\tau \right) d\sigma. \tag{1.3.17}$$

Using (1.3.17) in $u^p(t) \leq a(t) + b(t)\, z(t)$, we get the required inequality in (1.3.6).

Theorem 1.3.2. Let $u(t), a(t), b(t), g(t) \in C(R_+, R_+)$ and $p > 1$ be a real constant.

(b_1) Let $f : R_+^2 \to R_+$ be a continuous function such that

$$0 \leq f(t,x) - f(t,y) \leq m(t,y)(x-y), \tag{1.3.18}$$

for $t \in R_+$ and $x \geq y \geq 0$,where $m : R_+^2 \to R_+$ is a continuous function. If

$$u^p(t) \leq a(t) + b(t) \int_0^t f(s, u(s))\, ds, \tag{1.3.19}$$

for $t \in R_+$, then

$$u(t) \leq \left\{ a(t) + b(t) \int_0^t f\left(s, \frac{p-1}{p} + \frac{a(s)}{p}\right) \right.$$

$$\left. \times \exp \left(\int_s^t m\left(\sigma, \frac{p-1}{p} + \frac{a(\sigma)}{p}\right) \frac{b(\sigma)}{p} d\sigma \right) ds \right\}^{\frac{1}{p}}, \tag{1.3.20}$$

for $t \in R_+$.

(b_2) Let $f : R_+^2 \to R_+$ be a continuous function and $\phi : R_+ \to R_+$ be a continuous and strictly increasing function with $\phi(0) = 0$ such that

$$0 \leq f(t,x) - f(t,y) \leq m(t,y)\,\phi^{-1}(x-y), \tag{1.3.21}$$

for $t \in R_+$ and $x \geq y \geq 0$, where $m : R_+^2 \to R_+$ is a continuous function and ϕ^{-1} is the inverse function of ϕ and

$$\phi^{-1}(xy) \leq \phi^{-1}(x)\,\phi^{-1}(y), \tag{1.3.22}$$

for $x, y \in R_+$. If

$$u^p(t) \leq a(t) + b(t)\,\phi\left(\int_0^t f(s, u(s))\,ds\right), \tag{1.3.23}$$

for $t \in R_+$, then

$$u(t) \leq \left\{ a(t) + b(t)\,\phi\left(\int_0^t f\left(s, \frac{p-1}{p} + \frac{a(s)}{p}\right)\right)\right.$$
$$\left. \times \exp\left(\int_s^t m\left(\sigma, \frac{p-1}{p} + \frac{a(\sigma)}{p}\right)\phi^{-1}\left(\frac{b(\sigma)}{p}\right)d\sigma\right)ds\right\}^{\frac{1}{p}}, \tag{1.3.24}$$

for $t \in R_+$.

(b_3) Let $W(r)$ be a real-valued, continuous, nondecreasing, subadditive and submultiplative function defined on R_+ and $W(r) > 0$ on $(0, \infty)$. If

$$u^p(t) \leq a(t) + b(t)\int_0^t g(s)\,W(u(s))\,ds, \tag{1.3.25}$$

for $t \in R_+$, then for $0 \leq t \leq t_1$,

$$u(t) \leq \left\{ a(t) + b(t)\,G^{-1}\left[G(D(t)) + \int_0^t g(s)\,W\left(\frac{b(s)}{p}\right)ds\right]\right\}^{\frac{1}{p}}, \tag{1.3.26}$$

where for $t \in R_+$,

$$D(t) = \int_0^t g(s)\,W\left(\frac{p-1}{p} + \frac{a(s)}{p}\right)ds, \tag{1.3.27}$$

$$G(r) = \int_{r_0}^r \frac{ds}{W(s)}, r > 0, \tag{1.3.28}$$

$r_0 > 0$ is arbitrary, G^{-1} is the inverse function of G and $t_1 \in R_+$ is chosen so that

$$G(D(t)) + \int_0^t g(s) W\left(\frac{b(s)}{p}\right) ds \in Dom\left(G^{-1}\right),$$

for all $t \in R_+$ lying in the interval $0 \le t \le t_1$.

Proof. (b_1) Define a function $z(t)$ by

$$z(t) = \int_0^t f(s, u(s))\, ds. \tag{1.3.29}$$

Then as in the proof of Theorem 1.3.1, part (a_1), from (1.3.19) we see that the inequalities (1.3.10) and (1.3.12) hold. From (1.3.29), (1.3.12) and the condition (1.3.18) it follows that

$$\begin{aligned} z'(t) &= f(t, u(t)) \\ &\le f\left(t, \frac{p-1}{p} + \frac{a(t)}{p} + \frac{b(t)}{p} z(t)\right) - f\left(t, \frac{p-1}{p} + \frac{a(t)}{p}\right) \\ &+ f\left(t, \frac{p-1}{p} + \frac{a(t)}{p}\right) \\ &\le m\left(t, \frac{p-1}{p} + \frac{a(t)}{p}\right) \frac{b(t)}{p} z(t) + f\left(t, \frac{p-1}{p} + \frac{a(t)}{p}\right). \end{aligned} \tag{1.3.30}$$

The inequality (1.3.30) implies the estimate

$$\begin{aligned} z(t) &\le \int_0^t f\left(s, \frac{p-1}{p} + \frac{a(s)}{p}\right) \\ &\times \exp\left(\int_s^t m\left(\sigma, \frac{p-1}{p} + \frac{a(\sigma)}{p}\right) \frac{b(\sigma)}{p} d\sigma\right) ds. \end{aligned} \tag{1.3.31}$$

From (1.3.31) and (1.3.10) the desired inequality in (1.3.20) follows.

(b_2) Defining a function $z(t)$ by (1.3.29) and following the arguments as in the proof of Theorem 1.3.1, part (a_1) we see that corresponding to the inequalities (1.3.10) and (1.3.12) we get

$$u^p(t) \le a(t) + b(t)\phi(z(t)), \tag{1.3.32}$$

and

$$u(t) \le \frac{p-1}{p} + \frac{a(t)}{p} + \frac{b(t)}{p}\phi(z(t)). \tag{1.3.33}$$

From (1.3.29), (1.3.33) and the conditions (1.3.21), (1.3.22) it follows that

$$\begin{aligned} z'(t) &= f(t, u(t)) \\ &\le f\left(t, \frac{p-1}{p} + \frac{a(t)}{p} + \frac{b(t)}{p}\phi(z(t))\right) - f\left(t, \frac{p-1}{p} + \frac{a(t)}{p}\right) \\ &+ f\left(t, \frac{p-1}{p} + \frac{a(t)}{p}\right) \\ &\le m\left(t, \frac{p-1}{p} + \frac{a(t)}{p}\right)\phi^{-1}\left(\frac{b(t)}{p}\phi(z(t))\right) + f\left(t, \frac{p-1}{p} + \frac{a(t)}{p}\right) \\ &\le m\left(t, \frac{p-1}{p} + \frac{a(t)}{p}\right)\phi^{-1}\left(\frac{b(t)}{p}\right) z(t) + f\left(t, \frac{p-1}{p} + \frac{a(t)}{p}\right). \end{aligned} \quad (1.3.34)$$

The inequality (1.3.34) implies the estimate

$$\begin{aligned} z(t) \le &\int_0^t f\left(s, \frac{p-1}{p} + \frac{a(s)}{p}\right) \\ &\times \exp\left(\int_s^t m\left(\sigma, \frac{p-1}{p} + \frac{a(\sigma)}{p}\right)\phi^{-1}\left(\frac{b(\sigma)}{p}\right) d\sigma\right) ds. \end{aligned} \quad (1.3.35)$$

The required inequality (1.3.24) follows from (1.3.32) and (1.3.35).

(b_3) Define a function $z(t)$ by

$$z(t) = \int_0^t g(s) W(u(s))\, ds. \quad (1.3.36)$$

Then as in the proof of Theorem 1.3.1, part (a_1), from (1.3.25) we see that the inequalities (1.3.10) and (1.3.12) hold. From (1.3.36), (1.3.12) and the conditions on W it follows that

$$z(t) \le D(t) + \int_0^t g(s) W\left(\frac{b(s)}{p}\right) W(z(s))\, ds, \quad (1.3.37)$$

where $D(t)$ is defined by (1.3.27). The rest of the proof can be completed by closely looking at the proof of Theorem 2.4.2 given in [34, p.121]. We omit the further details.

Remark 1.3.1. We note that in the special cases when (i) $g = 0$, (ii) $g = 0, p = 2$ in Theorem 1.3.1, and (iii) $p = 2$ in Theorem 1.3.2, we get new inequalities which may be convenient in certain applications.

The following Bihari type inequality is proved by Pachpatte in [45].

Theorem 1.3.3. Let $u(t), f(t) \in C(R_+, R_+)$, $h(t,s) \in C(R_+^2, R_+)$, for $0 \le s \le t < \infty$ and $c \ge 0$, $p > 1$ are real constants. Let $g \in C(R_+, R_+)$ be a nondecreasing function, $g(u) > 0$ for $u > 0$ and

$$u^p(t) \le c + \int_0^t \left[f(s) g(u(s)) + \int_0^s h(s,\sigma) g(u(\sigma)) d\sigma \right] ds, \tag{1.3.38}$$

for $t \in R_+$, then for $0 \le t \le t_1$,

$$u(t) \le \left\{ H^{-1} \left[H(c) + E(t) \right] \right\}^{\frac{1}{p}}, \tag{1.3.39}$$

where

$$E(t) = \int_0^t \left[f(s) + \int_0^s h(s,\sigma) d\sigma \right] ds, \tag{1.3.40}$$

$$H(r) = \int_{r_0}^r \frac{ds}{g\left(s^{\frac{1}{p}}\right)}, r > 0, \tag{1.3.41}$$

$r_0 > 0$ is arbitrary, H^{-1} is the inverse function of H and $t_1 \in R_+$ is chosen so that

$$H(c) + E(t) \in Dom\left(H^{-1}\right),$$

for all $t \in R_+$ lying in the interval $0 \le t \le t_1$.

Proof. We first assume that $c > 0$ and define a function $z(t)$ by the right hand side of (1.3.38). Then $z(0) = c$, $u(t) \le (z(t))^{\frac{1}{p}}$, $z(t)$ is positive and nondecreasing for $t \in R_+$ and

$$z'(t) = f(t) g(u(t)) + \int_0^t h(t,\sigma) g(u(\sigma)) d\sigma$$

$$\le f(t) g\left((z(t))^{\frac{1}{p}}\right) + \int_0^t h(t,\sigma) g\left((z(\sigma))^{\frac{1}{p}}\right) d\sigma$$

$$\le g\left((z(t))^{\frac{1}{p}}\right) \left[f(t) + \int_0^t h(t,\sigma) d\sigma \right]. \tag{1.3.42}$$

From (1.3.41) and (1.3.42) we have

$$\frac{d}{dt}H\left(z\left(t\right)\right) = \frac{z'\left(t\right)}{g\left(\left(z\left(t\right)\right)^{\frac{1}{p}}\right)}$$

$$\leq \left[f\left(t\right) + \int_0^t h\left(t,\sigma\right)d\sigma\right]. \tag{1.3.43}$$

By setting $t = s$ in (1.3.43) and integrating it from 0 to t we have

$$H\left(z\left(t\right)\right) \leq H\left(c\right) + E\left(t\right). \tag{1.3.44}$$

Since H^{-1} is increasing, from (1.3.44) we have

$$z\left(t\right) \leq H^{-1}\left[H\left(c\right) + E\left(t\right)\right]. \tag{1.3.45}$$

Using (1.3.45) in $u\left(t\right) \leq \left(z\left(t\right)\right)^{\frac{1}{p}}$ we have the required inequality in (1.3.39). If c is nonngative, we carry out the above procedure with $c+\varepsilon$ instead of c, where $\varepsilon > 0$ is an arbitrary small constant, and by letting $\varepsilon \to 0$, we obtain (1.3.39). The subinterval $0 \leq t \leq t_1$ is obvious.

As an immediate consequence of Theorem 1.3.3 we have the following

Theorem 1.3.4. Let $u\left(t\right), f\left(t\right), h\left(t,s\right), c, p$ be as in Theorem 1.3.3. If

$$u^p\left(t\right) \leq c + \int_0^t \left[f\left(s\right)u\left(s\right) + \int_0^s h\left(s,\sigma\right)u\left(\sigma\right)d\sigma\right]ds, \tag{1.3.46}$$

for $t \in R_+$, then

$$u\left(t\right) \leq \left[c^{\frac{p-1}{p}} + \left(\frac{p-1}{p}\right)H\left(t\right)\right]^{\frac{1}{p-1}}, \tag{1.3.47}$$

for $t \in R_+$, where $E(t)$ is given by (1.3.40).

Proof. Let $g(u) = u$ in Theorem 1.3.3.Then (1.3.38) reduces to (1.3.46) and

$$H\left(r\right) = \frac{p}{p-1}\left[r^{\frac{p-1}{p}} - r_0^{\frac{p-1}{p}}\right],$$

$$H^{-1}\left(r\right) = \left[\frac{p-1}{p}r + r_0^{\frac{p-1}{p}}\right]^{\frac{p}{p-1}},$$

and consequently the bound obtained in (1.3.39) reduces to the bound in (1.3.47).

Remark 1.3.2. We note that the definition of the function H in (1.3.41) is motivated from Medveď [26]. If $\int_{r_0}^{\infty} \frac{ds}{g\left(s^{\frac{1}{p}}\right)} = \infty$, then $H(\infty) = \infty$ and the inequality in (1.3.39) is true for $t \in R_+$. In the special case when $p = 2$, the inequality given in Theorem 1.3.4 reduces to a variant of the inequality given in [34, p. 233]. We also note that by following the proof of Theorem 1.2.1, one can very easily obtain the bounds on the inequalities (1.3.38) and (1.3.46) by replacing the constant c by a function $a(t)$ as in Theorem 1.2.1.

1.4 Inequalities with iterated integrals

Integral inequalities with iterated integrals play a very important role in the qualitative theory of differential and integral equations. In this section we offer some fundamental iterated integral inequalities established by Bykov and Salpagarov in [9] and Pachpatte in [53,78].

Our first theorem deals with the inequalities established by Pachpatte in [53].

Theorem 1.4.1. Let $u(t), f(t), a(t) \in C(R_+, R_+)$, $k(t,s), \frac{\partial}{\partial t}k(t,s), C(D, R_+)$ and $c \geq 0$ is a constant, where $D = \left\{(t,s) \in R_+^2 : 0 \leq s \leq t < \infty\right\}$.

(a_1) If

$$u(t) \leq c + \int_0^t f(s)\left[u(s) + \int_0^s k(s,\sigma)\,u(\sigma)\,d\sigma\right] ds, \tag{1.4.1}$$

for $t \in R_+$, then

$$u(t) \leq c\left[1 + \int_0^t f(s)\exp\left(\int_0^s [f(\sigma) + g(\sigma)]d\sigma\right) ds\right], \tag{1.4.2}$$

for $t \in R_+$, where

$$A(t) = k(t,t) + \int_0^t \frac{\partial}{\partial t}k(t,\tau)\,d\tau. \tag{1.4.3}$$

(a_2) If

$$u(t) \leq a(t) + \int_0^t f(s)\left[u(s) + \int_0^s k(s,\sigma)\,u(\sigma)\,d\sigma\right] ds, \tag{1.4.4}$$

for $t \in R_+$, then

$$u(t) \leq a(t) + e(t)\left[1 + \int_0^t f(s)\exp\left(\int_0^s [f(\sigma) + A(\sigma)]d\sigma\right)ds\right], \quad (1.4.5)$$

for $t \in R_+$, where

$$e(t) = \int_0^t f(s)\left[a(s) + \int_0^s k(s,\sigma)a(\sigma)\,d\sigma\right]ds, \quad (1.4.6)$$

and $A(t)$ is defined by (1.4.3).

Proof. (a_1) Define a function $z(t)$ by the right hand side of (1.4.1). Then $z(0) = c$, $u(t) \leq z(t)$ and

$$z'(t) = f(t)\left[u(t) + \int_0^t k(t,\sigma)u(\sigma)\,d\sigma\right]$$

$$\leq f(t)\left[z(t) + \int_0^t k(t,\sigma)z(\sigma)\,d\sigma\right]. \quad (1.4.7)$$

Define a function $v(t)$ by

$$v(t) = z(t) + \int_0^t k(t,\sigma)z(\sigma)\,d\sigma. \quad (1.4.8)$$

Then $v(0) = z(0) = c$, $z(t) \leq v(t)$, $z'(t) \leq f(t)v(t)$ and $v(t)$ is nondecreasing for $t \in R_+$ and

$$v'(t) = z'(t) + k(t,t)z(t) + \int_0^t \frac{\partial}{\partial t}k(t,\sigma)z(\sigma)\,d\sigma$$

$$\leq f(t)v(t) + k(t,t)v(t) + \int_0^t \frac{\partial}{\partial t}k(t,\sigma)v(\sigma)\,d\sigma$$

$$\leq \left[f(t) + k(t,t) + \int_0^t \frac{\partial}{\partial t}k(t,\sigma)d\sigma\right]v(t)$$

$$= [f(t) + A(t)]\,v(t),$$

implying

$$v(t) \le c \exp\left(\int_0^t [f(\sigma) + A(\sigma)]\, d\sigma\right). \tag{1.4.9}$$

Using (1.4.9) in (1.4.7) and integrating the resulting inequality from 0 to t, $t \in R_+$, we get

$$z(t) \le c\left[1 + \int_0^t f(s) \exp\left(\int_0^s [f(\sigma) + A(\sigma)] d\sigma\right) ds\right]. \tag{1.4.10}$$

The desired inequality in (1.4.2) follows by using (1.4.10) in $u(t) \le z(t)$.

(a_2) Define a function $z(t)$ by

$$z(t) = \int_0^t f(s)\left[u(s) + \int_0^s k(s,\sigma)\, u(\sigma)\, d\sigma\right] ds. \tag{1.4.11}$$

Then from (1.4.4), $u(t) \le a(t) + z(t)$ and using this in (1.4.11) we get

$$z(t) \le \int_0^t f(s)\left[a(s) + z(s) + \int_0^s k(s,\sigma)\,(a(\sigma) + z(\sigma))\, d\sigma\right] ds$$

$$= e(t) + \int_0^t f(s)\left[z(s) + \int_0^s k(s,\sigma)\, z(\sigma)\, d\sigma\right] ds, \tag{1.4.12}$$

where $e(t)$ is defined by (1.4.6). Clearly $e(t)$ is nonnegative, continuous and nondecreasing for $t \in R_+$. First we assume that $e(t) > 0$ for $t \in R_+$. From (1.4.12) it is easy to observe that

$$\frac{z(t)}{e(t)} \le 1 + \int_0^t f(s)\left[\frac{z(s)}{e(s)} + \int_0^s k(s,\sigma)\,\frac{z(\sigma)}{e(\sigma)} d\sigma\right] ds. \tag{1.4.13}$$

Now, an application of the inequality in (a_1) to (1.4.13) we have

$$\frac{z(t)}{e(t)} \le \left[1 + \int_0^t f(s) \exp\left(\int_0^s [f(\sigma) + A(\sigma)] d\sigma\right) ds\right]. \tag{1.4.14}$$

The desired inequality in (1.4.5) follows from (1.4.14) and the fact that $u(t) \le a(t) + z(t)$. If $e(t) \ge 0$, we carry out the above procedure with $e(t) + \varepsilon$ instead of $e(t)$, where $\varepsilon > 0$ is an arbitrary small constant, and subsequently pass to the limit as $\varepsilon \to 0$ to obtain (1.4.5).

Remark 1.4.1. We note that in the special case when $k(t,s)=k(s)$, the inequality given in (a_1) reduces to the inequality established earlier by Pachpatte, see [34, p. 33]. For a number of inequalities involving iterated integrals and their applications, see [3,34].

In [9] Bykov and Salpagarov proved the inequalities in the following theorem.

Theorem 1.4.2. Let $u(t)\in C(R_+,R_+)$, $k(t,s)\in C(D,R_+)$, $h(t,s,\sigma)\in C(E,R_+)$ and $c\geq 0$ be a constant, where $D=\{(t,s)\in R_+^2 : 0\leq s\leq t<\infty\}$, $E=\{(t,s,\sigma)\in R_+^3 : 0\leq \sigma\leq s\leq t<\infty\}$.

(b_1) Let $b(t)\in C(R_+,R_+)$. If

$$u(t)\leq c+\int_0^t b(s)u(s)\,ds+\int_0^t\left(\int_0^s k(s,\tau)u(\tau)\,d\tau\right)ds$$

$$+\int_0^t\left(\int_0^s\left(\int_0^\tau h(s,\tau,\sigma)u(\sigma)\,d\sigma\right)d\tau\right)ds, \tag{1.4.15}$$

for $t\in R_+$, then

$$u(t)\leq c\exp\left(\int_0^t B(s)\,ds\right), \tag{1.4.16}$$

for $t\in R_+$, where

$$B(t)=b(t)+\int_0^t k(t,\tau)d\tau+\int_0^t\left(\int_0^\tau h(t,\tau,\sigma)d\sigma\right)d\tau. \tag{1.4.17}$$

(b_2) Let $\frac{\partial}{\partial t}k(t,s)\in C(D,R_+)$, $\frac{\partial}{\partial t}h(t,s,\sigma)\in C(E,R_+)$. If

$$u(t)\leq c+\int_0^t k(t,\tau)u(\tau)\,d\tau+\int_0^t\left(\int_0^s h(t,s,\sigma)u(\sigma)\,d\sigma\right)ds, \tag{1.4.18}$$

for $t\in R_+$, then

$$u(t)\leq c\exp\left(\int_0^t [R(s)+Q(s)]\,ds\right), \tag{1.4.19}$$

for $t\in R_+$, where

$$R(t)=k(t,t)+\int_0^t h(t,t,\sigma)\,d\sigma, \tag{1.4.20}$$

$$Q(t) = \int_0^t \frac{\partial}{\partial t} k(t,\sigma) d\sigma + \int_0^t \left(\int_0^s \frac{\partial}{\partial t} h(t,s,\sigma) d\sigma \right) ds. \tag{1.4.21}$$

The proof of this theorem follows as a consequence of the following more general theorem proved by Pachpatte in [78].

Theorem 1.4.3. Let $u(t), k(t,s), h(t,s,\sigma)$ be as in Theorem 1.4.2 and $a(t)$, $a'(t) \in C(R_+, R_+)$. Let $g \in C(R_+, R_+)$ be a nondecreasing function, $g(u) > 0$ on $(0, \infty)$.

(c_1) Let $b(t) \in C(R_+, R_+)$. If

$$u(t) \le a(t) + \int_0^t b(s) g(u(s)) ds + \int_0^t \left(\int_0^s k(s,\tau) g(u(\tau)) d\tau \right) ds$$

$$+ \int_0^t \left(\int_0^s \left(\int_0^\tau h(s,\tau,\sigma) g(u(\sigma)) d\sigma \right) d\tau \right) ds, \tag{1.4.22}$$

for $t \in R_+$,then for $t \in R_+$,

$$u(t) \le G^{-1} \left[G(a(t)) + \int_0^t B(s) ds \right], \tag{1.4.23}$$

where

$$G(r) = \int_{r_0}^r \frac{ds}{g(s)}, r > 0, \tag{1.4.24}$$

$r_0 > 0$ is arbitrary, G^{-1} is the inverse of G, $B(t)$ is given by (1.4.17) and $t_1 \in R_+$ is chosen so that

$$G(a(t)) + \int_0^t B(s) ds \in Dom\left(G^{-1}\right),$$

for all $t \in R_+$ lying in the interval $0 \le t \le t_1$.

(c_2) Let $\frac{\partial}{\partial t} k(t,s), \frac{\partial}{\partial t} h(t,s,\sigma)$ be as in Theorem 1.4.2, part (b_2). If

$$u(t) \le a(t) + \int_0^t k(t,s) g(u(s)) ds + \int_0^t \left(\int_0^s h(t,s,\sigma) g(u(\sigma)) d\sigma \right) ds, \tag{1.4.25}$$

for $t \in R_+$, then for $0 \le t \le t_2; t, t_2 \in R_+$,

$$u(t) \le G^{-1}\left[G(a(t)) + \int_0^t [R(s) + Q(s)]\,ds\right], \tag{1.4.26}$$

where G, G^{-1} are as defined in part (c_2), $R(t), Q(t)$ are given by (1.4.20), (1.4.21) and $t_2 \in R_+$ is chosen so that

$$G(a(t)) + \int_0^t [R(s) + Q(s)]\,ds \in Dom\left(G^{-1}\right),$$

for all $t \in R_+$ lying in the interval $0 \le t \le t_2$.

Proof. First we note that, since $a'(t) \ge 0$, the function $a(t)$ is monotonically increasing.

(c_1) Let $a(t) > 0$ for $t \in R_+$ and define a function $z(t)$ by the right hand side of (1.4.22). Then $z(t) > 0, z(0) = a(0)$, $u(t) \le z(t)$, $a(t) \le z(t)$, $z(t)$ is nondecreasing for $t \in R_+$ and

$$z'(t) = a'(t) + b(t)\,g(u(t)) + \int_0^t k(t,\tau)\,g(u(\tau))d\tau$$

$$+ \int_0^t \left(\int_0^\tau h(t,\tau,\sigma)\,g(u(\sigma))d\sigma\right)d\tau$$

$$\le a'(t) + b(t)\,g(z(t)) + \int_0^t k(t,\tau)\,g(z(\tau))d\tau$$

$$+ \int_0^t \left(\int_0^\tau h(t,\tau,\sigma)\,g(z(\sigma))d\sigma\right)d\tau$$

$$\le a'(t) + B(t)\,g(z(t)).$$

Now by following the same arguments as in the proof of Theorem 1.2.1 below the inequality (1.2.5) we get the required inequality in (1.4.23).

(c_2) Let $a(t) > 0$ for $t \in R_+$ and define a function $z(t)$ by the right hand side of (1.4.25). Then $z(t) > 0, z(0) = a(0)$, $u(t) \le z(t)$, $a(t) \le z(t)$. In view of the hypotheses, it is easy to observe that $z(t)$ is nondecreasing and

$$z'(t) = a'(t) + k(t,t)\,g(u(t)) + \int_0^t \frac{\partial}{\partial t}k(t,s)\,g(u(s))\,ds$$

$$+\int_0^t h(t,t,\sigma)\, g(u(\sigma))d\sigma + \int_0^t \left(\int_0^t \frac{\partial}{\partial t} h(t,s,\sigma)\, g(u(\sigma))d\sigma\right) ds$$

$$\leq a'(t) + k(t,t)\, g(z(t)) + \int_0^t \frac{\partial}{\partial t} k(t,s)\, g(z(s))\, ds$$

$$+\int_0^t h(t,t,\sigma)\, g(z(\sigma))d\sigma + \int_0^t \left(\int_0^t \frac{\partial}{\partial t} h(t,s,\sigma)\, g(z(\sigma))d\sigma\right) ds$$

$$\leq a'(t) + [R(t) + Q(t)]\, g(z(t)).$$

The remaining proof can be completed by following the proof of Theorem 1.2.1.

Remark 1.4.2. As a consequence of Theorem 1.4.3, if we take $g(u) = u$, then $G(r) = \log \frac{r}{r_0}, G^{-1}(r) = r_0 \exp(r)$ and the bounds obtained in (1.4.23) and (1.4.26) reduces respectively to

$$u(t) \leq a(t) \exp\left(\int_0^t B(s)\, ds\right), \tag{1.4.27}$$

and

$$u(t) \leq a(t) \exp\left(\int_0^t [R(s) + Q(s)]\, ds\right), \tag{1.4.28}$$

for $t \in R_+$. Furthermore, if we take $a(t) = c$, a nonnegative constant, then we get the inequalities in Theorem 1.4.2 established by Bykov and Salpagarov in [9].

Before giving the next result, we introduce some notations to simplify the details of presentation. Let $I = [0, \alpha)$ be the given subset of R and for $i = 1, ..., n$, let $I_i = \{(t_1, ..., t_i) : (t_1, ..., t_i) \in I^i\}$. For $i = 1, ..., n$ and any functions $w(t), a(t), b(t) \in C(I, R_+)$, $L_i(t_1, ..., t_i, w(t_i)), M_i(t_1, ..., t_i, a(t_i)) \in C(I_i \times R_+, R_+)$ and $t \in I$ we set

$$F_i[w](t) = \int_0^t \left(\int_0^{t_1} \cdots \left(\int_0^{t_{i-1}} L_i(t_1, ..., t_i, w(t_i))\, dt_i\right) \cdots\right) dt_1,$$

$$E(t) = L_1(t, a(t)) + \int_0^t L_2(t, t_2, a(t_2))\, dt_2 + \ldots$$

$$+\int_0^t\left(\int_0^{t_2}\cdots\left(\int_0^{t_{n-1}}L_n\left(t,t_2,...,t_n,a\left(t_n\right)\right)dt_n\right)dt_{n-1}\cdots\right)dt_2,$$

$$H\left(t\right)=M_1\left(t,a\left(t\right)\right)b\left(t\right)+\int_0^t M_2\left(t,t_2,a\left(t_2\right)\right)b\left(t_2\right)dt_2+...$$

$$+\int_0^t\left(\int_0^{t_2}\cdots\left(\int_0^{t_{n-1}}M_n\left(t,t_2,...,t_n,a\left(t_n\right)\right)b\left(t_n\right)dt_n\right)dt_{n-1}\cdots\right)dt_2.$$

The following theorem deals with the inequalities established by Pachpatte in [78].

Theorem 1.4.4. Let $u\left(t\right),a\left(t\right),b\left(t\right)\in C\left(I,R_+\right).$

(d_1) For $i=1,...,n$, let the functions $L_i\in C\left(I_i\times R_+,R_+\right)$ satisfy the conditions

$$0\leq L_i\left(t_1,...,t_i,x\left(t_i\right)\right)-L_i\left(t_1,...,t_i,y\left(t_i\right)\right)$$
$$\leq M_i\left(t_1,...,t_i,y\left(t_i\right)\right)\left(x\left(t_i\right)-y\left(t_i\right)\right), \tag{1.4.29}$$

for $\left(t_1,...,t_i\right)\in I_i$ and $x\left(t_i\right)\geq y\left(t_i\right)\geq 0$, where $M_i\in C\left(I_i\times R_+,R_+\right)$. If

$$u\left(t\right)\leq a\left(t\right)+b\left(t\right)\sum_{i=1}^n F_i\left[u\right]\left(t\right), \tag{1.4.30}$$

for $t\in I$, then

$$u\left(t\right)\leq a\left(t\right)+b\left(t\right)\int_0^t E\left(t_1\right)\exp\left(\int_{t_1}^t H\left(\sigma\right)d\sigma\right)dt_1, \tag{1.4.31}$$

for $t\in I$.

(d_2) Let $\psi\in C\left(R_+,R_+\right)$ be a strictly increasing function with $\psi\left(0\right)=0$. For $i=1,...,n$ let the functions $L_i\in C\left(I_i\times R_+,R_+\right)$ satisfy the conditions

$$0\leq L_i\left(t_1,...,t_i,x\left(t_i\right)\right)-L_i\left(t_1,...,t_i,y\left(t_i\right)\right)$$
$$\leq M_i\left(t_1,...,t_i,y\left(t_i\right)\right)\psi^{-1}\left(x\left(t_i\right)-y\left(t_i\right)\right), \tag{1.4.32}$$

for $\left(t_1,...,t_i\right)\in I_i$ and $x\left(t_i\right)\geq y\left(t_i\right)\geq 0$,where $M_i\in C\left(I_i\times R_+,R_+\right)$ and ψ^{-1} is the inverse function of ψ. If

$$u\left(t\right)\leq a\left(t\right)+\psi\left(b\left(t\right)\sum_{i=1}^n F_i\left[u\right]\left(t\right)\right), \tag{1.4.33}$$

for $t \in I$, then

$$u(t) \leq a(t) + \psi\left(b(t)\int_0^t E(t_1)\exp\left(\int_{t_1}^t H(\sigma)\,d\sigma\right)dt_1\right), \tag{1.4.34}$$

for $t \in I$.

(d_3) Let $L_i, M_i, \psi, \psi^{-1}$ be as in part (d_2)and the conditions in (1.4.32) hold. Suppose in addition that

$$\psi^{-1}(xy) \leq \psi^{-1}(x)\,\psi^{-1}(y), \tag{1.4.35}$$

for all $x, y \in R_+$. If

$$u(t) \leq a(t) + b(t)\,\psi\left(\sum_{i=1}^{n} F_i[u](t)\right), \tag{1.4.36}$$

for $t \in I$, then

$$u(t) \leq a(t) + b(t)\,\psi\left(\int_0^t E(t_1)\exp\left(\int_{t_1}^t H_1(\sigma)\,d\sigma\right)dt_1\right), \tag{1.4.37}$$

for $t \in I$, where $H_1(t)$ is obtained by replacing b by $\psi^{-1}(b)$ on the right hand side of the definition of $H(t)$.

(d_4) For $i = 1, ..., n$, let L_i, M_i be as in part (d_1) and the conditions (1.4.29) hold. Let $g \in C(R_+, R_+)$ be a nondecreasing function with $g(u) > 0$ for $u > 0$. If

$$u(t) \leq a(t) + b(t)\,g\left(\sum_{i=1}^{n} F_i[u](t)\right), \tag{1.4.38}$$

for $t \in I$, then for $0 \leq t \leq \bar{t}; t, \bar{t} \in I$,

$$u(t) \leq a(t) + b(t)\,g\left(G^{-1}\left[G\left(\bar{E}(t)\right) + \int_0^t H(t_1)\,dt_1\right]\right), \tag{1.4.39}$$

where

$$\bar{E}(t) = \int_0^t E(t_1)\,dt_1 \tag{1.4.40}$$

G, G^{-1} are as defined in Theorem 1.4.3, part (c_1) and $\bar{t} \in I$ is chosen so that

$$G\left(\bar{E}(t)\right) + \int_0^t H(t_1)\,dt_1 \in Dom\left(G^{-1}\right),$$

for all $t \in I$ lying in the interval $0 \leq t \leq \bar{t}$.

Proof. (d_1) Define a function $z(t)$ by

$$z(t) = \sum_{i=1}^{n} F_i[u](t)$$

$$= \int_0^t L_1(t_1, u(t_1))\, dt_1 + \int_0^t \left(\int_0^{t_1} L_2(t_1, t_2, u(t_2))\, dt_2 \right) dt_1 + \ldots$$

$$+ \int_0^t \left(\int_0^{t_1} \ldots \left(\int_0^{t_{n-1}} L_n(t_1, \ldots t_n, u(t_n))\, dt_n \right) dt_{n-1} \ldots \right) dt_1. \quad (1.4.41)$$

Then $z(0) = 0$, $z(t)$ is nondecreasing for $t \in I$ and (1.4.30) can be restated as

$$u(t) \leq a(t) + b(t) z(t). \quad (1.4.42)$$

From (1.4.41), (1.4.42) and the hypotheses we observe that

$$z'(t) = L_1(t, u(t)) + \int_0^t L_2(t, t_2, u(t_2)) dt_2 + \ldots$$

$$+ \int_0^t \left(\int_0^{t_2} \ldots \left(\int_0^{t_{n-1}} L_n(t, t_2, \ldots t_n, u(t_n))\, dt_n \right) dt_{n-1} \ldots \right) dt_2.$$

$$\leq \{L_1(t, a(t) + b(t) z(t)) - L_1(t, a(t))\} + L_1(t, a(t))$$

$$+ \int_0^t [\{L_2(t, t_2, a(t_2) + b(t_2) z(t_2)) - L_2(t, t_2, a(t_2))\}$$

$$+ L_2(t, t_2, a(t_2))]\, dt_2 + \ldots$$

$$+ \int_0^t \left(\int_0^{t_2} \ldots \left(\int_0^{t_{n-1}} [\{L_n(t, t_2, \ldots t_n, a(t_n) + b(t_n) z(t_n)) \right.\right.$$

$$\left.\left. - L_n(t, t_2, \ldots t_n, a(t_n))\} + L_n(t, t_2, \ldots t_n, a(t_n))]\, dt_n \right) dt_{n-1} \ldots \right) dt_2$$

$$\leq E(t) + M_1(t, a(t)) b(t) z(t)$$

$$+ \int_0^t M_2(t, t_2, a(t_2)) b(t_2) z(t_2)\, dt_2 + \ldots$$

$$+ \int_0^t \left(\int_0^{t_2} \ldots \left(\int_0^{t_{n-1}} M_n(t, t_2, \ldots t_n, a(t_n)) b(t_n) z(t_n) dt_n \right) dt_{n-1} \ldots \right) dt_2$$

$$\leq E(t) + H(t)\, z(t). \tag{1.4.43}$$

The inequality (1.4.43) yields

$$z(t) \leq \int_0^t E(t_1) \exp\left(\int_{t_1}^t H(s)\, ds\right) dt_1. \tag{1.4.44}$$

The desired inequality in (1.4.31) follows from (1.4.42) and (1.4.44).

(d_2) Define a function $z(t)$ by (1.4.41).Then $z(0) = 0, z(t)$ is nondecreasing for $t \in I$ and (1.4.33) can be restated as

$$u(t) \leq a(t) + \psi(b(t)\, z(t)). \tag{1.4.45}$$

By following a similar argument as in the proof of part (d_1) with suitable changes, see also [12,34] we obtain (1.4.44).Using (1.4.44) in (1.4.45) we get the required inequality in (1.4.34).

(d_3) Define a function $z(t)$ by (1.4.41). Then $z(0) = 0, z(t)$ is nondecreasing for $t \in I$ and (1.4.36) can be restated as

$$u(t) \leq a(t) + b(t)\, \psi(z(t)). \tag{1.4.46}$$

Now by following a similar argument as in the proof of part (d_1) with suitable modifications, we obtain

$$z(t) \leq \int_0^t E(t_1) \exp\left(\int_{t_1}^t H_1(s)\, ds\right) dt_1. \tag{1.4.47}$$

Using (1.4.47) in (1.4.46) we get (1.4.37).

(d_4) Define a function $z(t)$ by (1.4.41). Then $z(0) = 0, z(t)$ is nondecreasing for $t \in I$ and (1.4.38) can be restated as

$$u(t) \leq a(t) + b(t)\, g(z(t)). \tag{1.4.48}$$

From (1.4.41), (1.4.48), (1.4.29) and following the proof of part (d_1) we get

$$z'(t) \leq E(t) + H(t)\, g(z(t)),$$

which yields

$$z(t) \leq \bar{E}(t) + \int_0^t H(t_1)\, g(z(t_1))\, dt_1. \tag{1.4.49}$$

By following the same arguments as in the proof of Theorem 2.4.2 given in [34] we get

$$z(t) \leq G^{-1}\left[G\left(\bar{E}(t)\right) + \int_0^t H(t_1)\,dt_1\right]. \tag{1.4.50}$$

Using (1.4.50) in (1.4.48) we get (1.4.39). The subinterval $0 \leq t \leq \bar{t}$ is obvious.

Remark 1.4.3. If we take $L_1 = L, L_i = 0$ for $i = 1, ..., n$ and the interval $I = [\alpha, \beta)$ in Theorem 1.4.4, then we recapture the inequalities in Lemma 74, Theorem 81, Theorem 85, Theorem 91 given in [12] respectively. Here it is to be noted that, one can very easily obtain from Theorem 1.4.4 the corollaries similar to those of various corollaries of the corresponding results given in [12] which can be used in certain applications. We also note that, in view of the results given in Theorem 1.3.2, the inequalities in Theorem 1.4.4 can be extended when the function $u(t)$ on the left sides in (1.4.30), (1.4.33), (1.4.36), (1.4.38) is replaced by $u^p(t)$, where $p > 1$ is a real constant.

1.5 Bounds on certain integral inequalities

The classical integral inequalities which give explicit bounds for an unknown function have played a fundamental role in establishing the foundations of the theory of differential and integral equations. In this section we shall give explicit bounds on certain integral inequalities which will be equally important to achieve a diversity of desired goals in some applications. In what follows, $I = [\alpha, \beta]$ is a given subset of R and $D = \{(t, s) \in I^2 : \alpha \leq s \leq t \leq \beta\}$.

The following three theorems give the inequalities established by Pachpatte in [52,54,70,75].

Theorem 1.5.1. Let $u(t), a(t), b(t), f(t), g(t) \in C(I, R_+)$.

(a_1) Let $a(t)$ be continuously differentiable on I, $a'(t) \geq 0$ and

$$u(t) \leq a(t) + \int_\alpha^t b(s)\,u(s)\,ds + \int_\alpha^\beta c(s)\,u(s)\,ds, \tag{1.5.1}$$

for $t \in I$. If

$$p_1 = \int_\alpha^\beta c(s)\exp\left(\int_\alpha^s b(\sigma)\,d\sigma\right)ds < 1, \tag{1.5.2}$$

then

$$u(t) \leq M_1 \exp\left(\int_\alpha^t b(s)\,ds\right) + \int_\alpha^t a'(s) \exp\left(\int_s^t b(\sigma)\,d\sigma\right) ds, \tag{1.5.3}$$

for $t \in I$, where

$$M_1 = \frac{1}{1-p_1}\left[a(\alpha) + \int_\alpha^\beta c(s)\left(\int_\alpha^s a'(\tau)\exp\left(\int_\tau^s b(\sigma)\,d\sigma\right) d\tau\right) ds\right]. \tag{1.5.4}$$

(a_2) Suppose that

$$u(t) \leq a(t) + b(t)\int_\alpha^t f(s)\,u(s)\,ds + c(t)\int_\alpha^\beta g(s)\,u(s)\,ds. \tag{1.5.5}$$

for $t \in I$. If

$$p_2 = \int_\alpha^\beta g(s)\,K_2(s) ds < 1, \tag{1.5.6}$$

then

$$u(t) \leq K_1(t) + M_2 K_2(t), \tag{1.5.7}$$

for $t \in I$, where

$$K_1(t) = a(t) + b(t)\int_\alpha^t f(\tau) a(\tau) \exp\left(\int_\tau^t f(\sigma)\,b(\sigma)\,d\sigma\right) d\tau, \tag{1.5.8}$$

$$K_2(t) = c(t) + b(t)\int_\alpha^t f(\tau) c(\tau) \exp\left(\int_\tau^t f(\sigma)\,b(\sigma)\,d\sigma\right) d\tau, \tag{1.5.9}$$

and

$$M_2 = \frac{1}{1-p_2}\int_\alpha^\beta g(s)\,K_1(s)\,ds. \tag{1.5.10}$$

Proof. (a_1) Define a function $z(t)$ by the right hand side of (1.5.1). Then $u(t) \leq z(t)$,

$$z(\alpha) = a(\alpha) + \int_\alpha^\beta c(s)\,u(s)\,ds, \tag{1.5.11}$$

and

$$z'(t) = a'(t) + b(t)\,u(t) \le a'(t) + b(t)\,z(t),$$

which implies

$$u(t) \le z(t) \le z(\alpha)\exp\left(\int_\alpha^t b(\sigma)\,d\sigma\right) + \int_\alpha^t a'(s)\exp\left(\int_s^t b(\sigma)\,d\sigma\right)ds. \quad (1.5.12)$$

From (1.5.11) and (1.5.12)we have

$$z(\alpha) \le a(\alpha) + \int_\alpha^\beta c(s)\left\{z(\alpha)\exp\left(\int_\alpha^s b(\sigma)\,d\sigma\right) + \int_\alpha^s a'(\tau)\exp\left(\int_\tau^s b(\sigma)d\sigma\right)d\tau\right\}ds,$$

i.e.,

$$z(\alpha)\left\{1 - \int_\alpha^\beta c(s)\exp\left(\int_\alpha^s b(\sigma)d\sigma\right)ds\right\}$$

$$\le a(\alpha) + \int_\alpha^\beta c(s)\left(\int_\alpha^s a'(\tau)\exp\left(\int_\tau^s b(\sigma)\,d\sigma\right)d\tau\right)ds,$$

which implies

$$z(\alpha) \le M_1. \quad (1.5.13)$$

Using (1.5.13) in (1.5.12) we get the desired inequality in (1.5.3).

(a_2) Let

$$z(t) = \int_\alpha^t f(s)\,u(s)\,ds, \quad (1.5.14)$$

$$\lambda = \int_\alpha^\beta g(s)\,u(s)\,ds. \quad (1.5.15)$$

Then $z(0) = 0$, (1.5.5) can be restated as

$$u(t) \le a(t) + b(t)\,z(t) + c(t)\,\lambda, \quad (1.5.16)$$

and

$$z'(t) = f(t)\,u(t). \quad (1.5.17)$$

From (1.5.16) and (1.5.17) we have

$$z'(t) \le \{f(t)\,a(t) + \lambda f(t)\,c(t)\} + f(t)\,b(t)\,z(t),$$

which implies

$$z(t) \le \int_{\alpha}^{t} \{f(\tau)\,a(\tau) + \lambda f(\tau)\,c(\tau)\} \exp\left(\int_{\tau}^{t} f(\sigma)\,b(\sigma)\,d\sigma\right) d\tau. \qquad (1.5.18)$$

Using (1.5.18) in (1.5.16) we get

$$u(t) \le \{a(t) + \lambda c(t)\} + b(t) \int_{\alpha}^{t} \{f(\tau)\,a(\tau) + \lambda f(\tau)\,c(\tau)\}$$

$$\times \exp\left(\int_{\tau}^{t} f(\sigma)\,b(\sigma)\,d\sigma\right) d\tau$$

$$= K_1(t) + \lambda K_2(t). \qquad (1.5.19)$$

From (1.5.15) and (1.5.19) as in the proof of (a_1) it is easy to observe that

$$\lambda \le M_2. \qquad (1.5.20)$$

Using (1.5.20) in (1.5.19) we get 1.5.7).

Remark 1.5.1. If we take $a(t) = d$ (a constant) and hence $a'(t) = 0$, then the inequality given in (a_1) reduces to the special version of inequality given by Bainov and Simeonov in [3, p. 11] in case $u(t)$ and d therein are nonnegative. The inequality in (a_2) is a variant of the inequality given by Gamidov in [15, Lemma 1.2].

Theorem 1.5.2. Let $u(t), a(t), c(t) \in C(I, R_+)$

(b_1) Let $\frac{\partial}{\partial t} h(t,s) \in C(D, R_+)$ and

$$u(t) \le a(t) + \int_{\alpha}^{t} h(t,s)\,u(s)\,ds + \int_{\alpha}^{\beta} c(s)\,u(s)\,ds, \qquad (1.5.21)$$

for $t \in I$. If

$$p_3 = \int_{\alpha}^{\beta} c(s) \exp\left(\int_{\alpha}^{s} B(\sigma)\,d\sigma\right) ds < 1, \qquad (1.5.22)$$

then

$$u(t) \leq a(t) + M_3 \exp\left(\int_\alpha^t B(\sigma)\, d\sigma\right) + \int_\alpha^t A(s) \exp\left(\int_s^t B(\sigma)\, d\sigma\right) ds, \quad (1.5.23)$$

for $t \in I$, where

$$A(t) = h(t,t)\, a(t) + \int_\alpha^t \frac{\partial}{\partial t} h(t,s)\, a(s)\, ds, \quad (1.5.24)$$

$$B(t) = h(t,t) + \int_\alpha^t \frac{\partial}{\partial t} h(t,s)\, ds, \quad (1.5.25)$$

and

$$M_3 = \frac{1}{1-p_3} \int_\alpha^\beta c(s) \left[a(s) + \int_\alpha^s A(\tau) \exp\left(\int_\tau^s B(\sigma)\, d\sigma\right) d\tau\right] ds. \quad (1.5.26)$$

(b_2) Let $h(t,s), g(t,s) \in C(D, R_+)$ and be nondecreasing in $t \in I$, for each $s \in I$ and

$$u(t) \leq k + \int_\alpha^t h(t,s)\, u(s)\, ds + \int_\alpha^\beta g(t,s)\, u(s)\, ds, \quad (1.5.27)$$

for $t \in I$, where $k \geq 0$ is a constant. If

$$p(t) = \int_\alpha^\beta g(t,s) \exp\left(\int_\alpha^s h(s,\sigma)\, d\sigma\right) ds < 1, \quad (1.5.28)$$

for $t \in I$, then

$$u(t) \leq \frac{k}{1-p(t)} \exp\left(\int_\alpha^t h(t,s)\, ds\right), \quad (1.5.29)$$

for $t \in I$.

Proof. Define a function $z(t)$ by

$$z(t) = \int_\alpha^t h(t,s)\, u(s)\, ds + \int_\alpha^\beta c(s)\, u(s)\, ds. \quad (1.5.30)$$

Then $z(t)$ is nondecreasing for $t \in I$, (1.5.21) can be restated as

$$u(t) \leq a(t) + z(t), \tag{1.5.31}$$

$$z(\alpha) = \int_{\alpha}^{\beta} c(s)\, u(s)\, ds, \tag{1.5.32}$$

and

$$z'(t) = h(t,t)\, u(t) + \int_{0}^{t} \frac{\partial}{\partial t} h(t,s)\, u(s)\, ds$$

$$\leq h(t,t)\{a(t) + z(t)\} + \int_{0}^{t} \frac{\partial}{\partial t} h(t,s)\{a(s) + z(s)\}\, ds$$

$$\leq A(t) + B(t)\, z(t),$$

which implies

$$z(t) \leq z(\alpha) \exp\left(\int_{\alpha}^{t} B(\sigma)\, d\sigma\right) + \int_{\alpha}^{t} A(s) \exp\left(\int_{s}^{t} B(\sigma)\, d\sigma\right) ds. \tag{1.5.33}$$

The rest of the proof can be completed by following the proof of Theorem 1.5.1.

(b_2) Fix any T, $\alpha \leq T \leq \beta$, then for $\alpha \leq t \leq T$ we have

$$u(t) \leq k + \int_{\alpha}^{t} h(T,s)\, u(s)\, ds + \int_{\alpha}^{\beta} g(T,s)\, u(s)\, ds. \tag{1.5.34}$$

Define a function $z(t,T)$, $\alpha \leq t \leq T$ by the right hand side of (1.5.34). Then $u(t) \leq z(t,T), \alpha \leq t \leq T$,

$$z(\alpha, T) = k + \int_{\alpha}^{\beta} g(T,s)\, u(s)\, ds, \tag{1.5.35}$$

and

$$D_1 z(t,T) = h(T,t)\, u(t) \leq h(T,t)\, z(t), \tag{1.5.36}$$

for $\alpha \leq T$. By setting $t = \sigma$ in (1.5.36) and integrating it with respect to σ from α to T we get

$$z(T,T) \leq z(\alpha, T) \exp\left(\int_{\alpha}^{T} h(T,\sigma)\, d\sigma\right). \tag{1.5.37}$$

Since T is arbitrary,from (1.5.37) and (1.5.35) with T replaced by t and $u(t) \leq z(t,t)$ we have

$$u(t) \leq z(\alpha,t)\exp\left(\int_\alpha^t h(t,\sigma)\,d\sigma\right), \tag{1.5.38}$$

where

$$z(\alpha,t) = k + \int_\alpha^\beta g(t,s)\,u(s)\,ds. \tag{1.5.39}$$

Using (1.5.38) on the right hand side of (1.5.39) and in view of the condition (1.5.28) it is easy to observe that

$$z(\alpha,t) \leq \frac{k}{1-p(t)}. \tag{1.5.40}$$

Using (1.5.40) in (1.5.38) we get the desired inequality in (1.5.29).

Remark 1.5.2. In the special case when $c(t) = 0$, the inequality in (b_1) reduces to the inequality given in [3, Theorem 1.8, p. 11]. The inequality in (b_2) is a useful variant of the inequality given in [3, Theorem 1.7, p.11].

Theorem 1.5.3. Let $u(t) \in C(I,R_+)$ and $k \geq 0$ be a real constant.

(c_1) Let $a(t,s), b(t,s), c(t,s) \in C(D,R_+)$; $a(t,s), b(t,s)$ are nondecreasing in t for each $s \in I$ and

$$u(t) \leq k+\int_\alpha^t a(t,s)\left[u(s)+\int_\alpha^s c(s,\sigma)\,u(\sigma)\,d\sigma\right]ds+\int_\alpha^\beta b(t,s)\,u(s)\,ds \tag{1.5.41}$$

for $t \in I$. If

$$q(t) = \int_\alpha^\beta b(t,s)\exp\left(\int_\alpha^s E(s,\xi)\,d\xi\right)ds < 1, \tag{1.5.42}$$

for $t \in I$, where

$$E(t,\xi) = a(t,\xi)\left[1+\int_\alpha^\xi c(\xi,\sigma)\,d\sigma\right], \tag{1.5.43}$$

for $(t,\xi) \in D$, then

$$u(t) \leq \frac{k}{1-q(t)}\exp\left(\int_\alpha^t E(t,\xi)\,d\xi\right), \tag{1.5.44}$$

for $t \in I$.

(c_2) Let $f(t), g(t), h(t) \in C(I, R_+)$ and

$$u(t) \leq k + \int_\alpha^t f(s)\left[u(s) + \int_\alpha^s g(\sigma)u(\sigma)\,d\sigma + \int_\alpha^\beta h(\sigma)u(\sigma)\,d\sigma\right]ds, \quad (1.5.45)$$

for $t \in I$. If

$$r = \int_\alpha^\beta h(\sigma)\exp\left(\int_\alpha^s [f(\tau) + g(\tau)]d\tau\right)d\sigma < 1, \quad (1.5.46)$$

then

$$u(t) \leq \frac{k}{1-r}\exp\left(\int_\alpha^t [f(s) + g(s)]\,ds\right), \quad (1.5.47)$$

for $t \in I$.

Proof. (c_1) Let $k > 0$ and fix any $T \in I$, then for $\alpha \leq t \leq T$, from (1.5.41) we have

$$u(t) \leq k + \int_\alpha^t a(T,s)\left[u(s) + \int_\alpha^s c(s,\sigma)u(\sigma)\,d\sigma\right]ds$$

$$+ \int_\alpha^\beta b(T,s)u(s)\,ds. \quad (1.5.48)$$

Define a function $z(t,T)$, $t \in [\alpha, T]$ by the right hand side of (1.5.48). Then for $t \in [\alpha, T]$, $u(t) \leq z(t,T)$, $z(t,T) > 0$,

$$z(\alpha, T) = k + \int_\alpha^\beta b(T,s)u(s)\,ds, \quad (1.5.49)$$

and

$$D_1 z(t,T) = a(T,t)\left[u(t) + \int_\alpha^t c(t,\sigma)u(\sigma)\,d\sigma\right]$$

$$\leq a(T,t)\left[z(t) + \int_\alpha^t c(t,\sigma)z(\sigma,T)\,d\sigma\right]. \quad (1.5.50)$$

From (1.5.50) and using the fact that $z(t,T)$ is nondecreasing in t, it is easy to observe that

$$\frac{D_1 z(t,T)}{z(t,T)} \le a(T,t)\left[1+\int_{\alpha}^{t} c(t,\sigma)\,d\sigma\right], \tag{1.5.51}$$

for $t \in [\alpha, T]$. By setting $t=\xi$ in (1.5.51) and integrating it with respect to ξ from α to T we get

$$z(T,T) \le z(\alpha,T)\exp\left(\int_{\alpha}^{T} a(T,\xi)\left[1+\int_{\alpha}^{\xi} c(\xi,\sigma)\,d\sigma\right]d\xi\right). \tag{1.5.52}$$

Since T is arbitrary, from (1.5.52), (1.5.49) with T replaced by t we have for $t \in I$,

$$z(t,t) \le z(\alpha,t)\exp\left(\int_{\alpha}^{t} E(t,\xi)\,d\xi\right), \tag{1.5.53}$$

$$z(\alpha,t) = k + \int_{\alpha}^{\beta} b(t,s)\,u(s)\,ds. \tag{1.5.54}$$

Using (1.5.53) in $u(t) \le z(t)$ we get

$$u(t) \le z(\alpha,t)\exp\left(\int_{\alpha}^{t} E(t,\xi)\,d\xi\right), \tag{1.5.55}$$

for $t \in I$. Using (1.5.55) on the right hand side of (1.5.54) and in view of (1.5.42), it is easy to observe that

$$z(\alpha,t) \le \frac{k}{1-q(t)}. \tag{1.5.56}$$

The required inequality in (1.5.44) follows by using (1.5.56) in (1.5.55). If $k \ge 0$, we carry out the above procedure with $k+\varepsilon$ instead of k where $\varepsilon > 0$ is an arbitrary small constant, and subsequently pass to the limit as $\varepsilon \to 0$ to obtain (1.5.44).

(c_2) Define a function $z(t)$ by the right hand side of (1.5.45). Then $z(0)=0$, $u(t) \le z(t)$ and

$$z'(t) = f(t)\left[u(t) + \int_{\alpha}^{t} g(\sigma)\,u(\sigma)\,d\sigma + \int_{\alpha}^{\beta} h(\sigma)\,u(\sigma)\,d\sigma\right]$$

$$\leq f(t)\left[z(t)+\int_{\alpha}^{t} g(\sigma) z(\sigma) d\sigma+\int_{\alpha}^{\beta} h(\sigma) z(\sigma) d\sigma\right], \tag{1.5.57}$$

for $t \in I$. Define a function $v(t)$ by

$$v(t)=z(t)+\int_{\alpha}^{t} g(\sigma) z(\sigma) d\sigma+\int_{\alpha}^{\beta} h(\sigma) z(\sigma) d\sigma, \tag{1.5.58}$$

then $z(t) \leq v(t)$, $z'(t) \leq f(t) v(t)$,

$$v(\alpha)=k+\int_{\alpha}^{\beta} h(\sigma) z(\sigma) d\sigma,. \qquad (1.5.59))$$

and

$$v'(t)=z'(t)+g(t) z(t) \leq f(t) v(t)+g(t) z(t) \leq [f(t)+g(t)] v(t),$$

which implies

$$v(t) \leq v(\alpha) \exp\left(\int_{\alpha}^{t}[f(s)+g(s)] ds\right), \tag{1.5.60}$$

for $t \in I$. Using (1.5.60) in $z(t) \leq v(t)$ we get

$$z(t) \leq v(\alpha) \exp\left(\int_{\alpha}^{t}[f(s)+g(s)] ds\right), \tag{1.5.61}$$

for $t \in I$. Using (1.5.61) on the right hand side of (1.5.59) and in view of (1.5.46) it is easy to observe that

$$v(\alpha) \leq \frac{k}{1-r}. \tag{1.5.62}$$

Using (1.5.62) in (1.5.61) and the fact that $u(t) \leq z(t)$ we get the desired inequality in (1.5.47).

In the following theorem we present the inequalities established in [51] (see also [44]).

Theorem 1.5.4. Let $u(t), a(t), b(t) \in C(R_+, R_+)$.

(d_1) Let $a(t)$ be nonincreasing for $t \in R_+$. If

$$u(t) \leq a(t) + \int_t^\infty b(s) u(s) ds, \tag{1.5.63}$$

for $t \in R_+$, then

$$u(t) \leq a(t) \exp\left(\int_t^\infty b(s) ds\right), \tag{1.5.64}$$

for $t \in R_+$.

(d_2) Let $L \in C\left(R_+^2, R_+\right)$ and

$$0 \leq L(t, u) - L(t, v) \leq M(t, v)(u - v), \tag{1.5.65}$$

for $u \geq v \geq 0$, where $M \in C\left(R_+^2, R_+\right)$. If

$$u(t) \leq a(t) + \int_t^\infty b(s) u(s) ds + \int_t^\infty L(s, u(s)) ds, \tag{1.5.66}$$

for $t \in R_+$, then

$$u(t) \leq F(t)\left[a(t) + G(t) \exp\left(\int_t^\infty M(s, F(s) a(s)) F(s) ds\right)\right], \tag{1.5.67}$$

for $t \in R_+$, where

$$F(t) = \exp\left(\int_t^\infty b(s) ds\right), \tag{1.5.68}$$

$$G(t) = \int_t^\infty L(s, F(s) a(s)) ds, \tag{1.5.69}$$

for $t \in R_+$.

(d_3) Let L and M be as in (d_2). If

$$u(t) \leq a(t) + b(t) \int_t^\infty L(s, u(s)) ds, \tag{1.5.70}$$

for $t \in R_+$, then

$$u(t) \leq a(t) + b(t)\, e(t) \exp\left(\int_t^\infty M(s, a(s))\, b(s)\, ds \right), \tag{1.5.71}$$

for $t \in R_+$, where

$$e(t) = \int_t^\infty L(s, a(s))\, ds, \tag{1.5.72}$$

for $t \in R_+$.

Proof. (d_1) First we assume that $a(t) > 0$ for $t \in R_+$. From (1.5.63) it is easy to observe that

$$\frac{u(t)}{a(t)} \leq 1 + \int_t^\infty b(s) \frac{u(s)}{a(s)} ds. \tag{1.5.73}$$

Define a function $z(t)$ by the right hand side of (1.5.73), then $z(\infty) = 1$, $\frac{u(t)}{a(t)} \leq z(t)$ and

$$z'(t) = -b(t) \frac{u(t)}{a(t)} \geq -b(t)\, z(t). \tag{1.5.74}$$

The inequality (1.5.74) implies the estimate

$$z(t) \leq \exp\left(\int_t^\infty b(s)\, ds \right). \tag{1.5.75}$$

Using (1.5.75) in $\frac{u(t)}{a(t)} \leq z(t)$, we get the desired inequality in (1.5.64).

If $a(t)$ is nonnegative, we carry out the above procedure with $a(t) + \varepsilon$ instead of $a(t)$, where $\varepsilon > 0$ is an arbitrary small constant, and subsequently pass to the limit as $\varepsilon \to 0$ to obtain (1.5.64).

(d_2) Define a function $z(t)$ by

$$z(t) = \int_t^\infty L(s, u(s))\, ds, \tag{1.5.76}$$

then (1.5.66) can be restated as

$$u(t) \leq a(t) + z(t) + \int_t^\infty b(s)\, u(s)\, ds. \tag{1.5.77}$$

Since $a(t)+z(t)$ is nonnegative, continuous and nonincreasing for $t \in R_+$, by applying the inequality in part (d_1) to (1.5.77) we have

$$u(t) \le (a(t) + z(t))\, F(t). \tag{1.5.78}$$

From (1.5.76) and (1.5.78) and the hypotheses on L, we observe that

$$z(t) \le \int_t^\infty [L(s, F(s)\, a(s) + F(s)\, z(s)) - L(s, F(s)\, a(s)) + L(s, F(s)\, a(s))]\, ds$$

$$\le G(t) + \int_t^\infty M(s, F(s)\, a(s))\, F(s)\, z(s)\, ds. \tag{1.5.79}$$

Clearly, $G(t)$ is nonnegative, continuous and nonincreasing for $t \in R_+$. Now an application of the inequality in part (d_1) to (1.5.79) yields

$$z(t) \le G(t) \exp\left(\int_t^\infty M(s, F(s)\, a(s))\, F(s)\, ds\right). \tag{1.5.80}$$

Using (1.5.80) in (1.5.78) we get the required inequality in (1.5.67).

(d_3) Define a function $z(t)$ by (1.5.76). Then from (1.5.70) we have

$$u(t) \le a(t) + b(t)\, z(t). \tag{1.5.81}$$

From (1.5.76), (1.5.81) and the hypotheses on L, we observe that

$$z(t) \le \int_t^\infty [L(s, a(s) + b(s)\, z(s)) - L(s, a(s)) + L(s, a(s))] ds$$

$$\le e(t) + \int_t^\infty M(s, a(s))\, b(s)\, z(s)\, ds, \tag{1.5.82}$$

where $e(t)$ is defined by (1.5.72). Clearly $e(t)$ is real-valued, nonnegative, continuous and nonincreasing for $t \in R_+$. An application of the inequality in part (d_1) to (1.5.82) yields

$$z(t) \le e(t) \exp\left(\int_t^\infty M(s, a(s))\, b(s)\, ds\right). \tag{1.5.83}$$

The desired inequality in (1.5.71) follows from (1.5.81) and (1.5.83).

1.6 Applications

The study of various types of differential and integral equations has led to the investigation of a number of inequalities contained in earlier sections (see also [51,55,75,78]). In this section we present applications of some of the inequalities contained in sections 1.2-1.5 to study the qualitative properties of the solutions of certain differential and integral equations.

1.6.1 Nonlinear differential and integral equations

First,consider the nonlinear integral equation of the form

$$x^2(t) = f(t) + \int_0^t F(t, s, x(s))\, ds, \tag{1.6.1}$$

where $f \in C(R_+, R), F \in C(D \times R_+, R)$; D is as in Theorem 1.2.1. Here we assume that every solution $x(t)$ of (1.6.1) under discussion exists on R_+.

As an application of the inequality given in Theorem 1.2.2, part (a_1), we present the following theorem related to the solutions of equation (1.6.1) given in [55].

We list the following hypotheses on the functions f, F involved in (1.6.1):

$$|f(t)| \leq c, \ |F(t, s, x)| \leq k(t, s)|x|, \tag{1.6.2}$$

$$|f(t)| \leq ce^{-\alpha t}, \ |F(t, s, x)| \leq k(t, s)\, e^{-\alpha\left(t - \frac{1}{2}s\right)}|x|, \tag{1.6.3}$$

$$Q(t) = \sqrt{c} + \frac{1}{2}\int_0^t A(s)\, ds < \infty, \tag{1.6.4}$$

where $c, k(t, s), A(t)$ are as in Theorem 1.2.2 and $\alpha \geq 0$ is a real constant.

Theorem 1.6.1. (i) Suppose that the hypotheses (1.6.2), (1.6.4) are satisfied. Then all solutions of equation (1.6.1) are bounded for $t \in R_+$.

(ii) Suppose that the hypotheses (1.6.3), (1.6.4) are satisfied. Then all solutions of equation (1.6.1) approach zero as $t \to \infty$.

Proof. (i) Let $x(t)$, $t \in R_+$ be a solution of equation (1.6.1). From (1.6.1) and (1.6.2) we have

$$|x(t)|^2 \leq c + \int_0^t k(t,s)\, |x(s)|\, ds. \tag{1.6.5}$$

An application of the inequality given in Theorem 1.2.2 to (1.6.5) yields

$$|x(t)| \leq Q(t), \tag{1.6.6}$$

for $t \in R_+$. From the hypothesis (1.6.4), the estimation in (1.6.6) implies the boundedness of the solution $x(t)$ of equation (1.6.1) on R_+.

(ii) Let $x(t)$, $t \in R_+$ be a solution of equation (1.6.1). Then from (1.6.1), (1.6.3) we have

$$|x(t)|^2 \leq ce^{-\alpha t} + \int_0^t k(t,s)\, e^{-\alpha\left(t - \frac{1}{2}s\right)}\, |x(s)|\, ds. \tag{1.6.7}$$

From (1.6.7) it is easy to observe that

$$\left(|x(t)|\, e^{\frac{1}{2}\alpha t}\right)^2 \leq c + \int_0^t k(t,s) \left(|x(s)|\, e^{\frac{1}{2}\alpha s}\right) ds. \tag{1.6.8}$$

Now applying the inequality given in Theorem 1.2.2,part (a_1) to (1.6.8), and then multiplying the resulting inequality by $e^{-\frac{1}{2}\alpha t}$, we obtain

$$|x(t)| \leq Q(t)\, e^{-\frac{1}{2}\alpha t}, \tag{1.6.9}$$

for $t \in R_+$. In view of the hypothesis (1.6.4), the inequality (1.6.9) yields the desired result.

Next, we apply the inequality given in Theorem 1.3.3 (see [45]) to obtain a bound on the solution of the differential equation of the form

$$x^{p-1}(t)\, x'(t) = F(t, x(t)), \quad x(0) = x_0. \tag{1.6.10}$$

where x_0, $p > 1$ are constants and $F \in C(R_+ \times R_+, R)$.

Theorem 1.6.2. Assume that

$$|F(t,x)| \leq f(t)\, g(|x|), \tag{1.6.11}$$

for $t \in R_+$, where f and g are as in Theorem 1.3.3. Let $x(t)$ be a solution of equation (1.6.10) on R_+. Then

$$|x(t)| \leq \left\{ H^{-1} \left[H(|x_0|^p) + p \int_0^t f(s)\, ds \right] \right\}^{\frac{1}{p}}, \tag{1.6.12}$$

for $0 \leq t \leq t_1$; $t, t_1 \in R_+$, where H, H^{-1} are as in Theorem 1.3.3.

Proof. It is easy to see that the solution $x(t)$ of equation (1.6.10) satisfies the equivalent integral equation

$$\frac{x^p(t)}{p} - \frac{x_0^p}{p} = \int_0^t F(s, x(s))\, ds. \tag{1.6.13}$$

From (1.6.13) and (1.6.11) we observe that

$$|x(t)|^p \le |x_0|^p + p\int_0^t f(s)\, g(|x(s)|)\, ds. \tag{1.6.14}$$

Now a suitable application of Theorem 1.3.3 (when $h = 0$) to (1.6.14) yields the desired bound in (1.6.12).

1.6.2 Iterated Volterra integral equation

In this section, we present applications of the inequality given in Theorem 1.4.4, part (d_1) which provide estimates for the solutions of Volterra integral equation of the form

$$z(t) = f(t) + \sum_{i=1}^{n} G_i[z](t), \tag{1.6.15}$$

for $t \in I$, where

$$G_i[z](t) = \int_0^t \left(\int_0^{t_1} \cdots \left(\int_0^{t_{i-1}} k_i(t, t_1, ..., t_i, z(t_i))\, dt_i \right) \cdots \right) dt_1,$$

$f \in C(I, R), k_i \in C(I \times I_i \times R, R)$. Here we note that, our discussion uses the notations and definitions as used in Theorem 1.4.4.

Theorem 1.6.3. Suppose that the kernel functions k_i for $i = 1, ..., n$ satisfy

$$|k_i(t, t_1, ..., t_i, z(t_i))| \le b(t)\, L_i(t_1, ..., t_i, |z(t_i)|), \tag{1.6.16}$$

for $t \in I$, $(t_1, ..., t_i) \in I_i$,where $b \in C(I, R_+)$, L_i be as in Theorem 1.4.4, part (d_1) and verify the conditions in (1.4.29), M_i being the same as given therein. If $z(t) \in C(I, R)$ is any solution of equation (1.6.15), then

$$|z(t)| \le |f(t)| + b(t)\int_0^t \bar{E}(t_1) \exp\left(\int_{t_1}^t \bar{H}(\sigma) d\sigma \right) dt_1, \tag{1.6.17}$$

for $t \in I$, where $\bar{E}(t)$ and $\bar{H}(t)$ are respectively given by the right hand sides of the definitions of $E(t)$ and $H(t)$ given in Section 1.4, by replacing $a(t)$ by $|f(t)|$.

Proof. Let $z(t) \in C(I, R)$ be a solution of equation (1.6.15). Using the fact that $z(t)$ is a solution of (1.6.15) and (1.6.16) we observe that

$$|z(t)| \leq |f(t)| + \sum_{i=1}^{n} |G_i[z](t)|$$

$$\leq |f(t)| + b(t) \sum_{i=1}^{n} F_i[|z|](t). \tag{1.6.18}$$

Now a suitable application of the inequality given in Theorem 1.4.4, part (d_1) to (1.6.18) yields (1.6.17).

Theorem 1.6.4. Suppose that the kernel functions k_i for $i = 1, ..., n$ satisfy

$$|k_i(t, t_1, ..., t_i, x(t_i)) - k_i(t, t_1, ..., t_i, y(t_i))|$$

$$\leq b(t) L_i(t_1, ..., t_i, |x(t_i) - y(t_i)|), \tag{1.6.19}$$

for $t \in I$, $(t_1, ..., t_i) \in I_i$, where $b(t) \in C(I, R_+)$, L_i be as in Theorem 1.4.4, part (d_1) and verify the conditions in (1.4.29), M_i being the same functions as given therein. If $z(t) \in C(I, R)$ is any solution of equation (1.6.15), then

$$|z(t) - f(t)| \leq e(t) + b(t) \int_0^t E_0(t_1) \exp\left(\int_{t_1}^t H_0(\sigma)\, d\sigma\right) dt_1, \tag{1.6.20}$$

for $t \in I$, where

$$e(t) = \sum_{i=1}^{n} \int_0^t \left(\int_0^{t_1} \cdots \left(\int_0^{t_{i-1}} |k_i(t, t_1, ..., t_i, f(t_i))|\, dt_i\right) \cdots\right) dt_1,$$

$E_0(t)$ and $H_0(t)$ are respectively given by the right hand sides of the definitions of $E(t)$ and $H(t)$ given in Section 1.4, by replacing $a(t)$ by $e(t)$.

Proof. Let $z(t) \in C(I, R)$ be a solution of equation (1.6.15). Using the fact that $z(t)$ is a solution of (1.6.15) and (1.6.19) we observe that

$$|z(t) - f(t)| \leq \sum_{i=1}^{n} \int_0^t \left(\int_0^{t_1} \cdots \left(\int_0^{t_{i-1}} |k_i(t, t_i, ..., t_i, z(t_i))\right.\right.$$

$$\left.\left. - k_i(t, t_i, ..., t_i, f(t_i)) + k_i(t, t_i, ..., t_i, f(t_i))|\, dt_i\right) \cdots\right) dt_1$$

$$\leq e(t) + b(t) \sum_{i=1}^{n} F_i[|z - f|](t). \tag{1.6.21}$$

Now a suitable application of the inequality given in Theorem 1.4.4, part (d_1) to (1.6.21) yields (1.6.20).

1.6.3 General Volterra-Fredholm integral equation

Consider the following general Volterra-Fredholm integral equation

$$x(t) = f(t) + \int_{\alpha}^{t} F\left(t, s, x(s), \int_{\alpha}^{s} g(s, \sigma, x(\sigma))\, d\sigma\right) ds$$

$$+ \int_{\alpha}^{\beta} h(t, s, x(s))\, ds, \tag{1.6.22}$$

for $t \in I$, were $x(t)$ is an unknown function, $f \in C(I, R^n)$, $g, h \in C(D \times R^n, R^n)$, $F \in C(D \times R^n \times R^n, R^n)$, in which $I = [\alpha, \beta]$, $D = \{(t, s) \in I^2 : \alpha \leq s \leq t \leq \beta\}$ and R^n the n dimensional Euclidean space with norm $|.|$. Here we apply the inequality given in Theorem 1.5.3 to study certain properties of solutions of equation (1.6.22).

The following results are proved by Pachpatte in [75].

Theorem 1.6.5. Suppose that the functions f, g, h, F in equation (1.6.22) satisfy the conditions

$$|f(t)| \leq k, \tag{1.6.23}$$

$$|g(t, s, x)| \leq c(t, s)\, |x|, \tag{1.6.24}$$

$$|h(t, s, x)| \leq b(t, s)\, |x|, \tag{1.6.25}$$

$$|F(t, s, x, y)| \leq a(t, s)\, (|x| + |y|), \tag{1.6.26}$$

where $a(t, s), b(t, s), c(t, s)$ and k are as given in Theorem 1.5.3. Let $q(t)$ be as in (1.5.42). If $x(t)$, $t \in I$ is a solution of equation (1.6.22), then

$$|x(t)| \leq \frac{k}{1 - q(t)} \exp\left(\int_{\alpha}^{t} E(t, \xi) d\xi\right), \tag{1.6.27}$$

for $t \in I$, where $E(t, \xi)$ is defined by (1.5.43).

Proof. Using the fact that $x(t)$, $t \in I$ is a solution of equation (1.6.22) and the hypotheses (1.6.23)-(1.6.26) we have

$$|x(t)| \leq k + \int_{\alpha}^{t} a(t,s) \left(|x(s)| + \int_{\alpha}^{s} c(s,\sigma) |x(\sigma)| d\sigma \right) ds$$

$$+ \int_{\alpha}^{\beta} b(t,s) |x(s)| \, ds. \tag{1.6.28}$$

Now an application of Theorem 1.5.3 to (1.6.28) yields the required estimate in (1.6.27).

Theorem 1.6.6. Suppose that the functions f, g, h, F in equation (1.6.22) satisfy the conditions

$$|g(t,s,x) - g(t,s,y)| \leq c(t,s) |x - y|, \tag{1.6.29}$$

$$|h(t,s,x) - h(t,s,y)| \leq b(t,s) |x - y|, \tag{1.6.30}$$

$$|F(t,s,x,y) - F(t,s,\bar{x},\bar{y})| \leq a(t,s) (|x - \bar{x}| + |y - \bar{y}|), \tag{1.6.31}$$

where $a(t,s), b(t,s), c(t,s)$ are as in Theorem 1.5.3. Let $q(t)$ be as in (1.5.42). Then the equation (1.6.22) has at most one solution on I.

Proof. Let $u(t)$ and $v(t)$ be two solutions of equation (1.6.22) on I. Using these facts and hypotheses (1.6.29)-(1.6.31) we have

$$|u(t) - v(t)| \leq \int_{\alpha}^{t} a(t,s) \left(|u(s) - v(s)| + \int_{\alpha}^{s} c(s,\sigma) |u(\sigma) - v(\sigma)| d\sigma \right) ds$$

$$+ \int_{\alpha}^{\beta} b(t,s) |u(s) - v(s)| \, ds. \tag{1.6.32}$$

Now a suitable application of Theorem 1.5.3 to (1.6.32) yields $u(t) = v(t)$, $t \in I$ i.e., there is at most one solution of equation (1.6.22) on I.

1.6.4 Terminal value problem

In this section, we apply the inequality given in Theorem 1.5.4, part (d_1) to study certain properties of solutions of the following terminal value problem

$$u'(t) = f(t, u(t)) + p(t), \tag{1.6.33}$$

$$u(\infty) = u_{\infty}, \tag{1.6.34}$$

for $t \in R_+$, where $f \in C(R_+ \times R, R)$, $p \in C(R_+, R)$ and $u_{\infty} \in R$. For the exitence of solutions of problem (1.6.33)-(1.6.34) when $p(t) = 0$, see [3, p.80].

In the following theorems we present some results on the behavior of solutions of problem (1.6.33)-(1.6.34) given in [51].

Theorem 1.6.7. Suppose that

$$|f(t,u)| \le b(t)|u|, \tag{1.6.35}$$

$$|u_\infty - Q(t)| \le a(t), \tag{1.6.36}$$

where $a(t), b(t)$ are as in Theorem 1.5.4, part (d_1) and $Q(t) = \int_t^\infty p(s)\,ds$. If $u(t)$, $t \in R_+$ is a solution of the problem (1.6.33)-(1.6.34), then

$$|u(t)| \le a(t)\exp\left(\int_t^\infty b(s)\,ds\right), \tag{1.6.37}$$

for $t \in R_+$.

Proof. The solution $u(t)$ of the problem (1.6.33)-(1.6.34) can be written as (see [3, p. 80])

$$u(t) = u_\infty - \int_t^\infty [f(s,u(s)) + p(s)]\,ds, \tag{1.6.38}$$

for $t \in R_+$. From (1.6.38), (1.6.35), (1.6.36) we observe that

$$|u(t)| \le a(t) + \int_t^\infty b(s)|u(s)|\,ds. \tag{1.6.39}$$

Now an application of Theorem 1.5.4, part (d_1) to (1.6.39) yields the required estimate in (1.6.37).

Theorem 1.6.8. (i) Suppose that the function f in (1.6.33) satisfies the condition

$$|f(t,u) - f(t,v)| \le b(t)|u-v|, \tag{1.6.40}$$

where $b(t)$ is as defined in Theorem 1.5.4. Then the problem (1.6.33)-(1.6.34) has at most one solution on R_+.

(ii) Let $u_1(t)$ and $u_2(t)$, $t \in R_+$ be the solutions of (1.6.33) with the given terminal conditions

$$u_1(\infty) = u_{1\infty}, \tag{1.6.41}$$

and

$$u_2(\infty) = u_{2\infty}, \tag{1.6.42}$$

respectively, where $u_{1\infty}, u_{2\infty} \in R$. Suppose that the function f in (1.6.33) satisfies the condition (1.6.40). Then the solutions of (1.6.33) depends on terminal values and

$$|u_1(t) - u_2(t)| \leq |u_{1\infty} - u_{2\infty}| \exp\left(\int_t^\infty b(s)\,ds\right), \tag{1.6.43}$$

for $t \in R_+$.

Proof. (*i*) The problem (1.6.33)-(1.6.34) is equivalent to the integral equation (1.6.38). Let $u(t)$ and $v(t)$ be two solutions of (1.6.33)-(1.6.34) on R_+. Using the facts that $u(t)$ and $v(t)$ are the solutions of (1.6.38) and the condition (1.6.40) we have

$$|u(t) - v(t)| \leq \int_t^\infty b(s)\,|u(s) - v(s)|\,ds. \tag{1.6.44}$$

Now an application of Theorem 1.5.4, part (d_1) (when $a(t) = 0$) to (1.6.44) yields $u(t) = v(t)$ i.e., there is at most one solution to the problem (1.6.33)-(1.6.34) on R_+.

(*ii*) By using the facts that $u_1(t)$ and $u_2(t)$, $t \in R_+$ are the solutions of (1.6.33)-(1.6.41) and (1.6.33)-(1.6.42) respectively, we have

$$u_1(t) - u_2(t) = u_{1\infty} - u_{2\infty} - \int_t^\infty [f(s, u_1(s)) - f(s, u_2(s))]ds. \tag{1.6.45}$$

From (1.6.45) and (1.6.40) we have

$$|u_1(t) - u_2(t)| \leq |u_{1\infty} - u_{2\infty}| + \int_t^\infty b(s)\,|u_1(s) - u_2(s)|\,ds. \tag{1.6.46}$$

Now an application of Theorem 1.5.4, part (d_1) to (1.6.46) yields the required estimate in (1.6.43), which shows the dependency of solutions of (1.6.33) on terminal values.

1.7 Notes

The celebrated Gronwall's inequality [16,6] and its nonlinear generalization due to Bihari [8] have a profound and enduring influence on the development of the theory of differential and integral equations. Section 1.2 deals with some such basic nonlinear integral inequalities recently appeared in the literature. Theorem 1.2.1 is due to Pachpatte [68], which is a useful generalization of the well known Bihari's inequality [8], see also [34, p. 107]. The inequalities in Theorem 1.2.2 are taken from Pachpatte [55]. The results in Theorems 1.2.3-1.2.5 are recently established by Medveď in [24], which gives estimates on integral inequalities with weakly singular kernel. Indeed, the roots of such an inequality can be found in the work of Henry [17] who proved some results concerning linear integral inequalities with weakly singular kernel. Section 1.3 deals with some more nonlinear integral inequalities which claim their origin in the inequalities given by Ou-Iang [33] and Deformos [10]. The inequalities in Theorems 1.3.1-1.3.4 are due to Pachpatte and taken from [35,45]. Section 1.4 contains some useful integral inequalities involving iterated integrals.The inequalities in Theorem 1.4.1 are due to Pachpatte [53]. Theorem 1.4.2 is taken from Bykov and Salpagarov [9]. The results in Theorems 1.4.3 and 1.4.4 are the further generalizations of the inequalities in Theorem 1.4.2 and are due to Pachpatte [79,78]. The results given in Section 1.5 deals with some specific inequalities which are more convenient in certain situations. Theorems 1.5.1-1.5.3 are due to Pachpatte [52,54,70,75],while Theorem 1.5.4 is taken from Pachpatte and Pachpatte [51]. The material in Section 1.6 is taken from [51,55,75,78] and devoted to the applications of the inequalities given in earlier sections.

Chapter 2

Integral inequalities in two variables

2.1 Introduction

Inequalities involving functions of two or more independent variables, their partial derivatives and integrals play a fundamental role in the continuous development of the theory, methods and applications of partial differential and integral equations. In view of the wider applications, integral inequalities involving functions of two independent variables which furnish explicit known bounds have received considerable attention. Recently, different versions of such inequalities have been established, which are useful in the study of different classes of partial differential and integral equations. The main objective of this chapter is to present some useful integral inequalities in two independent variables recently appeared in the literature. These inequalities can be used as ready tools in the study of certain classes of partial differential and integral equations. We also give applications to convey the importance of some of these inequalities.

2.2 Some nonlinear integral inequalities

Integral inequalities involving functions of two and more independent variables which provide explicit bounds on unknown functions have played a fundamental role in the study of certain partial differential and integral equations. In this section we present some basic nonlinear integral inequalities in two variables which can be used as convenient tools in some applications.

The following theorem deals with a fairly general version of the inequality given by Pachpatte in [68].

Theorem 2.2.1. Let $u(x,y), a(x,y), D_1 a(x,y), D_2 a(x,y) \in C(R_+^2, R_+)$, $k(x,y,s,t), D_1 k(x,y,s,t), D_2 k(x,y,s,t), D_2 D_1 k(x,y,s,t) \in C(E, R_+)$, where $E = \{(x,y,s,t) \in R_+^4 : 0 \le s \le x < \infty, 0 \le t \le y < \infty\}$. Let $g(u)$ be a continuously differentiable function defined for $u \ge 0, g(u) > 0$ for $u > 0$ and $g'(u) \ge 0$ for $u \ge 0$. If

$$u(x,y) \le a(x,y) + \int_0^x \int_0^y k(x,y,\sigma,\tau)\, g(u(\sigma,\tau)) d\tau d\sigma, \tag{2.2.1}$$

for $x, y \in R_+$, then for $0 \le x \le x_1, 0 \le y \le y_1; x, x_1, y, y_1 \in R_+$,

$$u(x,y) \le G^{-1}\left[G(a(x,y)) + \int_0^x \int_0^y A(s,t)\, dt ds\right], \tag{2.2.2}$$

where

$$A(x,y) = k(x,y,x,y) + \int_0^x D_1 k(x,y,\sigma,y)\, d\sigma$$

$$+ \int_0^y D_2 k(x,y,x,\tau)\, d\tau + \int_0^x \int_0^y D_2 D_1 k(x,y,\sigma,\tau)\, d\sigma d\tau, \tag{2.2.3}$$

$$G(r) = \int_{r_0}^r \frac{ds}{g(s)}, r > 0, \tag{2.2.4}$$

$r_0 > 0$ is arbitrary and G^{-1} is the inverse of G and $x_1, y_1 \in R_+$ are chosen so that

$$G(a(x,y)) + \int_0^x \int_0^y A(s,t)\, dt ds \in Dom\left(G^{-1}\right),$$

for all x, y lying in $0 \le x \le x_1, 0 \le y \le y_1$ respectively.

Proof. From the hypotheses, it is easy to observe that the function $a(x,y)$ is monotonically increasing in both the variables x and y. We also note that since $g'(u) \ge 0$ on R_+, the function $g(u)$ is monotonically increasing on $(0,\infty)$. Let $a(x,y) > 0$ for $x, y \in R_+$ and define a function $z(x,y)$ by the right hand side of (2.2.1). Then $z(x,y)$ is positive and by hypotheses, it is nondecreasing in $x, y \in R_+$,$z(x,0) = a(x,0)$, $z(0,y) = a(0,y)$, $u(x,y) \le z(x,y)$ and

$$D_1 z(x,y) = D_1 a(x,y) + \int_0^y k(x,y,x,\tau)\, g(u(x,\tau))\, d\tau$$

$$+\int_0^x\int_0^y D_1k(x,y,\sigma,\tau)\,g(u(\sigma,\tau))d\tau d\sigma,$$

$$D_2z(x,y)=D_2a(x,y)+\int_0^x k(x,y,\sigma,y)\,g(u(\sigma,y))\,d\sigma$$

$$+\int_0^x\int_0^y D_2k(x,y,\sigma,\tau)\,g(u(\sigma,\tau))d\tau d\sigma,$$

$$D_2D_1z(x,y)=D_2D_1a(x,y)+k(x,y,x,y)\,g(u(x,y))$$

$$+\int_0^x D_1k(x,y,\sigma,y)\,g(u(\sigma,y))\,d\sigma+\int_0^y D_2k(x,y,x,\tau)\,g(u(x,\tau))\,d\tau$$

$$+\int_0^x\int_0^y D_2D_1k(x,y,\sigma,\tau)\,g(u(\sigma,\tau))d\tau d\sigma$$

$$\leq D_2D_1a(x,y)+k(x,y,x,y)\,g(z(x,y))$$

$$+\int_0^x D_1k(x,y,\sigma,y)\,g(z(\sigma,y))\,d\sigma+\int_0^y D_2k(x,y,x,\tau)\,g(z(x,\tau))\,d\tau$$

$$+\int_0^x\int_0^y D_2D_1k(x,y,\sigma,\tau)\,g(z(\sigma,\tau))d\tau d\sigma$$

$$\leq D_2D_1a(x,y)+A(x,y)\,g(z(x,y)). \tag{2.2.5}$$

It is easy to observe that

$$D_2D_1G(z(x,y))=G''(z(x,y))\,D_1z(x,y)\,D_2z(x,y)$$

$$+G'(z(x,y))\,D_2D_1z(x,y). \tag{2.2.6}$$

Since $a(x,y)\leq z(x,y)$ and $D_1z(x,y)\geq 0, D_2z(x,y)\geq 0$, $G'(z(x,y))=\frac{1}{g(z(x,y))}$ and $G''(z(x,y))\leq 0$, we obtain from (2.2.5) and (2.2.6)

$$D_2D_1G(z(x,y))\leq G'(z(x,y))\{D_2D_1a(x,y)+A(x,y)\,g(z(x,y))\}$$

$$=\frac{D_2D_1a(x,y)}{g(z(x,y))}+A(x,y)$$

$$\leq\frac{D_2D_1a(x,y)}{g(a(x,y))}+A(x,y). \tag{2.2.7}$$

On the other hand we observe that

$$D_2D_1G(a(x,y)) = D_2\left(D_1\left(\int_{r_0}^{a(x,y)}\frac{ds}{g(s)}\right)\right)$$

$$= D_2\left(\frac{D_1a(x,y)}{g(a(x,y))}\right)$$

$$= \frac{g(a(x,y))\,D_2D_1a(x,y) - D_1a(x,y)\,g'(a(x,y))\,D_2a(x,y)}{[g(a(x,y))]^2}$$

$$= \frac{D_2D_1a(x,y)}{g(a(x,y))} - \frac{g'(a(x,y))\,D_1a(x,y)\,D_2a(x,y)}{[g(a(x,y))]^2}$$

which implies

$$D_2D_1G(a(x,y)) \geq \frac{D_2D_1a(x,y)}{g(a(x,y))}. \tag{2.2.8}$$

From (2.2.7) and (2.2.8) we have

$$D_2D_1G(z(x,y)) \leq D_2D_1G(a(x,y)) + A(x,y),$$

and this yields

$$G(z(x,y)) \leq G(a(x,y)) + \int_0^x\int_0^y A(s,t)\,dtds,$$

which implies (see [34, Chapter 5])

$$u(x,y) \leq z(x,y) \leq G^{-1}\left[G(a(x,y)) + \int_0^x\int_0^y A(s,t)\,dtds\right].$$

If $a(x,y)$ is nonnegative, we carry out the above procedure with $a(x,y)+\varepsilon$ instead of $a(x,y)$, where $\varepsilon > 0$ is an arbitrary small constant, and subsequently pass to the limit $\varepsilon \to 0$ to obtain (2.2.2). The subdomain $0 \leq x \leq x_1, 0 \leq y \leq y_1$ is obvious.

Remark 2.2.1. If we take $g(u) = u$ in Theorem 2.2.1, then the bound obtained in (2.2.2) reduces to

$$u(x,y) \leq a(x,y)\exp\left(\int_0^x\int_0^y A(s,t)\,dtds\right),$$

for $x, y \in R_+$. In this case the inequality given in Theorem 2.2.1 is a generalization of the Wendroff's inequality given in [4, p. 154], see also [34].

In [55] Pachpatte has established the inequalities in the following theorem.

Theorem 2.2.2. Let $u(x,y), k(x,y,s,t)$, $D_1 k(x,y,s,t)$, $D_2 k(x,y,s,t)$, $D_2 D_1 k(x,y,s,t)$ be as in Theorem 2.2.1 and $c \geq 0$ is a constant.

(a_1) If

$$u^2(x,y) \leq c + \int_0^x \int_0^y k(x,y,\sigma,\tau) u(\sigma,\tau) d\tau d\sigma, \tag{2.2.9}$$

for $x, y \in R_+$, then

$$u(x,y) \leq \sqrt{c} + \frac{1}{2} \int_0^x \int_0^y A(s,t)\, dt ds, \tag{2.2.10}$$

for $x, y \in R_+$, where $A(x,y)$ is defined ny (2.2.3).

(a_2) Let $g(u)$ be as in Theorem 2.2.1. If

$$u^2(x,y) \leq c + \int_0^x \int_0^y k(x,y,\sigma,\tau) g(u(\sigma,\tau)) d\tau d\sigma, \tag{2.2.11}$$

for $x, y \in R_+$, then for $0 \leq x \leq x_2, 0 \leq y \leq y_2; x, x_2, y, y_2 \in R_+$

$$u(x,y) \leq G^{-1}\left[G\left(\sqrt{c}\right) + \frac{1}{2} \int_0^x \int_0^y A(s,t)\, dt ds\right] \tag{2.2.12}$$

where $A(x,y)$ is defined by (2.2.3), G, G^{-1} are as defined in Theorem 2.2.1 and $x_2, y_2 \in R_+$ are chosen so that

$$G\left(\sqrt{c}\right) + \frac{1}{2} \int_0^x \int_0^y A(s,t)\, dt ds \in Dom\left(G^{-1}\right),$$

for all $x, y \in R_+$ lying in $0 \leq x \leq x_2, 0 \leq y \leq y_2$.

Proof. (a_1) It is sufficient to assume that c is positive, since the standard limiting argument can be used to treat the remaining case. Let $c > 0$ and define a function $z(x,y)$ by the right hand side of (2.2.9). Then $z(0,y) = z(x,0) = c$, $u(x,y) \leq \sqrt{z(x,y)}$, $z(x,y)$ is positive and nondecreasing in $x, y \in R_+$ and

$$D_2 D_1 z(x,y) = k(x,y,x,y) u(x,y) + \int_0^x D_1 k(x,y,\sigma,y) u(\sigma,y) d\sigma$$

$$+\int_0^x D_2k(x,y,x,\tau)\,u(x,\tau)\,d\tau+\int_0^x\int_0^y D_2D_1k(x,y,\sigma,\tau)\,u(\sigma,\tau)\,d\tau d\sigma$$

$$\le k(x,y,x,y)\sqrt{z(x,y)}+\int_0^x D_1k(x,y,\sigma,y)\sqrt{z(\sigma,y)}d\sigma$$

$$+\int_0^x D_2k(x,y,x,\tau)\sqrt{z(x,\tau)}d\tau+\int_0^x\int_0^y D_2D_1k(x,y,\sigma,\tau)\sqrt{z(\sigma,\tau)}d\tau d\sigma$$

$$\le A(x,y)\sqrt{z(x,y)}. \tag{2.2.13}$$

Now by following the proof of Theorem 5.8.1 given in [34, p. 528], from (2.2.13) we get

$$\sqrt{z(x,y)}\le\sqrt{c}+\frac{1}{2}\int_0^x\int_0^y A(s,t)\,dtds. \tag{2.2.14}$$

Using (2.2.14) in $u(x,y)\le\sqrt{z(x,y)}$, we get the required inequality in (2.2.10).

(a_2) The proof follows by closely looking at the proof of Theorem 5.8.2 in [34].

Remark 2.2.2. If we take $k(x,y,s,t)=p(s,t)$ in Theorem 2.2.2, then the estimates obtained in (2.2.10), (2.2.12) reduces respectively to

$$u(x,y)\le\sqrt{c}+\frac{1}{2}\int_0^x\int_0^y p(s,t)\,dtds,$$

$$u(x,y)\le G^{-1}\left[G\left(\sqrt{c}\right)+\frac{1}{2}\int_0^x\int_0^y p(s,t)\,dtds\right].$$

In this case the inequalities in (a_1), (a_2) reduces respectively to the variants of the inequalities in Theorems 5.8.1, 5.8.2 given in [34]. We note that, by following the proof of Theorem 2.2.1 one can obtain the estimates on the inequalities (2.2.9), (2.2.11) when c is replaced by the function $a(x,y)$, where $a(x,y)$ is as in Theorem 2.2.1.

The inequalities in the following theorems are proved by Medveď in [26].

Theorem 2.2.3. Let $0 < T \leq \infty$ and $I = [0, T)$. Let $u(x,y), F(x,y), a(x,y), D_1 a(x,y), D_2 a(x,y), D_2 D_1 a(x,y) \subset C(I^2, R_+)$. Let $w(u)$ be a continuously differentiable function defined for $u \geq 0$, $w(u) > 0$ for $u > 0$ and $w'(u) \geq 0$ for $u \geq 0$. If

$$u(x,y) \leq a(x,y) + \int_0^x \int_0^y (x-s)^{\alpha-1} (y-t)^{\beta-1} F(s,t) w(u(s,t))\, dtds, \quad (2.2.15)$$

for $x, y \in I$, then the following assertions hold:

(i) Suppose $\alpha > \frac{1}{2}, \beta > \frac{1}{2}$ and w satisfies the condition (q) as given in Section 1.2 with $q = 2$. Then

$$u(x,y) \leq e^{x+y} \Phi(x,y), \quad (2.2.16)$$

for $x, y \in [0, T_1], T_1 \in I$, where

$$\Phi(x,y) = \left[\Omega^{-1}\left[\Omega\left(2a(x,y)^2\right) + 2K \int_0^x \int_0^y F(s,t)^2 R(s+t)\, dtds\right]\right]^{\frac{1}{2}}, \quad (2.2.17)$$

$$K = \frac{\Gamma(2\alpha - 1)\Gamma(2\beta - 1)}{4^{\alpha+\beta-1}}, \quad (2.2.18)$$

Γ is the Gamma function, $\Omega(r) = \int_{r_0}^{r} \frac{ds}{w(s)}, r > 0$ and $r_0 > 0$ is arbitrary, Ω^{-1} is the inverse of Ω and T_1 is chosen so that the argument of Ω^{-1} in (2.2.17) belongs to $Dom\left(\Omega^{-1}\right)$ for all $x, y \in [0, T_1]$.

(ii) Suppose $\alpha = \beta = \frac{1}{z+1}$ for some real number $z \geq 1$ and w satisfies the condition (q) in Section 1.2 with $q = z + 2$. Let Ω, Ω^{-1} be as in part (i). Then

$$u(x,y) \leq e^{x+y} \Psi(x,y) \quad (2.2.19)$$

for $x, y \in [0, T_2], T_2 \in I$, where

$$\Psi(x,y) = \left[\Omega^{-1}\left[\Omega\left(2^{q-1} a(x,y)^q\right) + 2^{q-1} M_z^q \int_0^x \int_0^y F(s,t)^q R(s+t)\, dtds\right]\right]^{\frac{1}{q}}, \quad (2.2.20)$$

$$M_z = \left[\frac{\Gamma(1-p\delta)}{p^{(1-p\delta)}}\right]^{\frac{2}{q}}, \delta = 1 - \beta = \frac{z}{z+1}, p = \frac{z+2}{z+1}, \quad (2.2.21)$$

T_2 is chosen so that the argument of Ω^{-1} in (2.2.20) belongs to $Dom\left(\Omega^{-1}\right)$ for all $x, y \in [0, T_2]$.

Proof. First we prove the assertion (i). Using the Cauchy-Schwarz inequality for double integrals we obtain from (2.2.15)

$$u(x,y) \le a(x,y) + \int_0^x \int_0^y (x-s)^{\alpha-1} e^s (y-t)^{\beta-1} e^t \left[e^{-(s+t)} F(s,t) w(u(s,t))\right] dtds$$

$$\le a(x,y) + \left[\int_0^x \int_0^y (x-s)^{2\alpha-2} e^{2s} (y-t)^{2\beta-2} e^{2t} dtds\right]^{\frac{1}{2}}$$

$$\times \left[\int_0^x \int_0^y e^{-2(s+t)} F(s,t)^2 w(u(s,t))^2 dtds\right]^{\frac{1}{2}}. \tag{2.2.22}$$

For the first integral in (2.2.22) we have the estimate

$$\int_0^x \int_0^y (x-s)^{2\alpha-2} e^{2s} (y-t)^{2\beta-2} e^{2t} dtds$$

$$= e^{2(x+y)} \int_0^x \sigma^{2\alpha-2} e^{-2\sigma} \int_0^y \eta^{2\beta-2} e^{-2\eta} d\eta d\sigma$$

$$= \frac{e^{2(x+y)}}{2^{2(\alpha+\beta)-2}} \int_0^{2x} \tau^{2\alpha-2} e^{-\tau} d\tau \int_0^{2y} \xi^{2\beta-2} e^{-\xi} d\xi$$

$$< \frac{e^{2(x+y)}}{2^{2(\alpha+\beta)-2}} \Gamma(2\alpha-1) \Gamma(2\beta-1).$$

Therefore we obtain from (2.2.22)

$$u(x,y) \le a(x,y) + e^{x+y} K^{\frac{1}{2}} \left[\int_0^x \int_0^y e^{-2(s+t)} F(s,t)^2 w(u(s,t))^2 dtds\right]^{\frac{1}{2}},$$

where K is as in (2.2.18). Using the inequality (1.2.23) with $n = 2, r = 2$ and the condition (q) in Section 1.2 we obtain

$$v(x,y) \le b(x,y) + 2K \int_0^x \int_0^y F(s,t)^2 R(s+t) w(v(s,t)) dtds, \tag{2.2.23}$$

where

$$v(x,y) = \left(e^{-(x+y)} u(x,y)\right)^2, b(x,y) = 2a(x,y)^2. \tag{2.2.24}$$

Now a suitable application of Theorem 2.2.1 to (2.2.23) yields

$$v(x,y) \leq \Omega^{-1}\left[\Omega(b(x,y)) + 2K\int_0^x\int_0^y F(s,t)^2 R(s+t)\,dtds\right]. \tag{2.2.25}$$

Using (2.2.24) in (2.2.25) we get the required inequality in (2.2.16).

Now we shall prove the assertion (ii). Let $p = \frac{z+2}{z+1}, q = z+2$. Then using the Hölder's integral inequality we obtain from (2.2.15)

$$u(x,y) \leq a(x,y) + \left[\int_0^x\int_0^y (x-s)^{-p\delta} e^{ps} (y-t)^{-p\delta} e^{pt}\,dtds\right]^{\frac{1}{p}}$$

$$\times\left[\int_0^x\int_0^y e^{-q(s+t)} F(s,t)^q\, w(u(s,t))^q\,dtds\right]^{\frac{1}{q}}. \tag{2.2.26}$$

For the first integral in (2.2.26) we have the estimate

$$\int_0^x\int_0^y (x-s)^{-p\delta} e^{ps} (y-t)^{-p\delta} e^{pt}\,dtds$$

$$= e^{p(x+y)}\int_0^x \sigma^{-p\delta} e^{-p\sigma}\int_0^y \eta^{-p\delta} e^{-p\eta}\,d\eta d\sigma$$

$$= \frac{e^{p(x+y)}}{p^{2(1-p\delta)}}\int_0^{px} \tau^{-p\delta} e^{-\tau}\,d\tau \int_0^{py} \xi^{-p\delta} e^{-\xi}\,d\xi$$

$$< \frac{e^{p(x+y)}}{p^{2(1-p\delta)}}\left\{\Gamma(1-p\delta)\right\}^2.$$

Thus (2.2.26) and the condition (q) yield

$$u(x,y) \leq a(x,y) + \left[\frac{e^{p(x+y)}}{p^{2(1-p\delta)}}\left\{\Gamma(1-p\delta)\right\}^2\right]^{\frac{1}{p}}$$

$$\times\left[\int_0^x\int_0^y F(s,t)^q R(s+t)\, w\left(e^{-q(s+t)} u(s,t)^q\right) dtds\right]^{\frac{1}{q}}. \tag{2.2.27}$$

From (2.2.27) and using the inequality (1.2.23) with $n = 2, r = q$ we obtain

$$v(x,y) \leq b(x,y) + 2^{q-1} M_z^q \int_0^x\int_0^y F(s,t)^q R(s+t)\, w(v(s,t))\,dtds, \tag{2.2.28}$$

where M_z is defined as in (2.2.21) and

$$v(x,y) = \left(e^{-(x+y)}u(x,y)\right)^q, b(x,y) = 2^{q-1}a(x,y)^q. \tag{2.2.29}$$

Now a suitable application of Theorem 2.2.1 to (2.2.28) yields

$$v(x,y) \le \Omega^{-1}\left[\Omega(b(x,y)) + 2^{q-1}M_z^q \int_0^x\int_0^y F(s,t)^q R(s+t)\,dtds\right]. \tag{2.2.30}$$

Using (2.2.29) in (2.2.30) we get the desired inequality in (2.2.19).

Theorem 2.2.4. Let $0 < T \le \infty$ and $I = [0,T)$. Let $u(x,y), F(x,y)$ and $w(u)$ be as in Theorem 2.2.3. If

$$u^2(x,y) \le a + \int_0^x\int_0^y (x-s)^{\alpha-1}(y-t)^{\beta-1} F(s,t)\,w(u(s,t))\,dtds, \tag{2.2.31}$$

for $x, y \in I$, where a is a positive constant. Then the following assertions hold:

(i) Suppose $\alpha > \frac{1}{2}, \beta > \frac{1}{2}$ and w satisfies the condition (q) as given in Section 1.2 with $q = 2$. Then

$$u(x,y) \le e^{x+y}L(x,y), \tag{2.2.32}$$

for $x, y \in [0, T_1], T_1 \in I$, where

$$L(x,y) = \left[\Lambda^{-1}\left[\Lambda(2a^2) + 2K\int_0^x\int_0^y F(s,t)^2 R(s+t)\,dtds\right]\right]^{\frac{1}{4}}, \tag{2.2.33}$$

K is defined as in (2.2.18) and $\Lambda(r) = \int_{r_0}^{r} \frac{ds}{w(\sqrt{s})}, r > 0$ and $r_0 > 0$ is arbitrary, T_1 is chosen so that the argument of Λ^{-1} in (2.2.33) belongs to $Dom\left(\Lambda^{-1}\right)$ for all $x, y \in [0, T_1]$.

(ii) Suppose $\alpha = \beta = \frac{1}{z+1}$ for some real number $z \ge 1$ and let $p = \frac{z+2}{z+1}, q = z+2$. Assume that w satisfies the condition (q) as given in Section 1.2 with $q = z+2$. Let Λ, Λ^{-1} be as in part (i). Then

$$u(x,y) \le e^{x+y}Q(x,y), \tag{2.2.34}$$

for $x, y \in [0, T_2], T_2 \in I$, where

$$Q(x,y) = \left[\Lambda^{-1}\left[\Lambda(2^{q-1}a^q) + 2^{q-1}M_z^q \int_0^x\int_0^y F(s,t)^q R(s+t)\,dtds\right]\right]^{\frac{1}{2q}}, \tag{2.2.35}$$

M_z is defined as in (2.2.21) and T_2 is chosen so that the argument of Λ^{-1} in (2.2.35) belongs to $Dom\left(\Lambda^{-1}\right)$ for all $x, y \in [0, T_2], T_2 \in I$.

Proof. First we prove the assertion (i). Following the proof of Theorem 2.2.3, part (i) one can show that

$$v^2(x,y) \leq c + 2K \int_0^x \int_0^y F(s,t)^2 R(s+t) w(v(s,t))\, dtds, \tag{2.2.36}$$

where

$$v(x,y) = \left(e^{-(x+y)} u(x,y)\right)^2, c = 2a^2. \tag{2.2.37}$$

Let $z(x,y)$ be the right hand side of (2.2.36). Then $z(x,y)$ is positive and nondecreasing for $x, y \in I$, $v(x,y) \leq \sqrt{z(x,y)}$, $z(x,0) = z(0,y) = c$,

$$D_2 D_1 z(x,y) = 2KF(x,y)^2 R(x+y) w(v(x,y))$$

$$\leq 2KF(x,y)^2 R(x+y) w\left(\sqrt{z(x,y)}\right), \tag{2.2.38}$$

and as in the proof of Theorem 2.2.1 we observe that

$$D_2 D_1 \Lambda(x,y) \leq \frac{D_2 D_1 z(x,y)}{w\left(\sqrt{z(x,y)}\right)}. \tag{2.2.39}$$

From (2.2.39) and (2.2.38) we have

$$D_2 D_1 \Lambda(x,y) \leq 2KF(x,y)^2 R(x+y),$$

and this yields

$$\Lambda(x,y) \leq \Lambda(c) + 2K \int_0^x \int_0^y F(s,t)^2 R(s+t)\, dtds. \tag{2.2.40}$$

Using (2.2.40), the fact that $v^2(x,y) \leq z(x,y)$ and (2.2.37) we get the desired inequality in (2.2.32).

The proof of the case (ii) can be completed by following the proof of case (i) given above and closely looking at the proof of Theorem 2.2.3 ,case (ii).

2.3 Further nonlinear integral inequalities

In view of the important applications of the integral inequalities which furnish explicit bounds on unknown functions, in the past few years, some new inequalities have been developed in the literature. In this section we give some integral inequalities involving functions of two variables established by Pachpatte in [46,40,45].

The inequalities in the following theorems are given in [46].

Theorem 2.3.1. Let $u(x,y), a(x,y), p(x,y), b(x,y) \in C(R_+^2, R_+)$. Let $a(x,y)$ be nondecreasing for $x, y \in R_+$ and $L \in C(R_+^3, R_+)$ satisfies the condition

$$0 \leq L(x,y,v_1) - L(x,y,v_2) \leq M(x,y,v_2)(v_1 - v_2), \tag{2.3.1}$$

for $v_1 \geq v_2 \geq 0$, where $M \in C(R_+^3, R_+)$.

(a_1) If

$$u(x,y) \leq a(x,y) + p(x,y) \int_0^x b(s,y) u(s,y) ds$$

$$+ \int_0^x \int_0^y L(s,t,u(s,t)) dt ds, \tag{2.3.2}$$

for $x, y \in R_+$, then

$$u(x,y) \leq f(x,y) [a(x,y) + e(x,y)$$

$$\times \exp\left(\int_0^x \int_0^y M(s,t,f(s,t) a(s,t)) f(s,t) dt ds\right)\Bigg], \tag{2.3.3}$$

for $x, y \in R_+$, where

$$f(x,y) = 1 + p(x,y) \int_0^x b(s,y) \exp\left(\int_s^x b(\sigma,y) p(\sigma,y) d\sigma\right) ds, \tag{2.3.4}$$

$$e(x,y) = \int_0^x \int_0^y L(s,t,f(s,t) a(s,t)) dt ds, \tag{2.3.5}$$

for $x, y \in R_+$.

(a_2) If

$$u(x,y) \leq a(x,y) + p(x,y) \int_0^y b(x,t) u(x,t) dt$$

$$+ \int_0^x \int_0^y L(s,t,u(s,t)) dt ds, \tag{2.3.6}$$

for $x, y \in R_+$, then

$$u(x,y) \leq \bar{f}(x,y)\left[a(x,y) + \bar{e}(x,y) \times \exp\left(\int_0^x \int_0^y M\left(s,t,\bar{f}(s,t)\,a(s,t)\right)\bar{f}(s,t)\,dtds\right)\right], \tag{2.3.7}$$

for $x, y \in R_+$, where

$$\bar{f}(x,y) = 1 + p(x,y)\int_0^y b(x,t)\exp\left(\int_t^y b(x,\tau)\,p(x,\tau)\,d\tau\right)dt, \tag{2.3.8}$$

$$\bar{e}(x,y) = \int_0^x \int_0^y L\left(s,t,\bar{f}(s,t)\,a(s,t)\right)dtds, \tag{2.3.9}$$

for $x, y \in R_+$.

Proof. (a_1) It is sufficient to assume that $a(x,y) > 0$ for $x, y \in R_+$, since the standard limiting argument can be used to treat the remaining case,see [34, p. 226]. Let $a(x,y) > 0$ for $x, y \in R_+$ and define a function $z(x,y)$ by

$$z(x,y) = a(x,y) + \int_0^x \int_0^y L(s,t,u(s,t))\,dtds. \tag{2.3.10}$$

Then (2.3.2) can be restated as

$$u(x,y) \leq z(x,y) + p(x,y)\int_0^y b(s,y)\,u(s,y)\,ds. \tag{2.3.11}$$

Clearly $z(x,y)$ is nonnegative and nondecreasing function for $x, y \in R_+$. Treating (2.3.11) as one-dimensional integral inequality for any fixed $y \in R_+$ and a suitable application of the inequality given in Theorem 1.3.3 in [34, p. 15] yields

$$u(x,y) \leq z(x,y,)\,f(x,y), \tag{2.3.12}$$

for $x, y \in R_+$, where $f(x,y)$ is defined by (2.3.4). From (2.3.10) and (2.3.12) we have

$$u(x,y) \leq f(x,y,)\left[a(x,y) + r(x,y)\right], \tag{2.3.13}$$

where

$$r(x,y) = \int_0^x \int_0^y L(s,t,u(s,t))\,dtds. \tag{2.3.14}$$

From (2.3.13), (2.3.14) and (2.3.1) we observe that

$$r(x,y) \le \int_0^x \int_0^y [L(s,t,f(s,t)[a(s,t)+r(s,t)])$$

$$- L(s,t,f(s,t)a(s,t)) + L(s,t,f(s,t)a(s,t))]\,dtds$$

$$\le e(x,y) + \int_0^x \int_0^y M(s,t,f(s,t)a(s,t))\,f(s,t)\,r(s,t)\,dtds, \qquad (2.3.15)$$

where $e(x,y)$ is defined by (2.3.5). Obviously, $e(x,y)$ is nonnegative and nondecreasing in each variable $x, y \in R_+$. A suitable application of Theorem 4.2.2 given in [34, p. 325] yields

$$r(x,y) \le e(x,y)\exp\left(\int_0^x \int_0^y M(s,t,f(s,t)a(s,t))\,f(s,t)dtds\right). \qquad (2.3.16)$$

Using (2.3.16) in (2.3.13) we get the required inequality in (2.3.3).

(a_2) The proof follows by a similar argument to that employed in (a_1). We omit the details.

Theorem 2.3.2. Let $u(x,y), a(x,y), g(x,y), h(x,y) \in C(R_+^2, R_+)$. Let $a(x,y)$, L and M be as in Theorem 2.3.1 and the condition (2.3.1) holds.

(b_1) If

$$u(x,y) \le a(x,y) + \int_0^x g(s,y)\left(u(s,y) + \int_0^s h(\sigma,y)\,u(\sigma,y)\,d\sigma\right)ds$$

$$+ \int_0^x \int_0^y L(s,t,u(s,t))\,dtds, \qquad (2.3.17)$$

for $x, y \in R_+$, then

$$u(x,y) \le k(x,y)\,[a(x,y) + E(x,y)$$

$$\times \exp\left(\int_0^x \int_0^y M(s,t,k(s,t)a(s,t))\,k(s,t)\,dtds\right)\Bigg], \qquad (2.3.18)$$

for $x, y \in R_+$, where

$$k(x,y) = 1 + \int_0^x g(s,y)\exp\left(\int_0^s [g(\sigma,y) + h(\sigma,y)]\,d\sigma\right)ds, \qquad (2.3.19)$$

$$E(x,y) = \int_0^x \int_0^y L(s,t,k(s,t)\,a(s,t))\,dtds, \tag{2.3.20}$$

for $x, y \in R_+$.

(b_2) If

$$u(x,y) \le a(x,y) + \int_0^y g(x,t) \left(u(x,t) + \int_0^t h(x,\tau)\,u(x,\tau)\,d\tau \right) dt$$

$$+ \int_0^x \int_0^y L(s,t,u(s,t))\,dtds, \tag{2.3.21}$$

for $x, y \in R_+$, then

$$u(x,y) \le \bar{k}(x,y) \Big[a(x,y) + \bar{E}(x,y)$$

$$\times \exp\left(\int_0^x \int_0^y M\big(s,t,\bar{k}(s,t)\,a(s,t)\big)\,\bar{k}(s,t)\,dtds \right) \Big], \tag{2.3.22}$$

for $x, y \in R_+$, where

$$\bar{k}(x,y) = 1 + \int_0^y g(x,t) \exp\left(\int_0^t [g(x,\tau) + h(x,\tau)]\,d\tau \right) dt, \tag{2.3.23}$$

$$\bar{E}(x,y) = \int_0^x \int_0^y L\big(s,t,\bar{k}(s,t)\,a(s,t)\big)\,dtds, \tag{2.3.24}$$

for $x, y \in R_+$.

Proof. As in the proof of Theorem 2.3.1, part (a_1) let $a(x,y) > 0$ for $x, y \in R_+$ and define a function $z(x,y)$ by (2.3.10). Then (2.3.17) can be written as

$$u(x,y) \le z(x,y) + \int_0^x g(s,y) \left(u(s,y) + \int_0^s h(\sigma,y)\,u(\sigma,y)\,d\sigma \right) ds. \tag{2.3.25}$$

Clearly $z(x,y)$ is nonnegative and nondecreasing function for $x, y \in R_+$. Treating (2.3.25) as an one-dimensional integral inequality for any fixed $y \in R_+$ and a suitable application of Theorem 1.7.4 given in [34, p. 39] yields

$$u(x,y) \le z(x,y)\,k(x,y), \tag{2.3.26}$$

where $k(x,y)$ is defined by (2.3.19). Now by following the proof of Theorem 2.3.1, part (a_1) with suitable modifications, we get the desired inequality in (2.3.18).

(b_2) The proof is similar to that of part (b_1) given above. We omit the details.

Remark 2.3.1. We note that from Theorems 2.3.1 and 2.3.2 one can obtain Corollaries similar to those of Corollaries of Lemma 74 discussed in [12, p. 43] which can be used in some applications.

The inequalities established in [40] are embodied in the following theorems.

Theorem 2.3.3. Let $u(x,y), a(x,y), b(x,y), g(x,y), h(x,y) \in C(R_+^2, R_+)$ and $p > 1$ be a real constant.

(c_1) If

$$u^p(x,y) \le a(x,y) + b(x,y) \int_0^x \int_0^y [g(s,t) u^p(s,t) + h(s,t) u(s,t)] dt ds, \quad (2.3.27)$$

for $x, y \in R_+$, then

$$u(x,y) \le \{a(x,y) + b(x,y) e(x,y)$$
$$\times \exp\left(\int_0^x \int_0^y \left[g(s,t) + \frac{h(s,t)}{p}\right] b(s,t) dt ds\right)\bigg\}^{\frac{1}{p}}, \quad (2.3.28)$$

for $x, y \in R_+$, where

$$e(x,y) = \int_0^x \int_0^y \left[g(s,t) a(s,t) + \left(\frac{p-1}{p} + \frac{a(s,t)}{p}\right) h(s,t)\right] dt ds, \quad (2.3.29)$$

for $x, y \in R_+$.

(c_2) Let $c(x,y)$ be a real-valued, continuous, positive and nondecreasing function defined for $x, y \in R_+$.If

$$u^p(x,y) \le c^p(x,y) + b(x,y)$$
$$\times \int_0^x \int_0^y [g(s,t) u^p(s,t) + h(s,t) u(s,t)] dt ds, \quad (2.3.30)$$

for $x, y \in R_+$, then

$$u(x,y) \le c(x,y) \{1 + b(x,y) e_0(x,y)$$
$$\times \exp\left(\int_0^x \int_0^y \left[g(s,t) + \frac{h(s,t)}{p} c^{1-p}(s,t)\right] b(s,t) dt ds\right)\bigg\}^{\frac{1}{p}}, \quad (2.3.31)$$

for $x, y \in R_+$, where

$$e_0(x,y) = \int_0^x \int_0^y \left[g(s,t) + h(s,t) c^{1-p}(s,t) \right] dtds, \tag{2.3.32}$$

for $x, y \in R_+$.

Proof. (c_1) Define a function $z(x,y)$ by

$$z(x,y) = \int_0^x \int_0^y \left[g(s,t) u^p(s,t) + h(s,t) u(s,t) \right] dtds, \tag{2.3.33}$$

then $z(0,y) = z(x,0) = 0$ and (2.3.27) can be written as

$$u^p(x,y) \le a(x,y) + b(x,y) z(x,y). \tag{2.3.34}$$

From (2.3.34) and using the elementary inequality (1.3.11) (see [30, p. 30]) we observe that

$$u(x,y) \le (a(x,y) + b(x,y) z(x,y))^{\frac{1}{p}} (1)^{1/(p/p-1)}$$

$$\le \frac{p-1}{p} + \frac{a(x,y)}{p} + \frac{b(x,y)}{p} z(x,y). \tag{2.3.35}$$

From (2.3.33)-(2.3.35) it is easy to observe that

$$z(x,y) \le e(x,y) + \int_0^x \int_0^y \left[g(s,t) + \frac{h(s,t)}{p} \right] b(s,t) z(s,t) dtds, \tag{2.3.36}$$

where $e(x,y)$ is defined by (2.3.29). Clearly $e(x,y)$ is nonnegative, continuous and nondecreasing for $x, y \in R_+$. A suitable application of Theorem 4.2.2 given in [34, p. 325] to (2.3.36) yields

$$z(x,y) \le e(x,y) \exp\left(\int_0^x \int_0^y \left[g(s,t) + \frac{h(s,t)}{p} \right] b(s,t) dtds \right), \tag{2.3.37}$$

for $x, y \in R_+$ The required inequality (2.3.28) follows from (2.3.34) and (2.3.37).

(c_2) Since $c(x,y)$ is positive, continuous and nondecreasing function for $x, y \in R_+$, from (2.3.30) we observe that

$$\left(\frac{u(x,y)}{c(x,y)} \right)^p \le 1 + \int_0^x \int_0^y \left[g(s,t) \left(\frac{u(s,t)}{c(s,t)} \right)^p \right.$$

$$\left. + h(s,t) c^{1-p}(s,t) \frac{u(s,t)}{c(s,t)} \right] dtds. \tag{2.3.38}$$

Now a suitable application of the inequality given in part (c_1) to (2.3.38) yields the desired inequality in (2.3.31).

Theorem 2.3.4. Let $u(x,y), a(x,y), b(x,y) \in C(R_+^2, R_+)$ and $p > 1$ be a real constant.

(d_1) Let $f \in C(R_+^3, R_+)$ satisfies the condition

$$0 \le f(x,y,u) - f(x,y,v) \le m(x,y,v)(u-v), \tag{2.3.39}$$

for $x, y \in R_+, u \ge v \ge 0$, where $m \in C(R_+^3, R_+)$.If

$$u^p(x,y) \le a(x,y) + b(x,y) \int_0^x \int_0^y f(s,t,u(s,t))\, dtds, \tag{2.3.40}$$

for $x, y \in R_+$, then

$$u(x,y) \le \{a(x,y) + b(x,y)\bar{e}(x,y)$$

$$\times \exp\left(\int_0^x \int_0^y m\left(s,t,\frac{p-1}{p} + \frac{a(s,t)}{p}\right)\frac{b(s,t)}{p} dtds\right)\Bigg\}^{\frac{1}{p}}, \tag{2.3.41}$$

for $x, y \in R_+$, where

$$\bar{e}(x,y) = \int_0^x \int_0^y f\left(s,t,\frac{p-1}{p} + \frac{a(s,t)}{p}\right) dtds, \tag{2.3.42}$$

for $x, y \in R_+$.

(d_2) Let $f \in C(R_+^3, R_+)$ and $\Phi \in C(R_+, R_+)$ be strictly increasing with $\Phi(0) = 0$ and

$$0 \le f(x,y,u) - f(x,y,v) \le m(x,y,v)\Phi^{-1}(u-v), \tag{2.3.43}$$

for $x, y \in R_+$, $u \ge v \ge 0$, where $m \in C(R_+^3, R_+)$ and Φ^{-1} is the inverse function of Φ and

$$\Phi^{-1}(uv) \le \Phi^{-1}(u)\Phi^{-1}(v), \tag{2.3.44}$$

for $u, v \in R_+$. If

$$u^p(x,y) \le a(x,y) + b(x,y)\Phi\left(\int_0^x \int_0^y f(s,t,u(s,t))\, dtds\right), \tag{2.3.45}$$

for $x, y \in R_+$, then

$$u(x,y) \le \{a(x,y) + b(x,y)\Phi(\bar{e}(x,y)$$

$$\times \exp\left(\int_0^x \int_0^y m\left(s,t,\frac{p-1}{p} + \frac{a(s,t)}{p}\right)\frac{b(s,t)}{p} dtds\right)\Bigg)\Bigg\}^{\frac{1}{p}}, \tag{2.4.46}$$

for $x, y \in R_+$, where $\bar{e}(x,y)$ is defined by (2.3.42).

(d_1) Define a function $z(x,y)$ by

$$z(x,y) = \int_0^x \int_0^y f(s,t,u(s,t))\,dtds, \tag{2.3.47}$$

then as in the proof of Theorem 2.3.3 part (c_1),from (2.3.40) we see that the inequalities (2.3.34), (2.3.35) hold. From (2.3.47), (2.3.35) and the assumptions on f, it follows that

$$z(x,y) \leq \int_0^x \int_0^y \left[f\left(s,t,\frac{p-1}{p} + \frac{a(s,t)}{p} + \frac{b(s,t)}{p} z(s,t)\right) \right.$$

$$\left. - f\left(s,t,\frac{p-1}{p} + \frac{a(s,t)}{p}\right) + f\left(s,t,\frac{p-1}{p} + \frac{a(s,t)}{p}\right)\right] dtds$$

$$\leq \bar{e}(x,y) + \int_0^x \int_0^y m\left(s,t,\frac{p-1}{p} + \frac{a(s,t)}{p}\right)\frac{b(s,t)}{p} z(s,t)\,dtds, \tag{2.3.48}$$

where $\bar{e}(x,y)$ is defined by (2.3.42). Clearly $\bar{e}(x,y)$ is nonnegative, continuous and nondecreasing function for $x, y \in R_+$. A suitable application of Theorem 4.2.2 given in [34, p. 325] to (2.3.48) yields

$$z(x,y) \leq \bar{e}(x,y) \exp\left(\int_0^x \int_0^y m\left(s,t,\frac{p-1}{p} + \frac{a(s,t)}{p}\right)\frac{b(s,t)}{p} dtds\right). \tag{2.3.49}$$

From (2.3.34) and (2.3.49) the desired inequality in (2.3.41) follows.

(d_2) Defining a function $z(x,y)$ by (2.3.47) and following the arguments as in the proof of Theorem 2.3.3, part (c_1) we see that, corresponding to the inequalities (2.3.34) and (2.3.35) we have the following inequalities

$$u^p(x,y) \leq a(x,y) + b(x,y)\Phi(z(x,y)), \tag{2.3.50}$$

and

$$u(x,y) \leq \frac{p-1}{p} + \frac{a(x,y)}{p} + \frac{b(x,y)}{p}\Phi(z(x,y)). \tag{2.3.51}$$

From (2.3.47), (2.3.51) and the assumptions on f and Φ we observe that

$$z(x,y) \leq \int_0^x \int_0^y \left[f\left(s,t,\frac{p-1}{p} + \frac{a(s,t)}{p} + \frac{b(s,t)}{p}\Phi(z(x,y))\right) \right.$$

$$\left. - f\left(s,t,\frac{p-1}{p} + \frac{a(s,t)}{p}\right) + f\left(s,t,\frac{p-1}{p} + \frac{a(s,t)}{p}\right)\right] dtds$$

$$\leq \bar{e}(x,y) + \int_0^x \int_0^y f\left(s,t,\frac{p-1}{p} + \frac{a(s,t)}{p}\right) \Phi^{-1}\left(\frac{b(s,t)}{p}\right) z(s,t)\, dtds, \quad (2.3.52)$$

where $\bar{e}(x,y)$ is defined by (2.3.42). Clearly $\bar{e}(x,y)$ is nonnegative, continuous and nondecreasing function for $x, y \in R_+$. A suitable application of Theorem 4.2.2 given in [34, p. 325] yields

$$z(x,y) \leq \bar{e}(x,y)$$
$$\times \exp\left(\int_0^x \int_0^y m\left(s,t,\frac{p-1}{p} + \frac{a(s,t)}{p}\right) \Phi^{-1}\left(\frac{b(s,t)}{p}\right) dtds\right). \quad (2.3.53)$$

The required inequality in (2.3.46) follows from (2.3.50) and (2.3.53).

In the following theorem we give the inequalities established in [45].

Theorem 2.3.5. Let $u(x,y), f(x,y) \in C(R_+^2, R_+)$, $h(x,y,s,t) \in C(R_+^4, R_+)$ for $0 \leq s \leq x < \infty$, $0 \leq t \leq y < \infty$ and $c \geq 0$, $p > 1$ are real constants.

(k_1) Let $g \in C(R_+, R_+)$ be a nondecreasing function, $g(u) > 0$ for $u > 0$. If

$$u^p(x,y) \leq c + \int_0^x \int_0^y \Big[f(s,t)\, g(u(s,t))$$
$$+ \int_0^s \int_0^t h(s,t,\sigma,\eta) g(u(\sigma,\eta))\, d\eta d\sigma\Big] dtds, \quad (2.3.54)$$

for $x, y \in R_+$, then for $0 \leq x \leq x_1, 0 \leq y \leq y_1; x, x_1, y, y_1 \in R_+$,

$$u(x,y) \leq \left\{G^{-1}[G(c) + A(x,y)]\right\}^{\frac{1}{p}}, \quad (2.3.55)$$

where

$$A(x,y) = \int_0^x \int_0^y \left[f(s,t) + \int_0^s \int_0^t h(s,t,\sigma,\eta)\, d\eta d\sigma\right] dtds, \quad (2.3.56)$$

$$G(r) = \int_{r_0}^r \frac{ds}{g\left(s^{\frac{1}{p}}\right)}, r > 0, \quad (2.3.57)$$

$r_0 > 0$ is arbitrary, G^{-1} is the inverse function of G and $x_1, y_1 \in R_+$ are chosen so that

$$G(c) + A(x,y) \in Dom\left(G^{-1}\right),$$

for all x, y lying in the intervals $0 \leq x \leq x_1, 0 \leq y \leq y_1$ of R_+.

(k_2) If

$$u^p(x,y) \le c + \int_0^x \int_0^y \left[f(s,t)\, u(s,t) \right.$$

$$\left. + \int_0^s \int_0^t h(s,t,\sigma,\eta) u(\sigma,\eta)\, d\eta d\sigma \right] dt ds, \tag{2.3.58}$$

for $x, y \in R_+$,, then

$$u(x,y) \le \left\{ c^{\frac{p-1}{p}} + \left(\frac{p-1}{p} \right) A(x,y) \right\}^{\frac{1}{p-1}}, \tag{2.3.59}$$

for $x, y \in R_+$, where $A(x,y)$ is defined by (2.3.56).

Proof. (k_1) Let $c > 0$ and define a function $z(x,y)$ by the right hand side of (2.3.54). Then $z(0,y) = z(x,0) = c$, $u(x,y) \le (z(x,y))^{\frac{1}{p}}$ and

$$D_1 z(x,y) = \int_0^y \left[f(x,t)\, g(u(x,t)) + \int_0^x \int_0^t h(x,t,\sigma,\eta)\, g(u(\sigma,\eta))\, d\eta d\sigma \right] dt$$

$$\le \int_0^y \left[f(x,t)\, g\left((z(x,t))^{\frac{1}{p}} \right) + \int_0^x \int_0^t h(x,t,\sigma,\eta)\, g\left((z(\sigma,\eta))^{\frac{1}{p}} \right) d\eta d\sigma \right] dt$$

$$\le g\left((z(x,y))^{\frac{1}{p}} \right) \int_0^y \left[f(x,t) + \int_0^x \int_0^t h(x,t,\sigma,\eta)\, d\eta d\sigma \right] dt. \tag{2.3.60}$$

From (2.3.57) and (2.3.60) we observe that

$$D_1 G(z(x,y)) = \frac{D_1 z(x,y)}{g\left((z(x,y))^{\frac{1}{p}} \right)}$$

$$\le \int_0^y \left[f(x,t) + \int_0^x \int_0^t h(x,t,\sigma,\eta)\, d\eta d\sigma \right] dt. \tag{2.3.61}$$

Keeping y fixed in (2.3.61), setting $x = s$ and integrating with respect to s from 0 to x and using the fact that $z(0,y) = c$, we have

$$G(z(x,y)) \le G(c) + A(x,y). \tag{2.3.62}$$

Now substituting the bound on $z(x,y)$ from (2.3.62) in $u(x,y) \le (z(x,y))^{\frac{1}{p}}$ we obtain the desired bound in (2.3.55). The proof of the case when $c \ge 0$ can be completed as mentioned in the proof of Theorem 1.3.3. The subdomain $0 \le x \le x_1, 0 \le y \le y_1$ is obvious.

(k_2) The proof is similar to that of given in Theorem 1.3.4 and we omit it here.

Remark 2.3.2. We note that the upper bound on the inequality (2.3.58) when $p = 1$ and $h = 0$ is first established by Wendroff, see [4, p. 154]. For various useful generalizations and variants of Wendroff's inequality, see [3,34,42].

2.4 Inequalities involving iterated integrals

During the past few years some useful integral inequalities in two independent variables which provide explicit bounds on unknown functions have appeared in the literature. In this section we shall deal with the inequalities involving iterated integrals established by Pachpatte in [53,72,78] which can be used as tools in certain applications.

First we introduce some notation to simplify the details of presentation. Let $I = [0, \alpha)\,, J = [0, \beta)$ are the given subsets of R and $\Delta = I \times J$. Let

$$D = \left\{(x, y, s, t) \in \Delta^2 : 0 \le s \le x < \alpha, 0 \le t \le y < \beta\right\},$$

and

$$E = \left\{(x, y, s, t, \sigma, \tau) \in \Delta^3 : 0 \le \sigma \le s \le x < \alpha, 0 \le \tau \le t \le y < \beta\right\}.$$

For any functions $k(x, y, s, t)$, $D_1k\,(x, y, s, t)\,, D_2k\,(x, y, s, t)$, $D_2D_1k\,(x, y, s, t) \in C\,(D, R_+)$ and $h\,(x, y, s, t, \sigma, \tau)\,, D_1h\,(x, y, s, t, \sigma, \tau)\,, D_2h\,(x, y, s, t, \sigma, \tau)\,, D_2D_1 h\,(x, y, s, t, \sigma, \tau) \in C\,(E, R_+)\,,$ we set

$$A\,(x, y) = k\,(x, y, x, y) + \int_0^x D_1k\,(x, y, \xi, y)\,d\xi$$

$$+ \int_0^y D_2k\,(x, y, x, \eta)\,d\eta + \int_0^x\int_0^y D_2D_1k\,(x, y, \xi, \eta)\,d\eta d\xi, \tag{2.4.1}$$

$$B\,(x, y) = \int_0^x\int_0^y h\,(x, y, x, y, \sigma, \tau)d\tau d\sigma$$

$$+ \int_0^x \left(\int_0^s\int_0^y D_1h\,(x, y, s, y, \sigma, \tau)d\tau d\sigma\right) ds$$

$$+ \int_0^y \left(\int_0^x\int_0^t D_2h\,(x, y, x, t, \sigma, \tau)d\tau d\sigma\right) dt$$

$$+\int_0^x\int_0^y\left(\int_0^s\int_0^t D_2D_1h\,(x,y,s,t,\sigma,\tau)d\tau d\sigma\right)dtds. \tag{2.4.2}$$

For $i = 1, ..., n$, let $I_i = \{(t_1, ..., t_i) : (t_1, ..., t_i) \in I^i\}$, $J_i = \{(s_1, ..., s_i) : (s_1, ..., s_i) \in J^i\}$ and $\Delta_i = I_i \times J_i$ and any functions $w(s,t), a(s,t), b(s,t) \in C(\Delta, R_+)$, and

$$L_i(t_1, ..., t_i, s_1, ..., s_i, w(t_i, s_i)), M_i(t_1, ..., t_i, s_1, ..., s_i, a(t_i, s_i)) \in C(\Delta_i \times R_+, R_+)$$

, we set

$$H_i[w](t,s)$$

$$=\int_0^t\int_0^s\left(\int_0^{t_1}\int_0^{s_1}\cdots\left(\int_0^{t_{i-1}}\int_0^{s_{i-1}} L_i(t_1,...,t_i,s_1,...,s_i,w(t_i,s_i))\,ds_idt_i\right)ds_{i-1}dt_{i-1}...\right)$$

$$\times ds_1dt_1, \tag{2.4.3}$$

$$P(t,s) = \int_0^t\int_0^s L_1(t_1,s_1,a(t_1,s_1))\,ds_1dt_1$$

$$+\int_0^t\int_0^s\left(\int_0^{t_1}\int_0^{s_1} L_2(t_1,t_2,s_1,s_2,a(t_2,s_2))\,ds_2dt_2\right)ds_1dt_1 + ...$$

$$+\int_0^t\int_0^s\left(\int_0^{t_1}\int_0^{s_1}\cdots\left(\int_0^{t_{n-1}}\int_0^{s_{n-1}} L_i(t_1,...,t_n,s_1,...,s_n,w(t_n,s_n))\,ds_ndt_n\right)ds_{n-1}dt_{n-1}...\right)$$

$$\times ds_1dt_1, \tag{2.4.4}$$

$$Q(s,t) = M_1(t,s,a(t,s))\,b(t,s) + \int_0^t\int_0^s M_2(t,t_2,s,s_2,a(t_2,s_2))\,b(t_2,s_2)\,ds_2dt_2 + ...$$

$$+\int_0^t\int_0^s\left(\int_0^{t_2}\int_0^{s_2}\cdots\left(\int_0^{t_{n-1}}\int_0^{s_{n-1}} M_n(t,t_2,...t_n,s,s_2,...s_n,a(t_n,s_n))\right.\right.$$

$$\times b(t_n,s_n)\,ds_ndt_n)\,ds_{n-1}dt_{n-1}...)\,ds_2dt_2. \tag{2.4.5}$$

Our first theorem deals with the inequalities proved in [53].

Theorem 2.4.1. Let $u(x,y), f(x,y), a(x,y) \in C(\Delta, R_+)$; $k(x,y,s,t)$, $D_1k(x,y,s,t)$, $D_2k(x,y,s,t)$, $D_2D_1k(x,y,s,t) \in C(D, R_+)$ and $c \geq 0$ be a real constant.

(a_1) If

$$u(x,y) \leq c + \int_0^x \int_0^y f(s,t) \left[u(s,t) + \int_0^s \int_0^t k(s,t,\sigma,\tau)\, u(\sigma,\tau)\, d\tau d\sigma \right] \times dtds, \tag{2.4.6}$$

for $(x,y) \in \Delta$, then

$$u(x,y) \leq c \left[1 + \int_0^x \int_0^y f(s,t) \times \exp\left(\int_0^s \int_0^t [f(\sigma,\tau) + A(\sigma,\tau)] d\tau d\sigma \right) dtds \right], \tag{2.4.7}$$

for $(x,y) \in \Delta$, where $A(x,y)$ is defined by (2.4.1).

(a_2) If

$$u(x,y) \leq a(x,y) + \int_0^x \int_0^y f(s,t) \left[u(s,t) + \int_0^s \int_0^t k(s,t,\sigma,\tau)\, u(\sigma,\tau)\, d\tau d\sigma \right] \times dtds, \tag{2.4.8}$$

for $(x,y) \in \Delta$, then

$$u(x,y) \leq e(x,y) \left[1 + \int_0^x \int_0^y f(s,t) \times \exp\left(\int_0^s \int_0^t [f(\sigma,\tau) + A(\sigma,\tau)] d\tau d\sigma \right) dtds \right], \tag{2.4.9}$$

for $(x,y) \in \Delta$, where

$$e(x,y) = \int_0^x \int_0^y f(s,t) \left[a(s,t) + \int_0^s \int_0^t k(s,t,\sigma,\tau)\, a(\sigma,\tau)\, d\tau d\sigma \right] dtds,$$

for $(x,y) \in \Delta$ and $A(x,y)$ is defined by (2.4.1).

Proof. (a_1) Let $c > 0$ and define a function $z(x, y)$ by the right hand side of (2.4.6). Then $z(x, y) > 0$, $z(0, y) = z(x, 0) = c$, $u(x, y) \leq z(x, y)$ and

$$D_2 D_1 z(x, y) = f(x, y)\left[u(x, y) + \int_0^x \int_0^y k(x, y, \sigma, \tau) u(\sigma, \tau) d\tau d\sigma\right]$$

$$\leq f(x, y)\left[z(x, y) + \int_0^x \int_0^y k(x, y, \sigma, \tau) z(\sigma, \tau) d\tau d\sigma\right]. \tag{2.4.10}$$

Define a function $v(x, y)$ by

$$v(x, y) = z(x, y) + \int_0^x \int_0^y k(x, y, \sigma, \tau) z(\sigma, \tau) d\tau d\sigma. \tag{2.4.11}$$

Then $v(x, y) > 0$, $v(0, y) = z(0, y) = c$, $v(x, 0) = z(x, 0) = c$, $z(x, y) \leq v(x, y)$, $D_2 D_1 z(x, y) \leq f(x, y) v(x, y)$, $v(x, y)$ is nondecreasing for $(x, y) \in \Delta$ and

$$D_2 D_1 v(x, y) = D_2 D_1 z(x, y) + k(x, y, x, y) z(x, y) + \int_0^x D_1 k(x, y, \sigma, y) z(\sigma, y) d\sigma$$

$$+ \int_0^y D_2 k(x, y, x, \tau) z(x, \tau) d\tau + \int_0^x \int_0^y D_2 D_1 k(x, y, \sigma, \tau) z(\sigma, \tau) d\tau d\sigma$$

$$\leq f(x, y) v(x, y) + k(x, y, x, y) v(x, y) + \int_0^x D_1 k(x, y, \sigma, y) v(\sigma, y) d\sigma$$

$$+ \int_0^y D_2 k(x, y, x, \tau) v(x, \tau) d\tau + \int_0^x \int_0^y D_2 D_1 k(x, y, \sigma, \tau) v(\sigma, \tau) d\tau d\sigma$$

$$\leq [f(x, y) + A(x, y)] v(x, y), \tag{2.4.12}$$

where $A(x, y)$ is defined by (2.4.1). Now by following the proof of Theorem 4.2.1 given in [34], the inequality (2.4.12) implies

$$v(x, y) \leq c \exp\left(\int_0^x \int_0^y [f(\sigma, \tau) + A(\sigma, \tau)] d\tau d\sigma\right). \tag{2.4.13}$$

Using (2.4.13) in (2.4.10) and integrating the resulting inequality first from 0 to y and then from 0 to x for $(x, y) \in \Delta$ we get

$$z(x, y) \leq c\left[1 + \int_0^x \int_0^y f(s, t)\right.$$

$$\times \exp\left(\int_0^s\int_0^t [f(\sigma,\tau)+A(\sigma,\tau)]d\tau d\sigma\right) dtds\Bigg]. \tag{2.4.14}$$

Using (2.4.14) in $u(x,y)\leq z(x,y)$, we get the required inequality in (2.4.7). If $c\geq 0$, we carry out the above procedure with $c+\varepsilon$ instead of c, where $\varepsilon>0$ is an arbitrary small constant, and then subsequently pass to the limit as $\varepsilon\to 0$ to obtain (2.4.7).

(a_2) The proof can be completed by closely looking at the proofs of Theorem 1.4.1, part (a_2) and (a_1) given above. Here we omit the details.

Remark 2.4.1. If we take $k(x,y,s,t)=k(s,t)$, then the inequality established in (a_1) reduces to the inequality given in [34, Remark 4.4.1]. For several other inequalities of the type given in Theorem 2.4.1, see [34].

The next two theorems are established in [71] which can be used in certain situations.

Theorem 2.4.2. Let $u(x,y)\in C(\Delta,R_+)$; $k(x,y,s,t)$, $D_1k(x,y,s,t)$, $D_2k(x,y,s,t)$, $D_2D_1k(x,y,s,t)\in C(D,R_+)$; $h(x,y,s,t,\sigma,\tau)$, $D_1h(x,y,s,t,\sigma,\tau)$, $D_2h(x,y,s,t,\sigma,\tau)$, $D_2D_1h(x,y,s,t)\sigma,\tau)\in C(E,R_+)$ and $c\geq 0$ be a real constant.

(b_1) If

$$u(x,y)\leq c+\int_0^x\int_0^y k(x,y,s,t)\,u(s,t)\,dtds$$

$$+\int_0^x\int_0^y\left(\int_0^s\int_0^t h(x,y,s,t,\sigma,\tau)\,u(\sigma,\tau)\,d\tau d\sigma\right)dtds, \tag{2.4.15}$$

for $(x,y)\in\Delta$, then

$$u(x,y)\leq c\exp\left(\int_0^x\int_0^y [A(m,n)+B(m,n)]\,dndm\right), \tag{2.4.16}$$

for $(x,y)\in\Delta$, where $A(x,y)$, $B(x,y)$ are given by (2.4.1), (2.4.2).

(b_2) Let $g(u)$ be continuously differentiable function defined for $u\geq 0$, $g(u)>0$ for $u>0$ and $g'(u)\geq 0$ for $u\geq 0$. If

$$u(x,y)\leq c+\int_0^x\int_0^y k(x,y,s,t)\,g(u(s,t))\,dtds$$

$$+\int_0^x\int_0^y\left(\int_0^s\int_0^t h(x,y,s,t,\sigma,\tau)\,g(u(\sigma,\tau))\,d\tau d\sigma\right)dtds, \tag{2.4.17}$$

for $(x,y)\in\Delta$, then for $0\le x\le x_1, 0\le y\le y_1; x,x_1\in I, y,y_1\in J$,

$$u(x,y)\le G^{-1}\left[G(c)+\int_0^x\int_0^y[A(m,n)+B(m,n)]\,dndm\right], \tag{2.4.18}$$

where $A(x,y), B(x,y)$ are given by (2.4.1), (2.4.2),

$$G(r)=\int_{r_0}^r\frac{ds}{g(w)}, r>0, \tag{2.4.19}$$

$r_0>0$ is arbitrary, G^{-1} is the inverse function of G and $x_1\in I, y_1\in J$ are chosen so that

$$G(c)+\int_0^x\int_0^y[A(m,n)+B(m,n)]\,dndm\in Dom\left(G^{-1}\right),$$

for $(x,y)\in\Delta$ such that $0\le x\le x_1, 0\le y\le y_1$.

Proof. We first assume that $c>0$ and define a function $z(x,y)$ by the right hand side of (2.4.15). Then $z(x,y)>0$, $z(0,y)=z(x,0)=c$, $u(x,y)\le z(x,y)$ and $z(x,y)$ is nondecreasing in both the variables $(x,y)\in\Delta$. It is easy to observe that (see [34, p. 328])

$$D_2D_1z(x,y)\le[A(x,y)+B(x,y)]\,z(x,y), \tag{2.4.20}$$

where $A(x,y), B(x,y)$ are given by (2.4.1), (2.4.2). Now by following the proof of Theorem 4.2.1 given in [34], from (2.4.20) we get

$$,z(x,y)\le c\exp\left(\int_0^x\int_0^y[A(m,n)+B(m,n)]\,dndm\right), \tag{2.4.21}$$

for $(x,y)\in\Delta$. Using (2.4.21) in $u(x,y)\le z(x,y)$, we get the required inequality in (2.4.16). If $c\ge 0$ we carry out the above procedure with $c+\varepsilon$ instead of c, where $\varepsilon>0$ is an arbitrary small constant, and subsequently pass to the limit as $\varepsilon\to 0$ to obtain (2.4.16).

(b_2) We note that since $g'(u)\ge 0$ on R_+, the function $g(u)$ is monotone increasing on $(0,\infty)$. Assume that $c>0$ and define a function $z(x,y)$ by the right hand side of (2.4.17). Then $z(x,y)>0$, $z(0,y)=z(x,0)=c$, $u(x,y)\le$

$z(x,y)$ and $z(x,y)$ is nondecreasing in both the variables $(x,y) \in \Delta$. It is easy to observe that

$$D_2 D_1 z(x,y) \leq [A(x,y) + B(x,y)]\, g(z(x,y)), \tag{2.4.22}$$

where $A(x,y), B(x,y)$ are given by (2.4.1), (2.4.2). The remaining proof can be completed by following the proof of Theorem 5.2.1 given in [34] . The proof of the case when $c \geq 0$ follows as mentioned in the proof of (b_1).

Theorem 2.4.3. Let $u(x,y), a(x,y) \in C(\Delta, R_+)$, $k(x,y,s,t) \in C(D, R_+)$, $h(x,y,s,t,\sigma,\tau) \in C(E, R_+)$ and $c \geq 0$ be a real constant.

(c_1) If

$$u(x,y) \leq c + \int_0^x \int_0^y a(s,t)\, u(s,t)\, dt ds$$

$$+ \int_0^x \int_0^y \left(\int_0^s \int_0^t k(s,t,\sigma,\tau)\, u(\sigma,\tau)\, d\tau d\sigma \right) dt ds$$

$$+ \int_0^x \int_0^y \left(\int_0^s \int_0^t \left(\int_0^\sigma \int_0^\tau h(s,t,\sigma,\tau,m,n) u(m,n)\, dn dm \right) d\tau d\sigma \right) dt ds, \tag{2.4.23}$$

for $(x,y) \in \Delta$, then

$$u(x,y) \leq c \exp\left(\int_0^x \int_0^y N(s,t)\, dt ds \right), \tag{2.4.24}$$

for $(x,y) \in \Delta$, where

$$N(x,y) = a(x,y) + \int_0^x \int_0^y k(x,y,\sigma,\tau) d\tau d\sigma$$

$$+ \int_0^x \int_0^y \left(\int_0^\sigma \int_0^\tau h(x,y,\sigma,\tau,m,n)\, dn dm \right) d\tau d\sigma. \tag{2.4.25}$$

(c_2) Let $g(u)$ be as in Theorem 2.4.2, part (b_2). If

$$u(x,y) \leq c + \int_0^x \int_0^y a(s,t)\, g(u(s,t))\, dt ds$$

$$+\int_0^x\int_0^y\left(\int_0^s\int_0^t k(s,t,\sigma,\tau)\,g(u(\sigma,\tau))\,d\tau d\sigma\right)dtds$$

$$+\int_0^x\int_0^y\left(\int_0^s\int_0^t\left(\int_0^\sigma\int_0^\tau h(s,t,\sigma,\tau,m,n)\,g(u(m,n))\,dndm\right)d\tau d\sigma\right)$$

$$\times dtds, \tag{2.4.26}$$

for $(x,y)\in\Delta$, then for $0\le x\le x_2, 0\le y\le y_2; x,x_2\in I, y,y_2\in J$,

$$u(x,y)\le G^{-1}\left[G(c)+\int_0^x\int_0^y N(s,t)\,dtds\right], \tag{2.4.27}$$

where $N(x,y)$ is given by (2.4.25), G, G^{-1} are as in Theorem 2.4.2, part (b_2) and $x_2\in I$, $y_2\in J$ are chosen so that

$$G(c)+\int_0^x\int_0^y N(s,t)\,dtds\in Dom\left(G^{-1}\right),$$

for all $(x,y)\in\Delta$ such that $0\le x\le x_2, 0\le y\le y_2$.

Proof. (c_1) Let $c>0$ and define a function $z(x,y)$ by the right hand side of (2.4.23).Then $z(x,y)>0$, $z(0,y)=z(x,0)=c$, $u(x,y)\le z(x,y)$ and $z(x,y)$ is nondecreasing in both the variables $(x,y)\in\Delta$ and

$$D_2D_1z(x,y)=a(x,y)\,u(x,y)+\int_0^x\int_0^y k(x,y,\sigma,\tau)\,u(\sigma,\tau)d\tau d\sigma$$

$$+\int_0^x\int_0^y\left(\int_0^\sigma\int_0^\tau h(x,y,\sigma,\tau,m,n)\,u(m,n)\,dndm\right)d\tau d\sigma$$

$$\le N(x,y)\,z(x,y), \tag{2.4.28}$$

where $N(x,y)$ is given by (2.4.25). Following the proof of Theorem 4.2.1 given in [34], from (2.4.28) we get

$$z(x,y)\le c\exp\left(\int_0^x\int_0^y N(s,t)\,dtds\right). \tag{2.4.29}$$

Using (2.4.29) in $u(x,y)\le z(x,y)$ we get the desired inequality in (2.4.24). The case when $c\ge 0$ follows as noted in the proof of Theorem 2.4.1, part (a_1).

(c_2) The proof can be completed by following the proof of (c_1) and closely looking at the proof of Theorem 5.2.1 given in [34]. Here we leave the details to the reader.

Remark 2.4.2. We note that, by following the proof of Theorem 2.2.1, one can very easily obtain the bounds on the inequalities in Theorems 2.4.2 and 2.4.3, when the constant c is replaced by the function $c(x,y)$ satisfying some suitable conditions.

The inequalities embodied in the following theorem are established in [78].

Theorem 2.4.4. Let $u(t,s), a(t,s), b(t,s) \in C(\Delta, R_+)$.

(d_1) For $i = 1, ..., n$ let the functions $L_i \in C(\Delta_i \times R_+, R_+)$ satisfy the conditions

$$0 \le L_i(t_1, ..., t_i, s_1, ..., s_i, x(t_i, s_i)) - L_i(t_1, ..., t_i, s_1, ..., s_i, y(t_i, s_i))$$

$$\le M_i(t_1, ..., t_i, s_1, ..., s_i, y(t_i, s_i))(x(t_i, s_i) - y(t_i, s_i)), \tag{2.4.30}$$

for $(t_1, ..., t_i, s_1, ..., s_i,) \in \Delta_i$ and $x(t_i, s_i) \ge y(t_i, s_i) \ge 0$, where $M_i \in C(\Delta_i \times R_+, R_+)$. If

$$u(s,t) \le a(s,t) + b(s,t) \sum_{i=1}^{n} H_i[u](t,s), \tag{2.4.31}$$

for $(t,s) \in \Delta$, then

$$u(s,t) \le a(s,t) + b(s,t) P(s,t) \exp\left(\int_0^t \int_0^s Q(t_1, s_1)\, ds_1 dt_1\right), \tag{2.4.32}$$

for $(t,s) \in \Delta$, where $P(s,t), Q(s,t)$ are given by (2.4.4),(2.4.5).

(d_2) Let $\Psi \in C(R_+, R_+)$ be strictly increasing function with $\Psi(0) = 0$ and Ψ^{-1} is the inverse function of Ψ. For $i = 1, ..., n$ let the functions $L_i \in C(\Delta_i \times R_+, R_+)$ satisfy the conditions

$$0 \le L_i(t_1, ..., t_i, s_1, ..., s_i, x(t_i, s_i)) - L_i(t_1, ..., t_i, s_1, ..., s_i, y(t_i, s_i))$$

$$\le M_i(t_1, ..., t_i, s_1, ..., s_i, y(t_i, s_i))\, \psi^{-1}(x(t_i, s_i) - y(t_i, s_i)), \tag{2.4.33}$$

for $(t_1, ..., t_i, s_1, ..., s_i,) \in \Delta_i$ and $x(t_i, s_i) \ge y(t_i, s_i) \ge 0$, where $M_i \in C(\Delta_i \times R_+, R_+)$. If

$$u(s,t) \le a(s,t) + \psi\left(b(s,t) \sum_{i=1}^{n} H_i[u](t,s)\right), \tag{2.4.34}$$

for $(x, y) \in \Delta$, then

$$u(t,s) \leq a(t,s) + \Psi\left(b(t,s)\,P(t,s)\exp\left(\int_0^t\int_0^s Q(t_1,s_1)\,ds_1dt_1\right)\right), \quad (2.4.35)$$

For $(t, s) \in \Delta$,, where $P(t, s), Q(t, s)$ are given by (2.4.4), (2.4.5).

(d_3) Let $L_i, M_i, \Psi, \psi^{-1}$ be as in part (d_2) and the conditions in (2.4.33) and

$$\psi^{-1}(xy) \leq \psi^{-1}(x)\,\psi^{-1}(y), \quad (2.4.36)$$

for all $x, y \in R_+$ hold. If

$$u(t,s) \leq a(t,s) + b(t,s)\,\Psi\left(\sum_{i=1}^{n} H_i[u](t,s)\right), \quad (2.4.37)$$

for $(t, s) \in \Delta$, then

$$u(t,s) \leq a(t,s) + b(t,s)\,\Psi\left(P(t,s)\exp\left(\int_0^t\int_0^s Q_1(t_1,s_1)\,ds_1dt_1\right)\right), \quad (2.4.38)$$

for $(t, s) \in \Delta$, where $P(t, s)$ is given by (2.4.4) and $Q_1(t, s)$ is obtained by replacing $b(t, s)$ by $\psi^{-1}(b(t, s))$ on the right hand side of (2.4.5).

(d_4) For $i = 1, ..., n$, let L_i, M_i be as in part (d_1) and (2.4.30) hold. Let $g(u)$ be as in Theorem 2.4.2, part (b_2). If

$$u(t,s) \leq a(t,s) + b(t,s)\,g\left(\sum_{i=1}^{n} H_i[u](t,s)\right), \quad (2.4.39)$$

for $(t, s) \in \Delta$, then for $0 \leq t \leq \bar{t}, 0 \leq s \leq \bar{s}; t, \bar{t} \in I, s, \bar{s} \in J$,

$$u(t,s) \leq a(t,s) + b(t,s)\,g\left(G^{-1}\left[G(P(t,s)) + \int_0^s\int_0^t Q(t_1,s_1)\,dt_1ds_1\right]\right), \quad (2.4.40)$$

where $P(s, t), Q(s, t)$ are given by (2.4.4),(2.4.5), G, G^{-1} be as in Theorem 2.4.2, part (b_2) and $(\bar{t}, \bar{s}) \in \Delta$ be chosen so that

$$G(P(t,s)) + \int_0^t\int_0^s Q(t_1,s_1)\,ds_1dt_1 \in Dom\left(G^{-1}\right),$$

for all $(t, s) \in \Delta$ such that $0 \leq t \leq \bar{t}, 0 \leq s \leq \bar{s}$.

Proof. (d_1) Define a function $z(t,s)$ by

$$z(t,s) = \sum_{i=1}^{n} H_i[u](t,s)$$

$$= \int_0^t \int_0^s L_1(t_1, s_1, u(t_1, s_1))\, ds_1 dt_1$$

$$+ \int_0^t \int_0^s \left(\int_0^{t_1} \int_0^{s_1} L_2(t_1, t_2, s_1, s_2, u(t_2, s_2))\, ds_2 dt_2 \right) ds_1 dt_1 + \ldots$$

$$+ \int_0^t \int_0^s \left(\int_0^{t_1} \int_0^{s_1} \ldots \left(\int_0^{t_{n-1}} \int_0^{s_{n-1}} L_n(t_1, \ldots t_n, s_1, \ldots, s_n, u(t_n, s_n))\, ds_n dt_n \right.\right.$$

$$\left.\left.\right) ds_{n-1} dt_{n-1} \ldots\right) ds_1 dt_1. \tag{2.4.41}$$

Then $z(t,0) = z(0,s) = 0$, $z(t,s)$ is nondecreasing for $(t,s) \in \Delta$ and (2.4.31) can be restated as

$$u(t,s) \le a(t,s) + b(t,s)\, z(t,s). \tag{2.4.42}$$

From (2.4.41), (2.4.42) and the hypotheses we observe that

$$z(s,t) \le \int_0^t \int_0^s [\{L_1(t_1, s_1, a(t_1, s_1) + b(t_1, s_1)\, z(t_1, s_1))$$

$$-L_1(t_1, s_1, a(t_1, s_1))\} + L_1(t_1, s_1, a(t_1, s_1))]\, ds_1 dt_1$$

$$+ \int_0^t \int_0^s \left(\int_0^{t_1} \int_0^{s_1} [\{L_2(t_1, t_2, s_1, s_2, a(t_2, s_2) + b(t_2, s_2)\, z(t_2, s_2)) \right.$$

$$\left. -L_2(t_1, t_2, s_1, s_2, a(t_2, s_2))\} + L_2(t_1, t_2, s_1, s_2, a(t_2, s_2))]\, ds_2 dt_2 \right) ds_1 dt_1 + \ldots$$

$$+ \int_0^t \int_0^s \left(\int_0^{t_1} \int_0^{s_1} \ldots \left(\int_0^{t_{n-1}} \int_0^{s_{n-1}} [\{L_n(t_1, \ldots, t_n, s_1, \ldots, s_n, \right.\right.$$

$$a(t_n, s_n) + b(t_n, s_n)\, z(t_n, s_n))$$

$$-L_n(t_1, \ldots, t_n, s_1, \ldots, s_n, a(t_n, s_n))\} + L_n(t_1, \ldots, t_n, s_1, \ldots, s_n, a(t_n, s_n))]$$

$$\left.\left. \times ds_n dt_n \right) ds_{n-1} dt_{n-1} \ldots\right) ds_1 dt_1$$

$$\le P(s,t) + \int_0^t \int_0^s M_1(t_1, s_1, a(t_1, s_1))\, b(t_1, s_1)\, z(t_1, s_1) ds_1 dt_1$$

$$+\int_0^t\int_0^s\left(\int_0^{t_1}\int_0^{s_1} M_2\left(t_1,t_2,s_1,s_2,a\left(t_2,s_2\right)\right)b\left(t_2,s_2\right)z\left(t_2,s_2\right)ds_2dt_2\right)ds_1dt_1+...$$

$$+\int_0^t\int_0^s\left(\int_0^{t_1}\int_0^{s_1}\cdots\left(\int_0^{t_{n-1}}\int_0^{s_{n-1}} M_n\left(t_1,...,t_n,s_1,...,s_n,a\left(t_n,s_n\right)\right)\right.\right.$$

$$\times b\left(t_n,s_n\right)z\left(t_n,s_n\right)ds_ndt_n\right)ds_{n-1}dt_{n-1}...\Big)ds_1dt_1$$

$$\leq P\left(t,s\right)+\int_0^t\int_0^s Q\left(t_1,s_1\right)z\left(t_1,s_1\right)ds_1dt_1. \tag{2.4.43}$$

Clearly $P(t,s)$ is continuous, nonnegative and nondecreasing in $(t,s)\in\Delta$. Now a suitable application of Theorem 4.2.2 given in [34, p. 325] to (2.4.43) yields

$$z\left(t,s\right)\leq P\left(t,s\right)\exp\left(\int_0^t\int_0^s Q\left(t_1,s_1\right)ds_1dt_1\right). \tag{2.4.44}$$

Using (2.4.44) in (2.4.42) we get (2.4.32).

The proofs of the remaining inequalities can be completed by following the proof of (d_1) and closely looking at the proof of Theorem 1.4.4, parts $(d_2)-(d_4)$ and the similar results given in [34]. Here we omit the details.

Remark 2.4.3. If we take $L_1=L$ and $L_i=0$ for $i=2,...,n$ in the inequality established in (d_1), then we get the inequality given in Theorem 5.3.1, part (i) in [34]. The inequalities in parts $(d_2)-(d_4)$ can be considered as further generalizations of the inequality in Theorem 5.3.1, part (i) given in [34]

2.5 Estimates on some integral inequalities

In the qualitative analysis of certain classes of differential,integral and integrodifferential equations some specific type of integral inequalities play a vital role. In this section we offer some such integral inequalities established by Pachpatte in [41,48,62,72,76] involving functions of two variables.

The following two theorems contain the inequalities investigated in [62] and [72] respectively, which can be used in some applications.

Theorem 2.5.1. Let $I = [0, \alpha]$, $J = [0, \beta]$ and $\Delta = I \times J$. Let $u(x, y)$, $p(x, y)$, $f(x, y)$, $g(x, y)$, $h(x, y) \in C(\Delta, R_+)$ and suppose that

$$u(x, y) \leq c + \int_0^x p(s, y) u(s, y) \, ds + \int_0^x \int_0^y f(s, t) \Big[u(s, t)$$

$$+ \int_0^s \int_0^t g(\sigma, \tau) u(\sigma, \tau) \, d\tau d\sigma + \int_0^\alpha \int_0^\beta h(\sigma, \tau) u(\sigma, \tau) \, d\tau d\sigma \Big] \, dt ds, \tag{2.5.1}$$

for $(x, y) \in \Delta$, where $c \geq 0$ is a real constant. If

$$k = \int_0^\alpha \int_0^\beta h(\sigma, \tau) A(\sigma, \tau) \exp\left(\int_0^\sigma \int_0^\tau A(s, t) [f(s, t) + g(s, t)] \, dt ds \right)$$

$$\times d\tau d\sigma < 1, \tag{2.5.2}$$

where

$$A(x, y) = \exp\left(\int_0^x p(s, y) \, ds \right), \tag{2.5.3}$$

for $(x, y) \in \Delta$, then

$$u(x, y) \leq \frac{c}{1 - k} A(x, y) \exp\left(\int_0^x \int_0^y A(s, t) [f(s, t) + g(s, t)] \, dt ds \right), \tag{2.5.4}$$

for $(x, y) \in \Delta$.

Proof. Let $c > 0$ and define a function $z(x, y)$ by

$$z(x, y) = \int_0^x \int_0^y f(s, t) \Big[u(s, t) + \int_0^s \int_0^t g(\sigma, \tau) u(\sigma, \tau) \, d\tau d\sigma$$

$$+ \int_0^\alpha \int_0^\beta h(\sigma, \tau) u(\sigma, \tau) \, d\tau d\sigma \Big] \, dt ds. \tag{2.5.5}$$

Then (2.5.1) can be restated as

$$u(x, y) \leq z(x, y) + \int_0^x p(s, y) u(s, y) \, ds. \tag{2.5.6}$$

It is easy to observe that $z(x,y)$ is positive, continuous and nondecreasing function for $(x,y) \in \Delta$. Treating y fixed in (2.5.6) and using Theorem 1.3.1 given in [34] to (2.5.6) we get

$$u(x,y) \le A(x,y)\, z(x,y), \tag{2.5.7}$$

for $(x,y) \in \Delta$, where $A(x,y)$ is defined by (2.5.3). From (2.5.5), (2.5.7) and the fact that $A(x,y) \ge 1$, we observe that

$$z(x,y) \le c + \int_0^x \int_0^y f(s,t) \Big[A(s,t)\, z(s,t) + \int_0^s \int_0^t g(\sigma,\tau) A(\sigma,\tau)\, z(\sigma,\tau)\, d\tau d\sigma$$

$$+ \int_0^\alpha \int_0^\beta h(\sigma,\tau) A(\sigma,\tau)\, z(\sigma,\tau)\, d\tau d\sigma \Big] dt ds. \tag{2.5.8}$$

Define a function $v(x,y)$ by the right hand side of (2.5.8). Then $v(x,y) > 0$, $v(0,y) = v(x,0) = c$, $z(x,y) \le v(x,y)$ and

$$D_2 D_1 v(x,y) = f(x,y)\, A(x,y) \Big[z(x,y) + \int_0^x \int_0^y g(\sigma,\tau)\, A(\sigma,\tau)\, z(\sigma,\tau)\, d\tau d\sigma$$

$$+ \int_0^\alpha \int_0^\beta h(\sigma,\tau) A(\sigma,\tau)\, z(\sigma,\tau)\, d\tau d\sigma \Big]$$

$$\le f(x,y)\, A(x,y) \Big[v(x,y) + \int_0^x \int_0^y g(\sigma,\tau)\, A(\sigma,\tau)\, v(\sigma,\tau)\, d\tau d\sigma$$

$$+ \int_0^\alpha \int_0^\beta h(\sigma,\tau) A(\sigma,\tau)\, v(\sigma,\tau)\, d\tau d\sigma \Big]. \tag{2.5.9}$$

Define a function $w(x,y)$ by

$$w(x,y) = v(x,y) + \int_0^x \int_0^y g(\sigma,\tau) A(\sigma,\tau)\, v(\sigma,\tau)\, d\tau d\sigma$$

$$+ \int_0^\alpha \int_0^\beta h(\sigma,\tau) A(\sigma,\tau)\, v(\sigma,\tau)\, d\tau d\sigma,$$

then $w(x,y) > 0$, $v(x,y) \le w(x,y)$, $D_2 D_1 v(x,y) \le f(x,y)\, A(x,y)\, w(x,y)$,

$$w(0,y) = w(x,0) = c + \int_0^\alpha \int_0^\beta h(\sigma,\tau) A(\sigma,\tau)\, v(\sigma,\tau)\, d\tau d\sigma = L\,(say), \tag{2.5.10}$$

and

$$D_2D_1w(x,y) = D_2D_1v(x,y) + g(x,y)A(x,y)v(x,y)$$
$$\leq f(x,y)A(x,y)w(x,y) + g(x,y)A(x,y)w(x,y)$$
$$= A(x,y)[f(x,y)+g(x,y)]w(x,y). \tag{2.5.11}$$

Now by following the proof of Theorem 4.2.1 given in [34], the inequality (2.5.11) implies the estimate

$$w(x,y) \leq L\exp\left(\int_0^x\int_0^y A(s,t)[f(s,t)+g(s,t)]\,dtds\right). \tag{2.5.12}$$

Using (2.5.12) in $v(x,y) \leq w(x,y)$ we get

$$v(x,y) \leq L\exp\left(\int_0^x\int_0^y A(s,t)[f(s,t)+g(s,t)]\,dtds\right). \tag{2.5.13}$$

From (2.5.10), (2.5.13) and (2.5.2) it is easy to observe that

$$L \leq \frac{c}{1-k}. \tag{2.5.14}$$

Using (2.5.14) in (2.5.13) and the facts that $z(x,y) \leq v(x,y)$, $u(x,y) \leq A(x,y)z(x,y)$ we get the desired inequality in (2.5.4). The proof of the case when $c \geq 0$ follows as mentioned in the proof of Theorem 2.4.2,part (b_1).

Remark 2.5.1. We note that, in the special cases when $(i)p = 0$, $(ii)g = 0$, $(iii)h = 0$, the inequality in Theorem 2.5.1 reduces to the new inequalities which can be used as tools in different applications.

Theorem 2.5.2. Let I, J, Δ be as in Theorem 2.5.1 and

$$D = \left\{(x,y,s,t) \in \Delta^2 : 0 \leq s \leq x \leq \alpha, 0 \leq t \leq y \leq \beta\right\},$$

$$E = \left\{(x,y,s,t,\sigma,\tau) \in \Delta^3 : 0 \leq \sigma \leq s \leq x \leq \alpha, 0 \leq \tau \leq t \leq y \leq \beta\right\}.$$

Let $u(x,y) \in C(\Delta, R_+)$ and $c \geq 0$ be a real constant.

(a_1) Let $k(x,y,s,t), e(x,y,s,t) \in C(D,R_+)$, $h(x,y,s,t,\sigma,\tau) \in C(E,R_+)$ be nondecreasing in $(x,y) \in \Delta$ for fixed $(s,t) \in \Delta$, $(s,t,\sigma,\tau) \in \Delta^2$ and suppose that

$$u(x,y) \leq c + \int_0^x\int_0^y k(x,y,s,t)u(s,t)\,dtds$$

$$+\int_0^x\int_0^y\left(\int_0^s\int_0^t h(x,y,s,t,\sigma,\tau)\,u(\sigma,\tau)\,d\tau d\sigma\right)dtds$$

$$+\int_0^\alpha\int_0^\beta e(x,y,s,t)\,u(s,t)\,dtds \tag{2.5.15}$$

for $(x,y)\in\Delta$. If

$$p(x,y)=\int_0^\alpha\int_0^\beta e(x,y,s,t)$$

$$\times\exp\left(\int_0^s\int_0^t [k(s,t,m,n)\right.$$

$$\left.+\int_0^m\int_0^n h(x,y,m,n,\sigma,\tau)\,d\tau d\sigma\right]dndm\Bigg)\,dtds<1, \tag{2.5.16}$$

for $(x,y)\in\Delta$, then

$$u(x,y)\le\frac{c}{1-p(x,y)}\exp\left(\int_0^s\int_0^t [k(s,t,m,n)\right.$$

$$\left.+\int_0^m\int_0^n h(x,y,m,n,\sigma,\tau)\,d\tau d\sigma\right]dndm\Bigg), \tag{2.5.17}$$

for $(x,y)\in\Delta$.

(a_2) Let $a(x,y), b(x,y)\in C(\Delta,R_+)$, $k(x,y,s,t)\in C(D,R_+)$, $h(x,y,s,t,\sigma,\tau)$ $\in C(E,R_+)$ and suppose that

$$u(x,y)\le c+\int_0^x\int_0^y a(s,t)\,u(s,t)\,dtds$$

$$+\int_0^x\int_0^y\left(\int_0^s\int_0^t k(s,t,\sigma,\tau)\,u(\sigma,\tau)d\tau d\sigma\right)dtds$$

$$+\int_0^x\int_0^y\left(\int_0^s\int_0^t\left(\int_0^\sigma\int_0^\tau h(s,t,\sigma,\tau,m,n)\,u(m,n)\,dndm\right)d\tau d\sigma\right)dtds$$

$$+\int_0^\alpha\int_0^\beta b(s,t)\,u(s,t)\,dtds, \tag{2.5.18}$$

for $(x,y)\in\Delta$. If

$$q=\int_0^\alpha\int_0^\beta b(s,t)\exp\left(\int_0^s\int_0^t Q(\xi,\eta)\,d\eta d\xi\right)dtds<1, \tag{2.5.19}$$

where

$$Q(x,y)=a(x,y)+\int_0^x\int_0^y k(x,y,\sigma,\tau)\,d\tau d\sigma$$

$$+\int_0^x\int_0^y\left(\int_0^\sigma\int_0^\tau h(x,y,\sigma,\tau,m,n)\,dndm\right)d\tau d\sigma, \tag{2.5.20}$$

for $(x,y)\in\Delta$, then

$$u(x,y)\le\frac{c}{1-q}\exp\left(\int_0^x\int_0^y Q(s,t)\,dtds\right), \tag{2.5.21}$$

for $(x,y)\in\Delta$.

Proof. (a_1) First assume that $c>0$ and fix any arbitrary element $(X,Y)\in\Delta$. Then for $0\le x\le X, 0\le y\le Y$ we have

$$u(x,y)\le c+\int_0^x\int_0^y k(X,Y,s,t)\,u(s,t)\,dtds$$

$$+\int_0^x\int_0^y\left(\int_0^s\int_0^t h(X,Y,s,t,\sigma,\tau)\,u(\sigma,\tau)\,d\tau d\sigma\right)dtds$$

$$+\int_0^\alpha\int_0^\beta e(X,Y,s,t)\,u(s,t)\,dtds. \tag{2.5.22}$$

Let

$$d(X,Y)=c+\int_0^\alpha\int_0^\beta e(X,Y,s,t)\,u(s,t)\,dtds, \tag{2.5.23}$$

then (2.5.22) can be restated as

$$u(x,y) \le d(X,Y) + \int_0^x \int_0^y k(X,Y,s,t)\, u(s,t)\, dt ds$$

$$+ \int_0^x \int_0^y \left(\int_0^s \int_0^t h(X,Y,s,t,\sigma,\tau)\, u(\sigma,\tau)\, d\tau d\sigma \right) dt ds, \tag{2.5.24}$$

for $0 \le x \le X, 0 \le y \le Y$. Define a function $z(x,y,X,Y)$ by the right hand side of (2.5.24). Then $z(x,y,X,Y) > 0$, $z(0,y,X,Y) = z(x,0,X,Y) = d(X,Y)$, $u(x,y) \le z(x,y,X,Y)$, $z(x,y,X,Y)$ is nondecreasing in both the variables x, y lying in $0 \le x \le X, 0 \le y \le Y$ and

$$D_2 D_1 z(x,y,X,Y) = k(X,Y,x,y)\, u(x,y) + \int_0^x \int_0^y h(X,Y,x,y,\sigma,\tau)\, u(\sigma,\tau)\, d\tau d\sigma$$

$$\le \left[k(X,Y,x,y) + \int_0^x \int_0^y h(X,Y,x,y,\sigma,\tau)\, d\tau d\sigma \right] z(x,y). \tag{2.5.25}$$

Now by following the proof of Theorem 4.2.1 given in [34] from (2.5.25) we get

$$z(x,y) \le d \exp\left(\int_0^x \int_0^y [k(X,Y,m,n) \right.$$

$$\left. + \int_0^m \int_0^n h(X,Y,m,n,\sigma,\tau)\, d\tau d\sigma \right] dn dm \Bigg), \tag{2.5.26}$$

for $0 \le x \le X, 0 \le y \le Y$. Since $(X,Y) \in \Delta$ is arbitrary, from (2.5.26), (2.5.23) with (X,Y) replaced by (x,y) and $u(x,y) \le z(x,y,x,y)$ we have

$$u(x,y) \le d \exp\left(\int_0^x \int_0^y [k(x,y,m,n) \right.$$

$$\left. + \int_0^m \int_0^n h(x,y,m,n,\sigma,\tau)\, d\tau d\sigma \right] dn dm \Bigg), \tag{2.5.27}$$

for $(x,y) \in \Delta$, where

$$d(x,y) = c + \int_0^\alpha \int_0^\beta e(x,y,s,t)\, u(s,t)\, dt ds, \tag{2.5.28}$$

for $(x,y) \in \Delta$. Using (2.5.27), in the integrand on the right hand side of (2.5.28) and in view of (2.5.16) we have

$$d(x,y) \leq \frac{c}{1-p(x,y)}. \tag{2.5.29}$$

Using (2.5.29) in (2.5.27) we get the required inequality in (2.5.17). The proof of the case when $c \geq 0$ follows as noted in the proof of Theorem 2.4.1, part (a_1).

(a_2) Let $c > 0$ and denote

$$d' = c + \int_0^\alpha \int_0^\beta b(s,t)\, u(s,t)\, dt ds. \tag{2.5.30}$$

Then (2.5.18) can be restated as

$$\begin{aligned} u(x,y) \leq d' &+ \int_0^x \int_0^y a(s,t)\, u(s,t)\, dt ds \\ &+ \int_0^x \int_0^y \left(\int_0^s \int_0^t k(s,t,\sigma,\tau)\, u(\sigma,\tau)\, d\tau d\sigma \right) dt ds \\ &+ \int_0^x \int_0^y \left(\int_0^s \int_0^t \left(\int_0^\sigma \int_0^\tau h(s,t,\sigma,\tau,m,n)\, u(m,n)\, dn dm \right) d\tau d\sigma \right) dt ds. \end{aligned} \tag{2.5.31}$$

Define a function $z(x,y)$ by the right hand side of (2.5.31). Then $z(x,y) > 0$, $z(0,y) = z(x,0) = d'$, $u(x,y) \leq z(x,y)$, $z(x,y)$ is nondecreasing in both the variables $(x,y) \in \Delta$ and

$$\begin{aligned} D_2 D_1 z(x,y) = a(x,y)\, u(x,y) &+ \int_0^s \int_0^t k(x,y,\sigma,\tau)\, u(\sigma,\tau)\, d\tau d\sigma \\ &+ \int_0^x \int_0^y \left(\int_0^\sigma \int_0^\tau h(x,y,\sigma,\tau,m,n)\, u(m,n)\, dn dm \right) d\tau d\sigma \\ &\leq Q(x,y)\, z(x,y), \end{aligned} \tag{2.5.32}$$

where $Q(x,y)$ is given by (2.5.20). The rest of the proof can be completed by following the proof of Theorem 4.2.1 given in [34] and closely looking at the proof of (a_1) given above.

Remark 2.5.2. In the various special cases the inequalities given in Theorem 2.5.2 reduces to different inequalities which can be used as tools in variety of applications.

The following three theorems contain the inequalities investigated in [41] which can be used more conveniently in certain applications.

Theorem 2.5.3. Let $u(x,y), a(x,y), b(x,y), c(x,y) \in C(R_+^2, R_+)$.

(b_1) If

$$u(x,y) \leq a(x,y) + b(x,y) \int_0^x \int_y^\infty c(s,t)\, u(s,t)\, dtds, \tag{2.5.33}$$

for $x, y \in R_+$, then

$$u(x,y) \leq a(x,y) + b(x,y)\, e(x,y) \exp\left(\int_0^x \int_y^\infty c(s,t)\, b(s,t)\, dtds\right) \tag{2.5.34}$$

for $x, y \in R_+$, where

$$e(x,y) = \int_0^x \int_y^\infty c(s,t)\, a(s,t)\, dtds, \tag{2.5.35}$$

for $x, y \in R_+$.

(b_2) If

$$u(x,y) \leq a(x,y) + b(x,y) \int_x^\infty \int_y^\infty c(s,t)\, u(s,t)\, dtds, \tag{2.5.36}$$

for $x, y \in R_+$, then

$$u(x,y) \leq a(x,y) + b(x,y)\, \bar{e}(x,y) \exp\left(\int_x^\infty \int_y^\infty c(s,t)\, b(s,t)\, dtds\right), \tag{2.5.37}$$

for $x, y \in R_+$, where

$$\bar{e}(x,y) = \int_x^\infty \int_y^\infty c(s,t)\, a(s,t)\, dtds, \tag{2.5.38}$$

for $x, y \in R_+$.

Theorem 2.5.4. Let $u(x,y), a(x,y), b(x,y), c(x,y) \in C(R_+^2, R_+)$.

(c_1) Assume that $a(x,y)$ is nondecreasing for $x \in R_+$. If

$$u(x,y) \leq a(x,y) + \int_0^x b(s,y)\,u(s,y)\,ds + \int_0^x \int_y^\infty c(s,t)\,u(s,t)\,dtds, \quad (2.5.39)$$

for $x, y \in R_+$, then

$$u(x,y) \leq p(x,y)\left[a(x,y) + A(x,y)\exp\left(\int_0^x \int_y^\infty c(s,t)\,p(s,t)\,dtds\right)\right], \quad (2.5.40)$$

for $x, y \in R_+$, where

$$p(x,y) = \exp\left(\int_0^x b(s,y)\,ds\right), \quad (2.5.41)$$

$$A(x,y) = \int_0^x \int_y^\infty c(s,t)\,p(s,t)\,a(s,t)\,dtds, \quad (2.5.42)$$

for $x, y \in R_+$.

(c_2) Assume that $a(x,y)$ is nonincreasing for $x \in R_+$. If

$$u(x,y) \leq a(x,y) + \int_x^\infty b(s,y)\,u(s,y)\,ds + \int_x^\infty \int_y^\infty c(s,t)\,u(s,t)\,dtds, \quad (2.5.43)$$

for $x, y \in R_+$, then

$$u(x,y) \leq \bar{p}(x,y)\left[a(x,y) + \bar{A}(x,y)\exp\left(\int_x^\infty \int_y^\infty c(s,t)\,\bar{p}(s,t)\,dtds\right)\right], \quad (2.5.44)$$

for $x, y \in R_+$, where

$$\bar{p}(x,y) = \exp\left(\int_x^\infty b(s,y)\,ds\right), \quad (2.5.45)$$

$$\bar{A}(x,y) = \int_x^\infty \int_y^\infty c(s,t)\,\bar{p}(s,t)\,a(s,t)\,dtds, \quad (2.5.46)$$

for $x, y \in R_+$.

Theorem 2.5.5. Let $u(x,y), a(x,y), b(x,y) \in C(R_+^2, R_+)$ and $F \in C(R_+^3, R_+)$ satisfies the condition

$$0 \le F(x,y,u) - F(x,y,v) \le K(x,y,v)(u-v), \tag{2.5.47}$$

for $u \ge v \ge 0$, where $K \in C(R_+^3, R_+)$.

(d_1) Assume that $a(x,y)$ is nondecreasing for $x \in R_+$. If

$$u(x,y) \le a(x,y) + \int_0^x b(s,y)u(s,y)\,ds + \int_0^x \int_y^\infty F(s,t,u(s,t))\,dtds, \tag{2.5.48}$$

for $x, y \in R_+$, then

$$u(x,y) \le p(x,y)\Big[a(x,y) + B(x,y)\exp\Big(\int_0^x \int_y^\infty K(s,t,p(s,t)a(s,t))p(s,t)\,dtds\Big)\Big], \tag{2.5.49}$$

for $x, y \in R_+$, where $p(x,y)$ is defined by (2.5.41) and

$$B(x,y) = \int_0^x \int_y^\infty F(s,t,p(s,t)a(s,t))\,dtds, \tag{2.5.50}$$

for $x, y \in R_+$.

(d_2) Assume that $a(x,y)$ is nonincreasing for $x \in R_+$.If

$$u(x,y) \le a(x,y) + \int_x^\infty b(s,y)u(s,y)\,ds + \int_x^\infty \int_y^\infty F(s,t,u(s,t))\,dtds, \tag{2.5.51}$$

for $x, y \in R_+$, then

$$u(x,y) \le \bar{p}(x,y)\Big[a(x,y) + \bar{B}(x,y)\exp\Big(\int_x^\infty \int_y^\infty K(s,t,\bar{p}(s,t)a(s,t))\bar{p}(s,t)\,dtds\Big)\Big], \tag{2.5.52}$$

for $x, y \in R_+$, where $\bar{p}(x,y)$ is defined by (2.5.45) and

$$\bar{B}(x,y) = \int_x^\infty \int_y^\infty F(s,t,\bar{p}(s,t)a(s,t))\,dtds, \tag{2.5.53}$$

for $x, y \in R_+$.

Proofs of Theorems 2.5.3-2.5.5. Since the proofs resemble one another, we give the details for $(b_1), (c_1)$ and (d_1); the proofs of $(b_2), (c_2)$ and (d_2) can be completed by following the proofs of the above mentioned results with suitable changes.

(b_1) Define a function $z(x,y)$ by

$$z(x,y) = \int_0^x \int_y^\infty c(s,t)\, u(s,t)\, dtds, \tag{2.5.54}$$

then (2.5.33) can be restated as

$$u(x,y) \le a(x,y) + b(x,y)\, z(x,y). \tag{2.5.55}$$

From (2.5.54) and (2.5.55) we have

$$z(x,y) \le \int_0^x \int_y^\infty c(s,t) \left[a(s,t) + b(s,t)\, z(s,t)\right] dtds,$$

$$= e(x,y) + \int_0^x \int_y^\infty c(s,t)\, b(s,t)\, z(s,t)\, dtds, \tag{2.5.56}$$

where $e(x,y)$ is defined by (2.5.35). Clearly $e(x,y)$ is nonnegative, continuous, nondecreasing in x and nonincreasing in y for $x, y \in R_+$. First we assume that $e(x,y) > 0$ for $x, y \in R_+$. From (2.5.56) it is easy to observe that

$$\frac{z(x,y)}{e(x,y)} \le 1 + \int_0^x \int_y^\infty c(s,t)\, b(s,t)\, \frac{z(s,t)}{e(s,t)} dtds. \tag{2.5.57}$$

Define a function $v(x,y)$ by the right hand side of (2.5.57), then $v(x,y) > 0$, $v(0,y) = v(x,\infty) = 1$, $\frac{z(x,y)}{e(x,y)} \le v(x,y)$, $v(x,y)$ is nonincreasing in y, $y \in R_+$ and

$$D_1 v(x,y) = \int_y^\infty c(x,t)\, b(x,t)\, \frac{z(x,t)}{e(x,t)} dt$$

$$\le \int_y^\infty c(x,t)\, b(x,t)\, v(x,t) dt$$

$$\le v(x,y) \int_y^\infty c(x,t)\, b(x,t) dt. \tag{2.5.58}$$

Treating y, $y \in R_+$ fixed in (2.5.58), dividing both sides of (2.5.58) by $v(x,y)$, setting $x = s$ and integrating the resulting inequality from 0 to x, $x \subset R_+$ we get

$$v(x,y) \leq \exp\left(\int_0^x \int_y^\infty c(s,t)\, b(s,t)\, dtds\right). \tag{2.5.59}$$

Using (2.5.59) in $\frac{z(x,y)}{e(x,y)} \leq v(x,y)$ we have

$$z(x,y) \leq e(x,y) \exp\left(\int_0^x \int_y^\infty c(s,t)\, b(s,t)\, dtds\right). \tag{2.5.60}$$

The desired inequality in (2.5.34) follows from (2.5.55) and (2.5.60). The proof of the case when $e(x,y) \geq 0$ follows as mentioned in the proof of Theorem 2.2.1.

(c_1) Define a function $z(x,y)$ by

$$z(x,y) = a(x,y) + \int_0^x \int_y^\infty c(s,t)\, u(s,t)\, dtds. \tag{2.5.61}$$

Then (2.5.39) can be restated as

$$u(x,y) \leq z(x,y) + \int_0^x b(s,y)\, u(s,y)\, ds. \tag{2.5.62}$$

Clearly $z(x,y)$ is nonnegative, continuous and nondecreasing function in $x, x \in R_+$. Treating y, $y \in R_+$ fixed in (2.5.62) and using the inequality given in Lemma 2, part (α_1) in [41], (see also [34]) to (2.5.62) we get

$$u(x,y) \leq z(x,y)p(x,y), \tag{2.5.63}$$

where $p(x,y)$ is defined by (2.5.41). From (2.5.63) and (2.5.61) we have

$$u(x,y) \leq p(x,y)\left[a(x,y) + v(x,y)\right], \tag{2.5.64}$$

where

$$v(x,y) = \int_0^x \int_y^\infty c(s,t)\, u(s,t)\, dtds. \tag{2.5.65}$$

From (2.5.64) and (2.5.65) we get

$$v(x,y) \leq \int_0^x \int_y^\infty c(s,t)\, p(s,t)\left[a(s,t) + v(s,t)\right] dtds$$

$$= A(x,y) + \int_0^x \int_y^\infty c(s,t)\,p(s,t)v(s,t)dtds,$$

where $A(x,y)$ is defined by (2.5.42). Clearly $A(x,y)$ is nonnegative, continuous, nondecreasing in x, $x \in R_+$ and nonincreasing in y, $y \in R_+$. Now by following the proof of (b_1) we obtain

$$v(x,y) \le A(x,y)\exp\left(\int_0^x \int_y^\infty c(s,t)\,p(s,t)\,dtds\right). \tag{2.5.66}$$

Using (2.5.66) in (2.5.64) we get the required inequality in (2.5.40).

(d_1) Define a function $z(x,y)$ by

$$z(x,y) = a(x,y) + \int_0^x \int_y^\infty F(s,t,u(s,t))\,dtds. \tag{2.5.67}$$

Then (2.5.48) can be restated as

$$u(x,y) \le z(x,y) + \int_0^x b(s,y)\,u(s,y)\,ds. \tag{2.5.68}$$

Clearly $z(x,y)$ is nonnegative, continuous and nondecreasing function in x, $x \in R_+$. Treating y, $y \in R_+$ fixed in (2.5.68) and using the inequality given in Lemma 2.1, part (α_1) in [41] (see also [34]) to (2.5.68) we obtain

$$u(x,y) \le z(x,y)p(x,y), \tag{2.5.69}$$

where $p(x,y)$ is defined by (2.5.41). From (2.5.69) and (2.5.67) we have

$$u(x,y) \le p(x,y)\left[a(x,y) + v(x,y)\right], \tag{2.5.70}$$

where

$$v(x,y) = \int_0^x \int_y^\infty F(s,t,u(s,t))\,dtds. \tag{2.5.71}$$

From (2.5.71), (2.5.70) and the hypotheses on F it follows that

$$v(x,y) \le \int_0^x \int_y^\infty [\{F(s,t,p(s,t)[a(s,t)+v(s,t)])$$

$$-F(s,t,p(s,t)\,a(s,t))\} + F(s,t,p(s,t)\,a(s,t))]\,dtds$$

$$\leq B(x,y) + \int_0^x \int_y^\infty K(s,t,p(s,t)a(s,t))\, p(s,t)v(s,t)dtds, \tag{2.5.72}$$

where $B(x,y)$ is defined by (2.5.50). Clearly $B(x,y)$ is nonnegative, continuous and nondecreasing in x and nonincreasing in y for $x,y \in R_+$. By following the proof of part (b_1) we get

$$v(x,y) \leq B(x,y) \exp\left(\int_0^x \int_y^\infty K(s,t,p(s,t)a(s,t))\, p(s,t)dtds\right). \tag{2.5.73}$$

The required inequality in (2.5.49) follows from (2.5.70) and (2.5.73).

The next theorem deals with the inequalities proved in [48].

Theorem 2.5.6. Let $u(x,y), a(x,y), b(x,y), c(x,y) \in C(R_+^2, R_+)$ and $p > 1$ be a real constant.

(k_1) If

$$u^p(x,y) \leq a(x,y) + b(x,y) \int_0^x \int_y^\infty c(s,t)\, u(s,t)\, dtds, \tag{2.5.74}$$

for $x,y \in R_+$, then

$$u(x,y) \leq \left[a(x,y) + b(x,y) A(x,y) \times \exp\left(\int_0^x \int_y^\infty c(s,t) \frac{b(s,t)}{p}\right) dtds\right]^{\frac{1}{p}}, \tag{2.5.75}$$

for $x,y \in R_+$, where

$$A(x,y) = \int_0^x \int_y^\infty c(s,t) \left(\frac{p-1}{p} + \frac{a(s,t)}{p}\right) dtds, \tag{2.5.76}$$

for $x,y \in R_+$.

(k_2) Let $L \in C(R_+^3, R_+)$ satisfies the condition

$$0 \leq L(x,y,u) - L(x,y,v) \leq G(x,y,v)(u-v), \tag{2.5.77}$$

for $u \geq v \geq 0$, where $G \in C(R_+^3, R_+)$. If

$$u^p(x,y) \leq a(x,y) + b(x,y) \int_0^x \int_y^\infty L(s,t,u(s,t))\, dtds, \tag{2.5.78}$$

for $x, y \in R_+$, then

$$u(x,y) \leq \Big[a(x,y) + b(x,y)\bar{A}(x,y)$$

$$\times \exp\left(\int_0^x \int_y^\infty G\left(s,t,\frac{p-1}{p} + \frac{a(s,t)}{p}\right)\frac{b(s,t)}{p}dtds\right)\Big]^{\frac{1}{p}}, \tag{2.5.79}$$

for $x, y \in R_+$, where

$$\bar{A}(x,y) = \int_0^x \int_y^\infty L\left(s,t,\frac{p-1}{p} + \frac{a(s,t)}{p}\right)dtds, \tag{2.5.80}$$

for $x, y \in R_+$.

Proof. (k_1) Define a function $z(x,y)$ by

$$z(x,y) = \int_0^x \int_y^\infty c(s,t)\,u(s,t)\,dtds, \tag{2.5.81}$$

then (2.5.74) can be restated as

$$u^p(x,y) \leq a(x,y) + b(x,y)\,u(x,y). \tag{2.5.82}$$

As in the proof of Theorem 2.3.3, part (c_1), from (2.5.82) and using the elementary inequality (1.3.11) (see [30, p. 30]) we get

$$u(x,y) \leq \frac{p-1}{p} + \frac{a(x,y)}{p} + \frac{b(x,y)}{p}z(x,y). \tag{2.5.83}$$

From (2.5.81) and (2.5.83) we have

$$z(x,y) \leq \int_0^x \int_y^\infty c(s,t)\left(\frac{p-1}{p} + \frac{a(s,t)}{p} + \frac{b(s,t)}{p}z(s,s)\right)dtds$$

$$= A(x,y) + \int_0^x \int_y^\infty c(s,t)\frac{b(s,t)}{p}z(s,s)\,dtds,$$

where $A(x,y)$ is defined by (2.5.76). The rest of the proof follows by the similar argument as in the proof of Theorem 2.5.3, part (b_1).

(k_2) The proof can be completed by closely looking at the proof of (k_1) given above and the proof of Theorem 2.3.4, part (d_1). Here we omit the details.

The inequalities in the following theorem are established in [76].

Theorem 2.5.7. Let $u(x,y), a(x,y), b(x,y), c(x,y), f(x,y), g(x,y) \in C(R_+^2, R_+)$.

(r_1) Suppose that

$$u(x,y) \le a(x,y) + b(x,y)\int_0^x\int_0^y f(s,t)\,u(s,t)\,dtds$$

$$+c(x,y)\int_0^\infty\int_0^\infty g(s,t)\,u(s,t)\,dtds, \tag{2.5.84}$$

for $x, y \in R_+$. If

$$p_1 = \int_0^\infty\int_0^\infty g(s,t)\,D_1(s,t)\,dtds < 1, \tag{2.5.85}$$

then

$$u(x,y) \le B_1(x,y) + M_1 D_1(x,y), \tag{2.5.86}$$

for $x, y \in R_+$, where

$$B_1(x,y) = a(x,y) + b(x,y)\,A_1(x,y)\int_0^x\int_0^y f(s,t)\,a(s,t)\,dtds, \tag{2.5.87}$$

$$D_1(x,y) = c(x,y) + b(x,y)\,A_1(x,y)\int_0^x\int_0^y f(s,t)\,c(s,t)\,dtds, \tag{2.5.88}$$

$$A_1(x,y) = \exp\left(\int_0^x\int_0^y f(s,t)\,b(s,t)\,dtds\right), \tag{2.5.89}$$

and

$$M_1 = \frac{1}{1-p_1}\int_0^\infty\int_0^\infty g(s,t)\,B_1(s,t)\,dtds. \tag{2.5.90}$$

(r_2) Suppose that

$$u(x,y) \le a(x,y) + b(x,y)\int_x^\infty\int_y^\infty f(s,t)\,u(s,t)\,dtds$$

$$+c(x,y)\int_0^\infty\int_0^\infty g(s,t)\,u(s,t)\,dtds, \tag{2.5.91}$$

for $x, y \in R_+$. If

$$p_2 = \int_0^\infty\int_0^\infty g(s,t)\,D_2(s,t)\,dtds < 1, \tag{2.5.92}$$

then

$$u(x,y) \le B_2(x,y) + M_2 D_2(x,y), \tag{2.5.93}$$

for $x, y \in R_+$, where

$$B_2(x,y) = a(x,y) + b(x,y)\,A_2(x,y)\int_x^\infty\int_y^\infty f(s,t)\,a(s,t)\,dtds, \tag{2.5.94}$$

$$D_2(x,y) = c(x,y) + b(x,y)\,A_2(x,y)\int_x^\infty\int_y^\infty f(s,t)\,c(s,t)\,dtds, \tag{2.5.95}$$

$$A_2(x,y) = \exp\left(\int_x^\infty\int_y^\infty f(s,t)\,b(s,t)\,dtds\right), \tag{2.5.96}$$

and

$$M_2 = \frac{1}{1-p_2}\int_x^\infty\int_y^\infty g(s,t)\,B_2(s,t)\,dtds. \tag{2.5.97}$$

(r_3) Suppose that

$$u(x,y) \le a(x,y) + b(x,y)\int_0^x\int_y^\infty f(s,t)\,u(s,t)\,dtds$$

$$+c(x,y)\int_0^\infty\int_0^\infty g(s,t)\,u(s,t)\,dtds, \tag{2.5.98}$$

for $x, y \in R_+$. If

$$p_3 = \int_0^\infty\int_0^\infty g(s,t)\,D_3(s,t)\,dtds < 1, \tag{2.5.99}$$

then

$$u(x,y) \leq B_3(x,y) + M_3 D_3(x,y), \tag{2.5.100}$$

for $x, y \in R_+$, where

$$B_3(x,y) = a(x,y) + b(x,y) A_3(x,y) \int_0^x \int_y^\infty f(s,t) a(s,t) \, dt ds, \tag{2.5.101}$$

$$D_3(x,y) = c(x,y) + b(x,y) A_3(x,y) \int_0^x \int_y^\infty f(s,t) c(s,t) \, dt ds, \tag{2.5.102}$$

$$A_3(x,y) = \exp\left(\int_0^x \int_y^\infty f(s,t) b(s,t) \right), \tag{2.5.103}$$

and

$$M_3 = \frac{1}{1-p_3} \int_0^\infty \int_0^\infty g(s,t) B_3(s,t) \, dt ds. \tag{2.5.104}$$

Proof. (r_1) Let

$$z(x,y) = \int_0^x \int_0^y f(s,t) u(s,t) \, dt ds, \tag{2.5.105}$$

$$k = \int_0^\infty \int_0^\infty g(s,t) u(s,t) \, dt ds. \tag{2.5.106}$$

Then (2.5.84) can be restated as

$$u(x,y) \leq a(x,y) + b(x,y) z(x,y) + c(x,y) k. \tag{2.5.107}$$

From (2.5.105) and (2.5.107) we have

$$z(x,y) \leq \int_0^x \int_0^y f(s,t) \left[a(s,t) + kc(s,t) + b(s,t) z(s,t) \right] dt ds$$

$$= e(x,y) + \int_0^x \int_0^y f(s,t) b(s,t) z(s,t) \, dt ds, \tag{2.5.108}$$

where

$$e(x,y) = \int_0^x \int_0^y \left[f(s,t) a(s,t) + kf(s,t) c(s,t) \right] dt ds.$$

Clearly $e(x, y)$ is continuous, nonnegative and nondecreasing in both the variables $x, y \in R_+$. Now an application of Theorem 4.2.2 given in [34] to (2.5.108) yields

$$z(x,y) \le e(x,y) A_1(x,y). \tag{2.5.109}$$

Using (2.5.109) in (2.5.107) we have

$$\begin{aligned} u(x,y) &\le a(x,y) + kc(x,y) + b(x,y) e(x,y) A_1(x,y) \\ &= B_1(x,y) + kD_1(x,y). \end{aligned} \tag{2.5.110}$$

Now, from (2.5.106), (2.5.110)and (2.5.85) we have

$$k \le \int_0^\infty \int_0^\infty g(s,t) \{B_1(s,t) + kD_1(s,t)\} \, dtds,$$

i.e.,

$$k \left\{ 1 - \int_0^\infty \int_0^\infty g(s,t) D_1(s,t) \, dtds \right\} \le \int_0^\infty \int_0^\infty g(s,t) B_1(s,t) \, dtds,$$

which implies

$$k \le M_1. \tag{2.5.111}$$

Using (2.5.111) in (2.5.110) we get (2.5.86).

(r_2) Let

$$z(x,y) = \int_x^\infty \int_y^\infty f(s,t) u(s,t) \, dtds, \tag{2.5.112}$$

and k be as in (2.5.106). The proof can be completed by following the proof of (r_1) and making use of the inequality in Theorem 1.2.3 given in [3, p. 110] (see also [34, p. 440]).

(r_3) Let

$$z(x,y) = \int_0^x \int_y^\infty f(s,t) u(s,t) \, dtds, \tag{2.5.113}$$

and k be as in (2.5.106). The proof follows by the similar arguments as in (r_1) and using the inequality in Theorem 1.2.4 given in [3, p. 110] (see also [34, p. 440]).

2.6 Applications

One of the main motivations for the discovery of different type of inequalities given in earlier sections was to apply them as tools in the study of various classes of partial differential, integrodifferential and integral equations. In this section we give applications of some of the inequalities which has been investigated during the past few years.

2.6.1 Nonlinear partial differential equation

Consider the partial differential equation

$$\frac{\partial}{\partial y}\left(u^{p-1}(x,y)\frac{\partial}{\partial x}u(x,y)\right)+F(x,y,u(x,y))=r(x,y), \tag{2.6.1}$$

with the given initial boundary conditions

$$u(x,0)=\sigma(x),u(0,y)=\tau(y),\sigma(0)=\tau(0)=0, \tag{2.6.2}$$

where $p>1$ is a real constant and $F\in C\left(R_+^2\times R,R\right), r\in C\left(R_+^2,R\right),\sigma,\tau\in C\left(R_+,R\right).$

As an application of Theorem2.3.5 we present the following result proved by Pachpatte in [45] which gives the bound on the solution of problem (2.6.1)-(2.6.2).

Theorem 2.6.1. Assume that

$$|F(x,y,u)|\leq f(x,y)\,g(|u|), \tag{2.6.3}$$

$$|\bar{a}(x,y)|\leq c, \tag{2.6.4}$$

where f,g,c are as in Theorem 2.3.5 and

$$\bar{a}(x,y)=\sigma^p(x)+\tau^p(y)+p\int_0^x\int_0^y r(s,t)\,dtds, \tag{2.6.5}$$

for $x,y\in R_+$.Let $u(x,y)$ be a solution of (2.6.1)-(2.6.2) for $x,y\in R_+$ Then for $0\leq x\leq x_1, 0\leq y\leq y_1; x,x_1,y,y_1\in R_+,$

$$|u(x,y)|\leq\left\{G^{-1}\left[G(c)+p\int_0^x\int_0^y f(s,t)\,dtds\right]\right\}^{\frac{1}{p}}, \tag{2.6.6}$$

where G, G^{-1} are as in Theorem 2.3.5, part (k_1) and $x_1, y_1 \in R_+$ are chosen so that

$$G(c) + p\int_0^x \int_0^y f(s,t)\, dtds \in Dom\left(G^{-1}\right),$$

for all x, y lying in $0 \le x \le x_1, 0 \le y \le y_1$.

Proof. It is easy to observe that the problem (2.6.1)-(2.6.2) is equivalent to the integral equation

$$\frac{u^p(x,y)}{p} - \frac{\sigma^p(x)}{p} - \frac{\tau^p(y)}{p} + \int_0^x \int_0^y F(s,t,u(s,t))\, dtds$$

$$= \int_0^x \int_0^y r(s,t)\, dtds. \tag{2.6.7}$$

From (2.6.7), (2.6.3) and (2.6.4) we observe that

$$|u(x,y)|^p \le c + p\int_0^x \int_0^y f(s,t)\, g(|u(s,t)|)\, dtds. \tag{2.6.8}$$

Now a suitable application of Theorem 2.3.5, part (k_1) (when $h = 0$) to (2.6.8) yields (2.6.6).

The next result proved by Pachpatte in [40] deals with the bound on the solution of (2.6.1)-(2.6.2) and is obtained by applying Theorem 2.3.3, part (c_1).

Theorem 2.6.2. Assume that

$$|F(x,y,u)| \le h(x,y)\, |u|, \tag{2.6.9}$$

where $h \in C\left(R_+^2, R_+\right)$. Let

$$a_0(x,y) = |\sigma(x)|^p + |\tau(y)|^p + p\int_0^x \int_0^y |r(s,t)|\, dtds, \tag{2.6.10}$$

and $u(x,y)$ be a solution of (2.6.1)-(2.6.2) for $x, y \in R_+$. Then

$$|u(x,y)| \le \left\{ a_0(x,y) + p e_0(x,y) \exp\left(\int_0^x \int_0^y h(s,t)\, dtds \right) \right\}^{\frac{1}{p}}, \tag{2.6.11}$$

for $x, y \in R_+$, where

$$e_0(x,y) = \int_0^x \int_0^y \left(\frac{p-1}{p} + \frac{a_0(s,t)}{p} \right) h(s,t)\, dtds, \tag{2.6.12}$$

for $x, y \in R_+$.

Proof. The solution $u(x,y)$ of (2.6.1)-(2.6.2) satisfies the equivalent integral equation (2.6.7). From (2.6.7), (2.6.9) and (2.6.10) we observe that

$$|u(x,y)|^p \le a_0(x,y) + p \int_0^x \int_0^y h(s,t) |u(s,t)|\, dtds. \tag{2.6.13}$$

Now a suitable application of Theorem 2.3.3, part (c_1) (with $a(x,y) = a_0(x,y)$, $b(x,y) = p$ and $g(x,y) = 0$) to (2.6.13) we get the required estimate in (2.6.11).

2.6.2 Hyperbolic partial differential equations with terminal values

Consider the hyperbolic partial differential equation

$$u_{xy}(x,y) = h(x,y,u(x,y)) + r(x,y), \tag{2.6.14}$$

with the given terminal value conditions

$$u(x,\infty) = \sigma_\infty(x), u(\infty,y) = \tau_\infty(y), u(\infty,\infty) = d, \tag{2.6.15}$$

where $h \in C(R_+^2 \times R, R_+), r \in C(R_+^2, R_+)$ $\sigma_\infty, \tau_\infty \in C(R_+, R_+)$ and d is a real constant.

As an application of Theorem 2.5.3, part (b_2) we present the results given by Pachpatte in [41] which deals with the estimate and uniqueness of solutions of (2.6.14)-(2.6.15).

Theorem 2.6.3. Suppose that

$$|h(x,y,u)| \le c(x,y)|u|, \tag{2.6.16}$$

$$\left| \sigma_\infty(x) + \tau_\infty(x) - d + \int_x^\infty \int_y^\infty r(s,t)\, dtds \right| \le a(x,y), \tag{2.6.17}$$

where $a(x,y), c(x,y) \in C(R_+^2, R_+)$. Let $u(x,y)$ be a solution of (2.6.14)-(2.6.15) for $x, y \in R_+$, then

$$|u(x,y)| \leq a(x,y) + \bar{e}(x,y) \exp\left(\int_x^\infty \int_y^\infty c(s,t)\, dtds\right), \tag{2.6.18}$$

for $x, y \in R_+$, where $\bar{e}(x,y)$ is defined by (2.5.38).

Proof. The solution $u(x,y)$ of (2.6.14)-(2.6.15) satisfies the following equivalent integral equation (see also [3, p. 80])

$$u(x,y) = \sigma_\infty(x) + \tau_\infty(y) - d + \int_x^\infty \int_y^\infty [h(s,t,u(s,t)) + r(s,t)]\, dtds, \tag{2.6.19}$$

for $x, y \in R_+$. From (2.6.19), (2.6.16), (2.6.17) we get

$$|u(x,y)| \leq a(x,y) + \int_x^\infty \int_y^\infty c(s,t)\, |u(s,t)|\, dtds. \tag{2.6.20}$$

Now a suitable application of Theorem 2.5.3, part (b_2) to (2.6.20) yields the required estimate in (2.6.18).

Theorem 2.6.4. Suppose that the function h in (2.6.14) satisfies the condition

$$|h(x,y,u) - h(x,y,v)| \leq c(x,y)\, |u - v|, \tag{2.6.21}$$

where $c(x,y) \in C(R_+^2, R_+)$.Then the problem (2.6.14)-(2.6.15) has at most one solution on R_+^2.

Proof. The problem (2.6.14)-(2.6.15) is equivalent to the integral equation (2.6.19). Let $u(x,y), v(x,y)$ be two solutions of (2.6.14)-(2.6.15). From (2.6.19), (2.6.21) we have

$$|u(x,y) - v(x,y)| \leq \int_x^\infty \int_y^\infty c(s,t)\, |u(s,t) - v(s,t)|\, dtds. \tag{2.6.22}$$

Now a suitable application of Theorem 2.6.3, part (b_2) yields $u(x,y) = v(x,y)$ i.e., there is at most one solution to the problem (2.6.14)-(2.6.15).

We note that the inequality given in Theorem 2.6.4, part (c_2) can be used to obtain the bound and uniqueness of the solutions of the following non-self-adjoint hyperbolic partial differential equation

$$u_{xy}(x,y) = (r(x,y)\, u(x,y))_x + h(x,y,u(x,y)), \tag{2.6.23}$$

with the given terminal value conditions given in (2.6.15), under some suitable conditions on the functions involved in proplem (2.6.23)-(2.6.15).

2.6.3 Non-self-adjoint Hyperbolic partial Fredholm integrodifferential equation

In this section we present some applications of the special version of the inequality in Theorem 2.5.1 to study certain properties of solutions of the initial boundary value problem (IBVP for short) for the following non-self-adjoint hyperbolic partial Fredholm integrodifferential equation

$$u_{xy}(x,y) = (p(x,y)\,u(x,y))_y$$

$$+F\left(x, y, u(x,y), \int_0^x \int_0^y k(x,y,\sigma,\tau,u(\sigma,\tau))\,d\tau d\sigma\right), \tag{2.6.24}$$

$$u(x,0) = \alpha(x),\, u(0,y) = \beta(y),\, \alpha(0) = \beta(0) = 0, \tag{2.6.25}$$

where $\alpha \in C(I_a, R),\, \beta \in C(I_b, R)$; for $0 \le \sigma \le x, 0 \le \tau \le y$, $k \in C(G^2 \times R, R)$, $F \in C(G \times R^2, R)$ and $p \in C(G, R)$ is differentiable with respect to y, in which $I_a = [0,a],\, I_b = [0,b]$ be subsets of R and $G = I_a \times I_b$.

The following theorems are given by Pachpatte in [62] which deals with the properties of solutions of IBVP (2.6.24)-(2.6.25).

Theorem 2.6.5. Assume that

$$|E(x,y)| \le c, \tag{2.6.26}$$

$$|k(x,y,s,t,u)| \le e(x,y)\,h(s,t)\,|u|, \tag{2.6.27}$$

$$|F(x,y,u,\bar{u})| \le f(x,y)\,(|u| + |\bar{u}|), \tag{2.6.28}$$

where

$$E(x,y) = \alpha(x) + \beta(y) - \int_0^x p(s,0)\,\alpha(s)\,ds, \tag{2.6.29}$$

f, h, c are as in Theorem 2.5.1 and $e(x,y) \in C(G, R_+)$ such that $e(x,y) \ge 1$. Let

$$q_0 = \int_0^a \int_0^b h(\sigma,\tau)\bar{A}(\sigma,\tau) \exp\left(\int_0^\sigma \int_0^\tau \bar{A}(s,t)\,f(s,t)\,e(s,t)\,dtds\right)$$

$$\times d\tau d\sigma < 1, \tag{2.6.30}$$

where

$$\bar{A}(x,y) = \exp\left(\int_0^x |p(s,y)|\, ds\right), \tag{2.6.31}$$

for $(x,y) \in G$. If $u(x,y)$ is any solution of IBVP (2.6.24)-(2.6.25) for $(x,y) \in G$, then

$$|u(x,y)| \leq \frac{c}{1-q_0}\bar{A}(x,y)\exp\left(\int_0^x\int_0^y \bar{A}(s,t)\, f(s,t)\, e(s,t)\, dtds\right), \tag{2.6.32}$$

for $(x,y) \in G$.

Proof. The solution $u(x,y)$ of IBVP (2.6.24)-(2.6.25) satisfies the equivalent integral equation

$$u(x,y) = E(x,y) + \int_0^x p(s,y)\, u(s,y)\, ds$$

$$+\int_0^x\int_0^y F\left(s,t,u(s,t),\int_0^a\int_0^b k(s,t,\sigma,\tau,u(\sigma,\tau))\, d\tau d\sigma\right) dtds, \tag{2.6.33}$$

where $E(x,y)$ is given by (2.6.29). Using (2.6.26)-(2.6.28) in (2.6.33) we have

$$|u(x,y)| \leq c + \int_0^x |p(s,y)|\, |u(s,y)|\, ds$$

$$+\int_0^x\int_0^y f(s,t)\left[|u(s,t)| + \int_0^a\int_0^b e(\sigma,\tau)h(\sigma,\tau)\, |u(\sigma,\tau)|\, d\tau d\sigma\right] dtds$$

$$\leq c + \int_0^x |p(s,y)|\, |u(s,y)|\, ds$$

$$+\int_0^x\int_0^y f(s,t)\, e(s,t)\left[|u(s,t)| + \int_0^a\int_0^b h(\sigma,\tau)\, |u(\sigma,\tau)| d\tau d\sigma\right] dtds. \tag{2.6.34}$$

Now a suitable application of Theorem 2.5.1 (with $g = 0$) to (2.6.34) yields (2.6.32).

Theorem 2.6.6. (i) Assume that

$$|k(x,y,s,t,u)-k(x,y,s,t,\bar{u})|\leq e(x,y)\,h(s,t)\,|u-\bar{u}|, \tag{2.6.35}$$

$$|F(x,y,u,\bar{u})-F(x,y,v,\bar{v})|\leq f(x,y)\,(|u-v|+|\bar{u}-\bar{v}|), \tag{2.6.36}$$

where e,h,f are as in Theorem 2.6.5. Let q_0 and $\bar{A}(x,y)$ be as in (2.6.30) and (2.6.31) respectively. Then the IBVP (2.6.24)-(2.6.25) has at most one solution on G.

(ii) Let $u(x,y)$ and $v(x,y)$ be the solutions of (2.6.24) with the initial boundary conditions (2.6.25) and

$$v(x,0)=\bar{\alpha}(x),\, v(0,y)=\bar{\beta}(y),\, \bar{\alpha}(0)=\bar{\beta}(0)=0, \tag{2.6.37}$$

respectively, where $\bar{\alpha}\in C(I_a,R),\bar{\beta}\in C(I_b,R)$. Suppose that the functions k and F in (2.6.24) satisfy the conditions (2.6.35) and (2.6.36) in part (i). Let $E(x,y)$ be given by (2.6.29),

$$\bar{E}(x,y)=\bar{\alpha}(x)+\bar{\beta}(y)-\int_0^x p(s,0)\,\bar{\alpha}(s)\,ds, \tag{2.6.38}$$

and

$$\left|E(x,y)-\bar{E}(x,y)\right|\leq c, \tag{2.6.39}$$

where c is as in Theorem 2.5.1. Let q_0 and $\bar{A}(x,y)$ be as in (2.6.30) and (2.6.31) respectively. Then the solutions of (2.6.24) depends on the initial boundry conditions and

$$|u(x,y)-v(x,y)|\leq\frac{c}{1-q_0}\bar{A}(x,y)$$

$$\times\exp\left(\int_0^x\int_0^y \bar{A}(s,t)\,f(s,t)\,e(s,t)\,dtds\right), \tag{2.6.40}$$

for $(x,y)\in G$.

Proof. Let $u(x,y)$ and $v(x,y)$ be two solutions of IBVP (2.6.24)-(2.6.25) on G, then we have

$$u(x,y)-v(x,y)=\int_0^x p(s,y)\{u(s,y)-v(s,y)\}\,ds$$

$$+\int_0^x\int_0^y\left\{F\left(s,t,u(s,t),\int_0^a\int_0^b k(s,t,\sigma,\tau,u(\sigma,\tau))\,d\tau d\sigma\right)\right.$$

$$-F\left(s,t,v\left(s,t\right),\int_0^a\int_0^b k\left(s,t,\sigma,\tau,v\left(\sigma,\tau\right)\right)d\tau d\sigma\right)\Bigg\}dtds. \qquad (2.6.41)$$

From (2.6.41), (2.6.35), (2.6.36) we obtain

$$|u\left(x,y\right)-v\left(x,y\right)|\leq\int_0^x p\left(s,y\right)|u\left(s,y\right)-v\left(s,y\right)|\,ds$$

$$+\int_0^x\int_0^y f\left(s,t\right)\left(|u\left(s,t\right)-v\left(s,t\right)|\right.$$

$$\left.+e\left(s,t\right)\int_0^a\int_0^b h\left(\sigma,\tau\right)|u\left(\sigma,\tau\right)-v\left(\sigma,\tau\right)|d\tau d\sigma\right)dtds. \qquad (2.6.42)$$

Rewtiting (2.6.42) in view of the fact that $e\left(x,y\right)\geq 1$ and a suitable application of Theorem 2.5.1 (with $c=0,\ g=0$) yields $|u\left(x,y\right)-v\left(x,y\right)|\leq 0$. Therefore $u(x,y)=v(x,y)$ i.e., there is at most one solution of IBVP (2.6.24)-(2.6.25) on G.

(*ii*) Since $u(x,y)$ and $v(x,y)$ are the solutions of IBVP (2.6.24)-(2.6.25) and (2.6.24)-(2.6.37) respectively, we have

$$u\left(x,y\right)-v\left(x,y\right)=E\left(x,y\right)-\bar{E}\left(x,y\right)+\int_0^x p\left(s,y\right)\left\{u\left(s,y\right)-v\left(s,y\right)\right\}ds$$

$$+\int_0^x\int_0^y\Bigg\{F\left(s,t,u\left(s,t\right),\int_0^a\int_0^b k\left(s,t,\sigma,\tau,u\left(\sigma,\tau\right)\right)d\tau d\sigma\right)$$

$$-F\left(s,t,v\left(s,t\right),\int_0^a\int_0^b k\left(s,t,\sigma,\tau,v\left(\sigma,\tau\right)\right)d\tau d\sigma\right)\Bigg\}dtds. \qquad (2.6.43)$$

From (2.6.43), (2.6.39), (2.6.36), (2.6.35), we have

$$|u\left(x,y\right)-v\left(x,y\right)|\leq c+\int_0^x p\left(s,y\right)|u\left(s,y\right)-v\left(s,y\right)|\,ds$$

$$+\int_0^x\int_0^y f\left(s,t\right)\left(|u\left(s,t\right)-v\left(s,t\right)|\right.$$

$$\left.+e\left(s,t\right)\int_0^a\int_0^b h\left(\sigma,\tau\right)|u\left(\sigma,\tau\right)-v\left(\sigma,\tau\right)|d\tau d\sigma\right)dtds$$

$$\leq c + \int_0^x p(s,y)\,|u(s,y) - v(s,y)|\,ds$$

$$+ \int_0^x \int_0^y f(s,t)\,e(s,t)\,(|u(s,t) - v(s,t)|$$

$$+ \int_0^a \int_0^b h(\sigma,\tau)\,|u(\sigma,\tau) - v(\sigma,\tau)|d\tau d\sigma \Bigg)\,dtds, \tag{2.6.44}$$

for $(x, y) \in G$. Now a suitable application of Theorem 2.5.1 (with $g = 0$) to (2.6.44) yields the estimate (2.6.40), which shows the dependency of solutions of (2.6.24) on given initial boundary data.

Here, we note that the inequality in Theorem 2.5.1 can be used to study similar properties as in Theorems 2.6.5 and 2.6.6 for solutions of the non-self-adjoint hyperbolic partial Volterra-Fredholm integrodifferential equation

$$u_{xy}(x,y) = (p(x,y)\,u(x,y))_y$$

$$+F\left(x, y, u(x,y), \int_0^x \int_0^y k_1(x,y,s,t,u(s,t))\,dtds,\right.$$

$$\left., \int_0^a \int_0^b k_2(x,y,s,t,u(s,t))\,dtds\right), \tag{2.6.45}$$

with the given initial boundary conditions (2.6.25) under some suitable conditions on the functions involved in IBVP (2.6.45)-(2.6.25). We omit the details.

2.6.4 Volterra-Fredholm integral equation

In this section we present applications of Theorem 2.5.7, part (r_1) to study certain properties of the solutions of Volterra-Fredholm integral equation of the form

$$z(x,y) = E(x,y) + \int_0^x \int_0^y F(x,y,s,t,z(s,t))\,dtds$$

$$+ \int_0^\infty \int_0^\infty H(x,y,s,t,z(s,t))\,dtds, \tag{2.6.46}$$

for $x, y \in R_+$, where $z(x,y)$ is an unknown function, $E \in C\left(R_+^2, R\right)$ and $F, H \in C\left(R_+^4 \times R, R\right)$.

In [76] Pachpatte has given the following theorems which deals with the properties of solutions of equation (2.6.46).

Theorem 2.6.7. Suppose that the functions E, F, H in equation (2.6.46) satisfy the conditions

$$|E(x,y)| \leq a(x,y), \tag{2.6.47}$$

$$|F(x,y,s,t,z)| \leq b(x,y) f(s,t) |z|, \tag{2.6.48}$$

$$|H(x,y,s,t,z)| \leq c(x,y) g(s,t) |z|, \tag{2.6.49}$$

where a, b, c, f, g are as in Theorem 2.5.7. Let p_1 be as in (2.5.85). If $z(x,y)$ is a solution of equation (2.6.46) on R_+^2,then

$$|z(x,y)| \leq B_1(x,y) + M_1 D_1(x,y), \tag{2.6.50}$$

where B_1, D_1, M_1 are as given in Theorem 2.5.7, part (r_1) .

Proof. Let $z(x,y)$ be a solution of equation (2.6.46) on R_+^2. Using the fact that $z(x,y)$ is a solution of equation (2.6.46) and (2.6.47)-(2.6.49) we observe that

$$|z(x,y)| \leq a(x,y) + b(x,y) \int_0^x \int_0^y f(s,t) |z(s,t)| \, dtds$$

$$+c(x,y) \int_0^\infty \int_0^\infty g(s,t) |z(s,t)| \, dtds. \tag{2.6.51}$$

Now an application of Theorem 2.5.7, part (r_1) to (2.6.51) yields the required estimate in (2.6.50).

Theorem 2.6.8. Suppose that the functions F, H in equation (2.6.46) satisfy the conditions

$$|F(x,y,s,t,z) - F(x,y,s,t,\bar{z})| \leq b(x,y) f(s,t) |z - \bar{z}|, \tag{2.6.52}$$

$$|H(x,y,s,t,z) - H(x,y,s,t,\bar{z})| \leq c(x,y) g(s,t) |z - \bar{z}|, \tag{2.6.53}$$

where b, c, f, g are as in Theorem 2.5.7. Let p_1 be as in (2.5.85).Then the equation (2.6.46) has at most one solution on R_+^2.

Proof. Let $u(x,y)$ and $v(x,y)$ be two solutions of equation (2.6.46) on R_+^2. Using the facts that $u(x,y)$ and $v(x,y)$ are the solutions of equation (2.6.46) and (2.6.52), (2.6.53) we have

$$|u(x,y) - v(x,y)| \leq b(x,y) \int_0^x \int_0^y f(s,t) |u(s,t) - v(s,t)| \, dtds$$

$$+c(x,y) \int_0^\infty \int_0^\infty g(s,t) |u(s,t) - v(s,t)| \, dtds. \tag{2.6.54}$$

Now an application of the inequality given in Theorem 2.5.7, part (r_1) (with $a(x,y) = 0$ which in fact implies $B_1(x,y) = 0, M_1 = 0$) to (2.6.54) yields $u(x,y) = v(x,y)$, i.e., there is at most one solution of equation (2.6.46) on R_+^2 .

Finally, we note that the applications presented here display the importance of some of the inequalities given in earlier sections.Most of the inequalities given here are recently developed and we hope that these inequalities will serve as a model for further investigation

2.7 Notes

The origin of the results included in this chapter can be traced back to the well known Wendroff's inequalities, see [4, p. 154]. Integral inequalities of Wendroff's type have proved to be very useful in the study of certain partial differential and integral equations. The material included in Section 2.2 contains some basic results on integral inequalities developed during the past few years. The inequalities in Theorems 2.2.1 and 2.2.2 are due to Pachpatte [68,55]. The inequalities in Theorems 2.2.3 and 2.2.4 are proved by Medveď [24], which yield estimates on nonlinear integral inequalities with singular kernels. Section 2.3 is dedicated to further nonlinear integral inequalities involving functions of two independent variables. The results given in Theorems 2.3.1-2.3.5 are due to Pachpatte [46,40,45]. Section 2.4 contains some integral inequalities in two independent variables involving iterated double integrals, which are adequate in new applications. The inequalities in Theorems 2.4.1-2.4.4 are all due to Pachpatte and taken from [53,72,78]. Section 2.5 is devoted to the inequalities which yield estimates on certain integral inequalities involving functions of two independent variables,which are mainly used when the earlier inequalities do not apply directly. All the results in this section are due to Pachpatte and taken from [62,72,41,48,76]. Section 2.6 is devoted to the applications of some of the inequalities given in this chapter, to study various aspects of certain partial differential and integral equations. The literature concerning such inequalities is rich and for earlier work we refer the reader to the books by Bainov and

Simeonov [3] Martinjuk and Gutowski [23] and Pachpatte [34] which contains many references on this topic.

Chapter 3

Retarded integral inequalities

3.1 Introduction

Differential equations with retarded arguments have been studied by many investigators and various methods and ideas have been proposed for the study of their different aspects. The fundamental role played by the integral inequalities which provide explicit bounds on unknown functions in the development of the theory of differential and integral equations is well known, see [3,6,12,14,17,19, 23,83,84]. It is natural to expect that some new generalizations and variants of such inequalities would also be equally important in certain new applications. Motivated by a desire to apply integral inequalities which provide explicit bounds on unknown functions, in the development of the theory of differential and integral equations with retarded arguments, recently some new inequalities have been developed to achive a diversity of desired goals, see [21,22,43,47,58-61,69,74,77]. This chapter deals with some basic retarded integral inequalities involving functions of one and two independent variables, which can be used as tools in the study of differential and integral equations involving retarded arguments. Applications of some of the inequalities are also given.

3.2 Basic retarded integral inequalities in one variable

Motivated by the needs of diverse applications in different branchas of differential and integral equations,various investigators have discovered many useful integral inequalities in the literature. In this section, we present some basic retarded integral inequalities established by Pachpatte in [43,60,69,74] which can be used as handy tools in the study of certain new classes of retarded differential and integral equations.

The following theorems contains some useful inequalities proved in [43].

Theorem 3.2.1. Let $I = [t_0, T) \subset R$ (the set of real numbers), $a(t), b(t) \in C(I, R_+)$, $\alpha(t) \in C^1(I, I)$ be nondecreasing with $\alpha(t) \le t$ on I and $k \ge 0$, $c \ge 1$ and $p > 1$ are real constants.

(a_1) If $u(t) \in C(I, R_+)$ and

$$u(t) \le k + \int_{t_0}^{t} a(s) u(s) \, ds + \int_{\alpha(t_0)}^{\alpha(t)} b(s) u(s) \, ds, \tag{3.2.1}$$

for $t \in I$, then

$$u(t) \le k \exp(A(t) + B(t)), \tag{3.2.2}$$

for $t \in I$, where

$$A(t) = \int_{t_0}^{t} a(s) \, ds, \tag{3.2.3}$$

$$B(t) = \int_{\alpha(t_0)}^{\alpha(t)} b(s) \, ds, \tag{3.2.4}$$

for $t \in I$.

(a_2) Let $R_1 = [1, \infty)$. If $u(t) \in C(I, R_1)$ and

$$u(t) \le c + \int_{t_0}^{t} a(s) u(s) \log u(s) \, ds + \int_{\alpha(t_0)}^{\alpha(t)} b(s) u(s) \log u(s) \, ds, \tag{3.2.5}$$

for $t \in I$, then

$$u(t) \le c^{\exp(A(t) + B(t))}, \tag{3.2.6}$$

for $t \in I$ where $A(t)$ and $B(t)$ are defined by (3.2.3) and (3.2.4).

(a_3) If $u(t) \in C(I, R_+)$ and

$$u^p(t) \le k + \int_{t_0}^{t} a(s)\,u(s)\,ds + \int_{\alpha(t_0)}^{\alpha(t)} b(s)\,u(s)\,ds, \tag{3.2.7}$$

for $t \in I$, then

$$u(t) \le \left[k^{\frac{p-1}{p}} + \left(\frac{p-1}{p}\right)(A(s) + B(s))\right]^{\frac{1}{p-1}}, \tag{3.2.8}$$

for $t \in I$, where $A(t)$ and $B(t)$ are defined by (3.2.3) and (3.2.4).

Proof. From the hypotheses we observe that $\alpha'(t) \ge 0$ for $t \in I$.

(a_1) Let $k > 0$ and define a function $z(t)$ by the right hand side of (3.2.1). Then $z(t_0) = k, u(t) \le z(t)$, $z(t)$ is positive, nondecreasing for $t \in I$ and

$$\begin{aligned} z'(t) &= a(t)\,u(t) + b(\alpha(t))\,u(\alpha(t))\,\alpha'(t) \\ &\le a(t)\,u(t) + b(\alpha(t))\,u(\alpha(t))\,\alpha'(t) \\ &\le a(t)\,z(t) + b(\alpha(t))\,z(t)\,\alpha'(t) \end{aligned}$$

i.e.,

$$\frac{z'(t)}{z(t)} \le a(t) + b(\alpha(t))\,\alpha'(t). \tag{3.2.9}$$

Integrating (3.2.9) from t_0 to t, $t \in I$, and the change of variable yield

$$z(t) \le k \exp(A(t) + B(t)), \tag{3.2.10}$$

for $t \in I$. Using (3.2.10) in $u(t) \le z(t)$ we get the inequality (3.2.2). If $k \ge 0$, we carry out the above procedure with $k + \varepsilon$ instead of k, where $\varepsilon > 0$ is an arbitrary small constant, and subsequently pass the limit as $\varepsilon \to 0$ to obtain (3.2.2).

(a_2) Define a function $z(t)$ by the right hand side of (3.2.5). Then $z(t_0) = c$, $u(t) \le z(t)$, $z(t)$ is positive and nondecreasing for $t \in I$ and as in the proof of part (a_1) we get

$$\frac{z'(t)}{z(t)} \le a(t)\log z(t) + b(\alpha(t))\log z(\alpha(t))\,\alpha'(t). \tag{3.2.11}$$

Integrating (3.2.11) from t_0 to t, $t \in I$, and the change of variable yield

$$\log z(t) \le \log c + \int_{t_0}^{t} a(s)\log z(s)\,ds + \int_{\alpha(t_0)}^{\alpha(t)} b(s)\log z(s)\,ds. \tag{3.2.12}$$

Now by a suitable application of the inequality given in (a_1) to (3.2.12) we get

$$\log z(t) \le (\log c)\exp(A(t)+B(t))$$
$$= \log c^{\exp(A(t)+B(t))}. \qquad (3.2.13)$$

From (3.2.13) we observe that

$$z(t) \le c^{\exp(A(t)+B(t))}. \qquad (3.2.14)$$

Using (3.2.14) in $u(t) \le z(t)$ we get the required inequality in (3.2.6).

(a_3) Let $k > 0$ and define a function $z(t)$ by the right hand side of (3.2.7). Then $z(t_0) = k, u(t) \le \{z(t)\}^{\frac{1}{p}}$, $z(t)$ is positive and nondecreasing for $t \in I$ and as in the proof of part (a_1) we have

$$\{z(t)\}^{\frac{1}{p}} z'(t) \le a(t) + b(\alpha(t))\,\alpha'(t). \qquad (3.2.15)$$

Integrating (3.2.15) from t_0 to t, $t \in I$, and the change of variable gives

$$z(t) \le \left[k^{\frac{p-1}{p}} + \left(\frac{p-1}{p}\right)(A(t)+B(t))\right]^{\frac{p}{p-1}}. \qquad (3.2.16)$$

The desired inequality in (3.2.8) follows by using (3.2.16) in $u(t) \le \{z(t)\}^{\frac{1}{p}}$. The case $k \ge 0$ can be completed as mentioned in the proof of part (a_1).

Remark 3.2.1. If we take $a(t) = 0$ in part (a_1), then we get the inequality given by Lipovan in [21, Corollary, p. 391] which in turn contains as a special case, the celebrated Gronwall-Bellman inequality, see [34, p.11] and in this special case, the inequality in (a_2) reduces to the further extension of the inequality given in [34, Theorem 3.8.2, p. 268]. The inequality in (a_3) can be considered as a generalization of the inequality given in [34, Theorem 4.3.1, p. 233].

Theorem 3.2.2. Let $a(t), b(t), \alpha(t), k, c, p$ be as in Theorem 3.2.1. For $i = 1,2$, let $g_i \in C(R_+, R_+)$ be nondecreasing functions with $g_i(u) > 0$ for $u > 0$.

(b_1) If $u(t) \in C(I, R_+)$ and for $t \in I$,

$$u(t) \le k + \int_{t_0}^{t} a(s)\, g_1(u(s))\, ds + \int_{\alpha(t_0)}^{\alpha(t)} b(s)\, g_2(u(s))\, ds, \qquad (3.2.17)$$

then for $t_0 \le t \le t_1; t, t_1 \in I$,

(*i*) in case $g_2(u) \le g_1(u)$,

$$u(t) \le G_1^{-1}[G_1(k) + A(t) + B(t)]; \tag{3.2.18}$$

(*ii*) in case $g_1(u) \le g_2(u)$,

$$u(t) \le G_2^{-1}[G_2(k) + A(t) + B(t)]; \tag{3.2.19}$$

where $A(t)$ and $B(t)$ are given by (3.2.3) and (3.2.4) and for $i = 1, 2$, G_i^{-1} are the inverse functions of

$$G_i(r) = \int_{r_0}^{r} \frac{ds}{g_i(s)}, r > 0, \tag{3.2.20}$$

$r_0 > 0$ is arbitrary and $t_1 \in I$ is chosen so that

$$G_i(k) + A(t) + B(t) \in Dom\left(G_i^{-1}\right),$$

respectively, for all t lying in the interval $[t_0, t_1]$.

(b_2) If $u(t)$ be as in Theorem 3.2.1, part(a_2) and for $t \in I$,

$$u(t) \le c + \int_{t_0}^{t} a(s)\, u(s)\, g_1(\log u(s))\, ds + \int_{\alpha(t_0)}^{\alpha(t)} b(s)\, u(s)\, g_2(\log u(s))\, ds, \tag{3.2.21}$$

then for $t_0 \le t \le t_2; t, t_2 \in I$,

(*i*) in case $g_2(u) \le g_1(u)$,

$$u(t) \le \exp\left(G_1^{-1}[G_1(\log c) + A(t) + B(t)]\right); \tag{3.2.22}$$

(*ii*) in case $g_1(u) \le g_2(u)$,

$$u(t) \le \exp\left(G_2^{-1}[G_2(\log c) + A(t) + B(t)]\right); \tag{3.2.23}$$

where G_i, G_i^{-1}, $A(t), B(t)$ are as in (b_1) and $t_2 \in I$ is chosen so that for $i = 1, 2$

$$G_i(\log c) + A(t) + B(t) \in Dom\left(G_i^{-1}\right),$$

respectively, for all t lying in the interval $[t_0, t_2]$.

(b_3) If $u(t) \in C(I, R_+)$ and for $t \in I$,

$$u^p(t) \le k + \int_{t_0}^{t} a(s)\, g_1(u(s))\, ds + \int_{\alpha(t_0)}^{\alpha(t)} b(s)\, g_2(u(s))\, ds, \tag{3.2.24}$$

then for $t_0 \le t \le t_3; t, t_3 \in I$,

(i) in case $g_2(u) \leq g_1(u)$,

$$u(t) \leq \{H_1^{-1}[H_1(k) + A(t) + B(t)]\}^{\frac{1}{p}}, \tag{3.2.25}$$

(ii) in case $g_1(u) \leq g_2(u)$,

$$u(t) \leq \{H_2^{-1}[H_2(k) + A(t) + B(t)]\}^{\frac{1}{p}}, \tag{3.2.26}$$

where $A(t), B(t)$ are given as in (b_1) and for $i = 1, 2$, H_i^{-1} are the inverse functions of

$$H_i(r) = \int_{r_0}^{r} \frac{ds}{g_i\left(s^{\frac{1}{p}}\right)}, r > 0, \tag{3.2.27}$$

$r_0 > 0$ is arbitrary and $t_3 \in I$ is chosen so that

$$H_i(k) + A(t) + B(t) \in Dom\left(H_i^{-1}\right),$$

respectively, for all t lying in the interval $[t_0, t_3]$.

Proof. (b_1) From the hypotheses we observe that $\alpha'(t) \geq 0$ for $t \in I$. Let $k > 0$ and define a function $z(t)$ by the right hand side of (3.2.17). Then $z(t_0) = k$, $u(t) \leq z(t)$, $z(t)$ is positive and nondecreasing for $t \in I$ and as in the proof of Theorem 3.2.1, part (a_1) we get

$$z'(t) \leq a(t)g_1(z(t)) + b(\alpha(t))g_2(z(\alpha(t)))\alpha'(t). \tag{3.2.28}$$

(i) when $g_2(u) \leq g_1(u)$, then from (3.2.28) we observe that

$$z'(t) \leq g_1(z(t))[a(t) + b(\alpha(t))\alpha'(t)]. \tag{3.2.29}$$

From (3.2.20) and (3.2.29) we have

$$\frac{d}{dt}G_1(z(t)) = \frac{z'(t)}{g_1(z(t))} \leq a(t) + b(\alpha(t))\alpha'(t). \tag{3.2.30}$$

Integrating (3.2.30) from t_0 to t, $t \in I$, and by making the change of variable, we have

$$G_1(z(t)) \leq G_1(k) + A(t) + B(t). \tag{3.2.31}$$

Since G_1^{-1} is increasing, from (3.2.31) we have

$$z(t) \leq G_1^{-1}[G_1(k) + A(t) + B(t)]. \tag{3.2.32}$$

Using (3.2.32) in $u(t) \leq z(t)$ gives the required inequality in (3.2.18). The case $k \geq 0$ can be completed as mentioned in the proof of Theorem 3.2.1, part (a_1). The proof of the case when $g_1(u) \leq g_2(u)$ can be completed similarly. The subinterval $t_0 \leq t \leq t_1$ is obvious.

(i) in case $g_2(u) \leq g_1(u)$,

$$u(t) \leq G_1^{-1}[G_1(k) + A(t) + B(t)]; \tag{3.2.18}$$

(ii) in case $g_1(u) \leq g_2(u)$,

$$u(t) \leq G_2^{-1}[G_2(k) + A(t) + B(t)]; \tag{3.2.19}$$

where $A(t)$ and $B(t)$ are given by (3.2.3) and (3.2.4) and for $i = 1, 2$, G_i^{-1} are the inverse functions of

$$G_i(r) = \int_{r_0}^{r} \frac{ds}{g_i(s)}, r > 0, \tag{3.2.20}$$

$r_0 > 0$ is arbitrary and $t_1 \in I$ is chosen so that

$$G_i(k) + A(t) + B(t) \in Dom\left(G_i^{-1}\right),$$

respectively, for all t lying in the interval $[t_0, t_1]$.

(b_2) If $u(t)$ be as in Theorem 3.2.1, part(a_2) and for $t \in I$,

$$u(t) \leq c + \int_{t_0}^{t} a(s)\, u(s)\, g_1(\log u(s))\, ds + \int_{\alpha(t_0)}^{\alpha(t)} b(s)\, u(s)\, g_2(\log u(s))\, ds, \tag{3.2.21}$$

then for $t_0 \leq t \leq t_2; t, t_2 \in I$,

(i) in case $g_2(u) \leq g_1(u)$,

$$u(t) \leq \exp\left(G_1^{-1}[G_1(\log c) + A(t) + B(t)]\right); \tag{3.2.22}$$

(ii) in case $g_1(u) \leq g_2(u)$,

$$u(t) \leq \exp\left(G_2^{-1}[G_2(\log c) + A(t) + B(t)]\right); \tag{3.2.23}$$

where G_i, G_i^{-1}, $A(t), B(t)$ are as in (b_1) and $t_2 \in I$ is chosen so that for $i = 1, 2$

$$G_i(\log c) + A(t) + B(t) \in Dom\left(G_i^{-1}\right),$$

respectively, for all t lying in the interval $[t_0, t_2]$.

(b_3) If $u(t) \in C(I, R_+)$ and for $t \in I$,

$$u^p(t) \leq k + \int_{t_0}^{t} a(s)\, g_1(u(s))\, ds + \int_{\alpha(t_0)}^{\alpha(t)} b(s)\, g_2(u(s))\, ds, \tag{3.2.24}$$

then for $t_0 \leq t \leq t_3; t, t_3 \in I$,

(i) in case $g_2(u) \le g_1(u)$,

$$u(t) \le \left\{H_1^{-1}\left[H_1(k) + A(t) + B(t)\right]\right\}^{\frac{1}{p}}, \tag{3.2.25}$$

(ii) in case $g_1(u) \le g_2(u)$,

$$u(t) \le \left\{H_2^{-1}\left[H_2(k) + A(t) + B(t)\right]\right\}^{\frac{1}{p}}, \tag{3.2.26}$$

where $A(t), B(t)$ are given as in (b_1) and for $i = 1, 2$, H_i^{-1} are the inverse functions of

$$H_i(r) = \int_{r_0}^{r} \frac{ds}{g_i\left(s^{\frac{1}{p}}\right)}, r > 0, \tag{3.2.27}$$

$r_0 > 0$ is arbitrary and $t_3 \in I$ is chosen so that

$$H_i(k) + A(t) + B(t) \in Dom\left(H_i^{-1}\right),$$

respectively, for all t lying in the interval $[t_0, t_3]$.

Proof. (b_1) From the hypotheses we observe that $\alpha'(t) \ge 0$ for $t \in I$. Let $k > 0$ and define a function $z(t)$ by the right hand side of (3.2.17). Then $z(t_0) = k$, $u(t) \le z(t)$, $z(t)$ is positive and nondecreasing for $t \in I$ and as in the proof of Theorem 3.2.1, part (a_1) we get

$$z'(t) \le a(t) g_1(z(t)) + b(\alpha(t)) g_2(z(\alpha(t))) \alpha'(t). \tag{3.2.28}$$

(i) when $g_2(u) \le g_1(u)$, then from (3.2.28) we observe that

$$z'(t) \le g_1(z(t)) \left[a(t) + b(\alpha(t)) \alpha'(t)\right]. \tag{3.2.29}$$

From (3.2.20) and (3.2.29) we have

$$\frac{d}{dt} G_1(z(t)) = \frac{z'(t)}{g_1(z(t))} \le a(t) + b(\alpha(t)) \alpha'(t). \tag{3.2.30}$$

Integrating (3.2.30) from t_0 to t, $t \in I$, and by making the change of variable, we have

$$G_1(z(t)) \le G_1(k) + A(t) + B(t). \tag{3.2.31}$$

Since G_1^{-1} is increasing, from (3.2.31) we have

$$z(t) \le G_1^{-1}\left[G_1(k) + A(t) + B(t)\right]. \tag{3.2.32}$$

Using (3.2.32) in $u(t) \le z(t)$ gives the required inequality in (3.2.18). The case $k \ge 0$ can be completed as mentioned in the proof of Theorem 3.2.1, part (a_1). The proof of the case when $g_1(u) \le g_2(u)$ can be completed similarly. The subinterval $t_0 \le t \le t_1$ is obvious.

The proofs of (b_2) and (b_3) can be completed by following the proof of (b_1) and closely looking at the proofs of similar inequalities given in [34]. We omit the details.

Remark 3.2.2. We note that the inequalities in Theorem 3.2.2 parts (b_1)–(b_3) can be considered as further generalizations of the inequalities given in Theorems 2.3.1, 3.9.1, 3.4.1 in [34] respectively. We also note that the definitions of the functions H_i in (3.2.27) are motivated from the work of Medveď [26].

The following useful generalization of the inequality (3.2.17) is proved in [69].

For suitable functions defined on the respective domains of their definitions, first we give the following notation used to simplify the details of presentation:

$$H\left[t, m; \phi_1, a_1, p_1; \phi_2, a_2, p_2\right] = \int_{\phi_1(t_0)}^{\phi_1(t)} a_1(s)\, p_1(m(s))\, ds + \int_{\phi_2(t_0)}^{\phi_2(t)} a_2(s)\, p_2(m(s))\, ds.$$

Theorem 3.2.3. Let $u(t), f(t), b(t), a_1(t), a_2(t) \in C(I, R_+)$; $\phi_1(t), \phi_2(t) \in C^1(I, I)$ be nondecreasing with $\phi_1(t) \le t, \phi_2(t) \le t$ on $I = [t_0, T)$. For $i = 1, 2$, let $g_i(t) \in C(R_+, R_+)$ be nondecreasing, subadditive and submultiplicative functions with $g_i(u) > 0$ for $u > 0$ and for $t \in I$,

$$u(t) \le f(t) + b(t)\, H\left[t, u; \phi_1, a_1, g_1; \phi_2, a_2, g_2\right], \tag{3.2.33}$$

then for $t_0 \le t \le t_1; t, t_1 \in I$,

(i) in case $g_2(u) \le g_1(u)$,

$$u(t) \le f(t) + b(t)\, G_1^{-1}\left[G_1(e(t)) + H\left[t, b; \phi_1, a_1, g_1; \phi_2, a_2, g_2\right]\right], \tag{3.2.34}$$

(ii) in case $g_1(u) \le g_2(u)$,

$$u(t) \le f(t) + b(t)\, G_2^{-1}\left[G_2(e(t)) + H\left[t, b; \phi_1, a_1, g_1; \phi_2, a_2, g_2\right]\right], \tag{3.2.35}$$

where

$$e(t) = H\left[t, f; \phi_1, a_1, g_1; \phi_2, a_2, g_2\right], \tag{3.2.36}$$

G_i, G_i^{-1} are as in Theorem 3.2.2, part (b_1) and $t_1 \in I$ is chosen so that

$$G_i(e(t)) + H\left[t, b; \phi_1, a_1, g_1; \phi_2, a_2, g_2\right] \in Dom\left(G_i^{-1}\right),$$

respectively, for all t lying in the interval $[t_0, t_1]$.

Proof. From the hypotheses we observe that $\phi_1'(t) \geq 0, \phi_2'(t) \geq 0$ for $t \in I$. Define a function $z(t)$ by

$$z(t) = H[t, u; \phi_1, a_1, g_1; \phi_2, a_2, g_2]$$

$$= \int_{\phi_1(t_0)}^{\phi_1(t)} a_1(s) g_1(u(s)) ds + \int_{\phi_2(t_0)}^{\phi_2(t)} a_2(s) g_2(u(s)) ds. \tag{3.2.37}$$

Then $z(t_0) = 0$ and (3.2.33) can be restated as

$$u(t) \leq f(t) + b(t) z(t). \tag{3.2.38}$$

Using (3.2.38) in (3.2.37) and the hypotheses on g_1, g_2 we have

$$z(t) \leq \int_{\phi_1(t_0)}^{\phi_1(t)} a_1(s) g_1(f(s) + b(s) z(s)) ds$$

$$+ \int_{\phi_2(t_0)}^{\phi_2(t)} a_2(s) g_2(f(s) + b(s) z(s)) ds$$

$$\leq e(t) + \int_{\phi_1(t_0)}^{\phi_1(t)} a_1(s) g_1(b(s)) g_1(z(s)) ds$$

$$+ \int_{\phi_2(t_0)}^{\phi_2(t)} a_2(s) g_2(b(s)) g_2(z(s)) ds. \tag{3.2.39}$$

Let $\beta \in I$ be an arbitrary number. From (3.2.39), for $t_0 \leq t \leq \beta$ we have

$$z(t) \leq e(\beta) + \int_{\phi_1(t_0)}^{\phi_1(t)} a_1(s) g_1(b(s)) g_1(z(s)) ds$$

$$+ \int_{\phi_2(t_0)}^{\phi_2(t)} a_2(s) g_2(b(s)) g_2(z(s)) ds. \tag{3.2.40}$$

Now assume that $e(\beta) > 0$ and let $g_2(u) \leq g_1(u)$. Define a function $v(t)$ by the right hand side of (3.2.40). Then $v(t_0) = e(\beta)$ $z(t) \leq v(t)$, $v(t)$ is positive and nondecreasing for $t_0 \leq t \leq \beta$ and

$$v'(t) = a_1(\phi_1(t)) g_1(b(\phi_1(t))) g_1(z(\phi_1(t))) \phi_1'(t)$$

$$+ a_2(\phi_2(t)) g_2(b(\phi_2(t))) g_2(z(\phi_2(t))) \phi_2'(t)$$

$$\leq a_1(\phi_1(t))\, g_1(b(\phi_1(t)))\, g_1(v(\phi_1(t)))\, \phi_1'(t)$$

$$+a_2(\phi_2(t))\, g_2(b(\phi_2(t)))\, g_2(v(\phi_2(t)))\, \phi_2'(t)$$

$$\leq a_1(\phi_1(t))\, g_1(b(\phi_1(t)))\, g_1(v(t))\, \phi_1'(t)$$

$$+a_2(\phi_2(t))\, g_2(b(\phi_2(t)))\, g_2(v(t))\, \phi_2'(t)$$

$$\leq [a_1(\phi_1(t))\, g_1(b(\phi_1(t)))\, \phi_1'(t)$$

$$+a_2(\phi_2(t))\, g_2(b(\phi_2(t)))\, \phi_2'(t)]\, g_1(v(t)). \tag{3.2.41}$$

From (3.2.20) and (3.2.41) we have

$$\frac{d}{dt} G_1(v(t)) = \frac{v'(t)}{g_1(v(t))}$$

$$\leq [a_1(\phi_1(t))\, g_1(b(\phi_1(t)))\, \phi_1'(t)$$

$$+a_2(\phi_2(t))\, g_2(b(\phi_2(t)))\, \phi_2'(t)]. \tag{3.2.42}$$

By taking $t = s$ in (3.2.42) and integrating it with respect to s from t_0 to t for $t_0 \leq t \leq \beta$ we get

$$G_1(v(t)) \leq G_1(e(\beta)) + \int_{t_0}^{t} [a_1(\phi_1(s))\, g_1(b(\phi_1(s)))\, \phi_1'(s)$$

$$+a_2(\phi_2(s))\, g_2(b(\phi_2(s)))\, \phi_2'(s)]\, ds, \tag{3.2.43}$$

for $t_0 \leq t \leq \beta$. Since $z(t) \leq v(t)$ for $t_0 \leq t \leq \beta$ and $\beta \in I$ is arbitrary, from (3.2.43) we have

$$z(t) \leq G_1^{-1}\left[G_1(e(\beta)) + \int_{t_0}^{t} [a_1(\phi_1(s))\, g_1(b(\phi_1(s)))\, \phi_1'(s)\right.$$

$$\left. +a_2(\phi_2(s))\, g_2(b(\phi_2(s)))\, \phi_2'(s)]\, ds\right], \tag{3.2.44}$$

for $t_0 \leq t \leq t_1$. By making the change of variable in the integral on the right hand side in (3.2.44) we have

$$z(t) \leq G_1^{-1}\left[G_1(e(t)) + H[t, b; \phi_1, a_1, g_1; \phi_2, a_2, g_2]\right], \tag{3.2.45}$$

for $t_0 \leq t \leq t_1$. The conclusion (3.2.34) follows from (3.2.38) and (3.2.45). If $e(\beta)$ in (3.2.40) is nonnegative, we carry out the above procedure with $e(\beta) + \varepsilon$ instead of $e(\beta)$, where $\varepsilon > 0$ is an arbitrary small constant, and subsequently pass to the limit $\varepsilon \to 0$ to obtain (3.2.34). The subinterval $t_0 \leq t \leq t_1$ is obvious. The proof of the case when $g_1(u) \leq g_2(u)$ can be completed similarly.

Remark 3.2.3. We note that in [2] the authors have given the upper bound on (3.2.33) (when $b(t) = 1$), which depends on the continuous solution of a certain initial value problem for first order differential equation. The bound obtained on (3.2.33) in Theorem 3.2.3 is explicit and it is more convenient in applications. Here, it is to be noted that the conditions required on the functions involved in (3.2.33) are different from those of given in [2].

Next, we shall give the inequalities established in [74] which can be used more conveniently in certain situations.

Theorem 3.2.4. Let $u(t), a(t), b_i(t) \in C(I, R_+)$; $\alpha_i(t) \in C^1(I, I)$ be nondecreasing with $\alpha_i(t) \leq t$ on $I = [t_0, T)$ for $i = 1, ..., n$ and $k \geq 0$ be a real constant.

(c_1) If

$$u(t) \leq k + \sum_{i=1}^{n} \int_{\alpha_i(t_0)}^{\alpha_i(t)} b_i(s) u(s) ds, \tag{3.2.46}$$

for $t \in I$, then

$$u(t) \leq k \exp(E(t)), \tag{3.2.47}$$

for $t \in I$, where

$$E(t) = \sum_{i=1}^{n} \int_{\alpha_i(t_0)}^{\alpha_i(t)} b_i(\sigma) d\sigma, \tag{3.2.48}$$

for $t \in I$.

(c_2) If $a(t)$ is nondecreasing for $t \in I$ and

$$u(t) \leq a(t) + \sum_{i=1}^{n} \int_{\alpha_i(t_0)}^{\alpha_i(t)} b_i(s) u(s) ds, \tag{3.2.49}$$

for $t \in I$, then

$$u(t) \leq a(t) \exp(E(t)), \tag{3.2.50}$$

for $t \in I$, where $E(t)$ is given by (3.2.48).

Proof. From the hypotheses on $\alpha_i(t)$ we observe that $\alpha_i'(t) \geq 0$ for $t \in I$ and $i = 1, ..., n$.

(c_1) Let $k > 0$ and define a function $z(t)$ by the right hand side of (3.2.46). Then $z(t_0) = k$, $u(t) \leq z(t)$, $z(t) > 0$ and

$$z'(t) = \sum_{i=1}^{n} b_i(\alpha_i(t))\, u(\alpha_i(t))\, \alpha_i'(t)$$

$$\leq \sum_{i=1}^{n} b_i(\alpha_i(t))\, z(\alpha_i(t))\, \alpha_i'(t)$$

$$\leq \sum_{i=1}^{n} b_i(\alpha_i(t))\, z(t)\, \alpha_i'(t)$$

i.e.,

$$\frac{z'(t)}{z(t)} \leq \sum_{i=1}^{n} b_i(\alpha_i(t))\, \alpha_i'(t). \tag{3.2.51}$$

Integrating (3.2.51) from t_0 to t; $t \in I$ and then the change of variables yields

$$z(t) \leq k \exp(E(t)), \tag{3.2.52}$$

for $t \in I$. Using (3.2.52) in $u(t) \leq z(t)$ we get the inequality in (3.2.47). If $k \geq 0$ we carry out the above procedure with $k + \varepsilon$ insteead of k, where $\varepsilon > 0$ is an arbitrary small constant, and subsequently pass the limit $\varepsilon \to 0$ to obtain (3.2.47).

(c_2) First we assume that $a(t) > 0$ for $t \in I$. From the hypotheses, for $s \leq \alpha_i(t) \leq t$, we have $a(s) \leq a(\alpha_i(t)) \leq a(t)$. In view of this, from (3.2.49) we observe that

$$\frac{u(t)}{a(t)} \leq 1 + \sum_{i=1}^{n} \int_{\alpha_i(t_0)}^{\alpha_i(t)} b_i(s) \frac{u(s)}{a(s)} ds. \tag{3.2.53}$$

Now an application of the inequality in part (c_1) to (3.2.53) yields the required inequality in (3.2.50). If $a(t) = 0$, then from (3.2.49) we observe that

$$u(t) \leq \varepsilon + \sum_{i=1}^{n} \int_{\alpha_i(t_0)}^{\alpha_i(t)} b_i(s)\, u(s)\, ds, \tag{3.2.54}$$

where $\varepsilon > 0$ is an arbitrary small constant. An application of the inequality in part (c_1) to (3.2.54) yields

$$u(t) \leq \varepsilon \exp(E(t)). \tag{3.2.55}$$

Now by letting $\varepsilon \to 0$ in (3.2.55) we have $u(t) = 0$ and hence (3.2.50) holds.

Theorem 3.2.5. Let $u(t), b_i(t), \alpha_i(t)$ be as in Theorem 3.2.4. Let $k \geq 0, p > 1$ be real constants. Let $g \in C(R_+, R_+)$ be nondecreasing function with $g(u) > 0$ for $u > 0$.

(d_1) If for $t \in I$,

$$u(t) \leq k + \sum_{i=1}^{n} \int_{\alpha_i(t_0)}^{\alpha_i(t)} b_i(s)\, g(u(s))\, ds, \tag{3.2.56}$$

then for $t_0 \leq t \leq t_1; t, t_1 \in I$,

$$u(t) \leq G^{-1}\left[G(k) + E(t)\right], \tag{3.2.57}$$

where $E(t)$ is given by (3.2.48) and G^{-1} is the inverse function of

$$G(r) = \int_{r_0}^{r} \frac{ds}{g(s)}, r > 0, \tag{3.2.58}$$

$r_0 > 0$ is arbitrary and $t_1 \in I$ is chosen so that

$$G(k) + E(t) \in Dom\left(G^{-1}\right),$$

for all t lying in the interval $[t_0, t_1]$.

(d_2) If for $t \in I$,

$$u^p(t) \leq k + \sum_{i=1}^{n} \int_{\alpha_i(t_0)}^{\alpha_i(t)} b_i(s)\, g(u(s))\, ds, \tag{3.2.59}$$

then for $t_0 \leq t \leq t_2; t, t_2 \in I$,

$$u(t) \leq \left\{H^{-1}\left[H(k) + E(t)\right]\right\}^{\frac{1}{p}}, \tag{3.2.60}$$

where $E(t)$ is given by (3.2.48) and H^{-1} is the inverse function of

$$H(r) = \int_{r_0}^{r} \frac{ds}{g\left(s^{\frac{1}{p}}\right)}, r > 0, \tag{3.2.61}$$

$r_0 > 0$ is arbitrary and $t_2 \in I$ is chosen so that

$$H(k) + E(t) \in Dom\left(H^{-1}\right),$$

for all t lying in the interval $[t_0, t_2]$.

Proof. From the hypotheses on $\alpha_i(t)$ we observe that $\alpha_i'(t) \geq 0$ for $t \in I$ and $i = 1, ..., n$.

(d_1) Let $k > 0$ and define a function $z(t)$ by the right hand side of (3.2.56). Then $z(t_0) = k, u(t) \leq z(t)$, $z(t)$ is positive and nondecreasing for $t \in I$ and following the proof of Theorem 3.2.4, part (c_1) we have

$$\frac{z'(t)}{g(z(t))} \leq \sum_{i=1}^{n} b_i(\alpha_i(t))\,\alpha_i'(t). \tag{3.2.62}$$

From (3.2.58) and (3.2.62) we have

$$\frac{d}{dt}G(z(t)) = \frac{z'(t)}{g(z(t))} \leq \sum_{i=1}^{n} b_i(\alpha_i(t))\,\alpha_i'(t). \tag{3.2.63}$$

Integrating (3.2.63) from t_0 to t; $t \in I$, and making the change of variables, we get

$$G(z(t)) \leq G(k) + E(t),$$

which implies

$$z(t) \leq G^{-1}[G(k) + E(t)]. \tag{3.2.64}$$

Using (3.2.64) in $u(t) \leq z(t)$ gives the required inequality in (3.2.57). The case $k \geq 0$ can be completed as mentioned in the proof of Theorem 3.2.4, part (c_1). The subinterval $t_0 \leq t \leq t_1$ is obvious.

(d_2) The proof can be completed by following the proof of part (d_1) given above with suitable changes. Here we omit the details.

The inequalities established in [60] are embodied in the following theorem.

Theorem 3.2.6. Let $u(t), a_i(t), b_i(t) \in C(I, R_+)$ and $\alpha_i(t) \in C^1(I, I)$ be nondecreasing with $\alpha_i(t) \leq t$ on $I = [t_0, T)$ for $i = 1, ..., n$. Let $p > 1$ and $c \geq 0$ be real constants.

(q_1) If

$$u^p(t) \leq c + p\sum_{i=1}^{n} \int_{\alpha_i(t_0)}^{\alpha_i(t)} [a_i(s)\,u^p(s) + b_i(s)\,u(s)]\,ds, \tag{3.2.65}$$

for $t \in I$, then

$$u(t) \leq \left\{ M(t)\exp\left((p-1)\sum_{i=1}^{n} \int_{\alpha_i(t_0)}^{\alpha_i(t)} a_i(\sigma)d\sigma \right)\right\}^{\frac{1}{p-1}}, \tag{3.2.66}$$

for $t \in I$, where

$$M(t) = \{c\}^{\frac{p-1}{p}} + (p-1)\sum_{i=1}^{n} \int_{\alpha_i(t_0)}^{\alpha_i(t)} b_i(\sigma)d\sigma, \tag{3.2.67}$$

for $t \in I$.

(q_2) Let $w \in C(R_+, R_+)$ be nondecreasing function with $w(u) > 0$ on $(0, \infty)$. If for $t \in I$,

$$u^p(t) \le c + p\sum_{i=1}^{n} \int_{\alpha_i(t_0)}^{\alpha_i(t)} [a_i(s)\,u(s)\,w(u(s)) + b_i(s)\,u(s)]\,ds, \tag{3.2.68}$$

then for $t_0 \le t \le t_1; t, t_1 \in I$,

$$u(t) \le \left\{F^{-1}\left[F(M(t)) + (p-1)\sum_{i=1}^{n} \int_{\alpha_i(t_0)}^{\alpha_i(t)} a_i(\sigma)d\sigma\right]\right\}^{\frac{1}{p-1}}, \tag{3.2.69}$$

where $M(t)$ is given by (3.2.67), F^{-1} is the inverse function of

$$F(r) = \int_{r_0}^{r} \frac{ds}{w\left(s^{\frac{1}{p-1}}\right)}, r > 0, \tag{3.2.70}$$

$r_0 > 0$ is arbitrary and $t_1 \in I$ is chosen so that

$$F(M(t)) + (p-1)\sum_{i=1}^{n} \int_{\alpha_i(t_0)}^{\alpha_i(t)} a_i(\sigma)d\sigma \in Dom\left(F^{-1}\right),$$

for all t lying in the interval $t_0 \le t \le t_1$.

Proof. From the hypotheses on $\alpha_i(t)$ we observe that $\alpha_i'(t) \ge 0$ for $t \in I$ and $i = 1, ..., n$.

(q_1) Let $c > 0$ and define a function $z(t)$ by the right hand side of (3.2.65). Then $z(t_0) = c, u(t) \le \{z(t)\}^{\frac{1}{p}}$, $z(t)$ is positive and nondecreasing for $t \in I$ and

$$z'(t) = p\sum_{i=1}^{n} [a_i(\alpha_i(t))\,u^p(\alpha_i(t)) + b_i(\alpha_i(t))\,u(\alpha_i(t))]\alpha_i'(t)$$

$$\le p\sum_{i=1}^{n} \left[a_i(\alpha_i(t))\,z(\alpha_i(t)) + b_i(\alpha_i(t))\,\{z(\alpha_i(t))\}^{\frac{1}{p}}\right]\alpha_i'(t)$$

$$= p\sum_{i=1}^{n}\left[a_i\left(\alpha_i\left(t\right)\right)\left\{z\left(\alpha_i\left(t\right)\right)\right\}^{1-\frac{1}{p}} + b_i\left(\alpha_i\left(t\right)\right)\right]\left\{z\left(\alpha_i\left(t\right)\right)\right\}^{\frac{1}{p}}\alpha_i'\left(t\right)$$

$$\leq p\sum_{i=1}^{n}\left[a_i\left(\alpha_i\left(t\right)\right)\left\{z\left(\alpha_i\left(t\right)\right)\right\}^{\frac{p-1}{p}} + b_i\left(\alpha_i\left(t\right)\right)\right]\left\{z\left(t\right)\right\}^{\frac{1}{p}}\alpha_i'\left(t\right)$$

i.e.,

$$\frac{z'\left(t\right)}{\left\{z\left(t\right)\right\}^{\frac{1}{p}}} \leq p\sum_{i=1}^{n}\left[a_i\left(\alpha_i\left(t\right)\right)\left\{z\left(\alpha_i\left(t\right)\right)\right\}^{\frac{p-1}{p}} + b_i\left(\alpha_i\left(t\right)\right)\right]\alpha_i'\left(t\right). \tag{3.2.71}$$

By taking $t = s$ in (3.2.71) and integrating it with respect to s from t_0 to t we get

$$\left\{z\left(t\right)\right\}^{\frac{p-1}{p}} \leq c^{\frac{p-1}{p}} + (p-1)$$
$$\times\int_{t_0}^{t}\sum_{i=1}^{n}\left[a_i\left(\alpha_i\left(s\right)\right)\left\{z\left(\alpha_i\left(s\right)\right)\right\}^{\frac{p-1}{p}} + b_i\left(\alpha_i\left(s\right)\right)\right]\alpha_i'\left(s\right)ds. \tag{3.2.72}$$

Making the change of variables on the right hand side of (3.2.72) and rewriting we get

$$\left\{z\left(t\right)\right\}^{\frac{p-1}{p}} \leq M\left(t\right) + (p-1)\int_{\alpha_i(t_0)}^{\alpha_i(t)}\sum_{i=1}^{n}a_i\left(\sigma\right)\left\{z\left(\sigma\right)\right\}^{\frac{p-1}{p}}d\sigma. \tag{3.2.73}$$

Clearly $M(t)$ is continuous,positive and nondecreasing function for $t \in I$. Now by following the idea used in the proof of Theorem 1 in [22] (see also [43]) we get

$$\left\{z\left(t\right)\right\}^{\frac{p-1}{p}} \leq M\left(t\right)\exp\left((p-1)\sum_{i=1}^{n}\int_{\alpha_i(t_0)}^{\alpha_i(t)}a_i\left(\sigma\right)d\sigma\right). \tag{3.2.74}$$

Using (3.2.74) in $u\left(t\right) \leq \left\{z\left(t\right)\right\}^{\frac{1}{p}}$ we get the desired inequality in (3.2.66). The case $c \geq 0$ can be completed as mentioned in the proof of Theorem 3.2.4, part (c_1).

(q_2) Let $c > 0$ and define a function $z(t)$ by the right hand side of (3.2.68). Then $z\left(t_0\right) = c, u\left(t\right) \leq \left\{z\left(t\right)\right\}^{\frac{1}{p}}$, $z(t)$ is positive and nondecreasing for $t \in I$ and by following the proof of (q_1) given above upto (3.2.73) with suitable changes we get

$$\left\{z\left(t\right)\right\}^{\frac{p-1}{p}} \leq M\left(t\right) + (p-1)\sum_{i=1}^{n}\int_{\alpha_i(t_0)}^{\alpha_i(t)}a_i\left(\sigma\right)w\left(\left\{z\left(\sigma\right)\right\}^{\frac{1}{p}}\right)d\sigma. \tag{3.2.75}$$

Now fix $\lambda \in I$ such that $t_0 \leq t \leq \lambda \leq t_1$. Then from (3.2.75) we observe that

$$\{z(t)\}^{\frac{p-1}{p}} \leq M(\lambda) + (p-1)\sum_{i=1}^{n} \int_{\alpha_i(t_0)}^{\alpha_i(t)} a_i(\sigma)\, w\left(\{z(\sigma)\}^{\frac{1}{p}}\right) d\sigma, \tag{3.2.76}$$

for $t_0 \leq t \leq \lambda$. Define a function $v(t)$ by the right hand side of (3.2.76). Then $v(t_0) = M(\lambda), \{z(t)\}^{\frac{p-1}{p}} \leq v(t)$, $v(t)$ is positive and nondecreasing for $t_0 \leq t \leq \lambda$ and

$$v(t) \leq M(\lambda) + (p-1)\sum_{i=1}^{n} \int_{\alpha_i(t_0)}^{\alpha_i(t)} a_i(\sigma)\, w\left(\{v(\sigma)\}^{\frac{1}{p-1}}\right) d\sigma,$$

for $t_0 \leq t \leq \lambda$. The rest of the proof can be completed by following the proof of Theorem 3.2.5, part (d_1) with suitable changes (see also [43]). We omit the details.

3.3 Further retarded integral inequalities in one variable

In view to widen the scope of applications of the inequalities of the type given earlier section, in [58,61,64,74,77] Pachpatte has established a number of such inequalities. In this section we offer some of the inequalities given in the above references, which are more adequate in certain situations.

First we shall give the following theorems which deals with the inequalities proved in [61].

Theorem 3.3.1. Let $u(t), a(t) \in C(I, R_+)$, $k_i(t,s), \frac{\partial}{\partial t}k_i(t,s) \in C(I^2, R_+)$ for $t_0 \leq s \leq t < T$ and $\alpha_i(t) \in C^1(I, I)$ be nondecreasing with $\alpha_i(t) \leq t$ on $I = [t_0, T)$ for $i = 1, ..., n$.

(a_1) If $c \geq 0$ is a real constant and

$$u(t) \leq c + \sum_{i=1}^{n} \int_{\alpha_i(t_0)}^{\alpha_i(t)} k_i(t,s)\, u(s)\, ds, \tag{3.3.1}$$

for $t \in I$, then

$$u(t) \leq c \exp\left(\int_{t_0}^{t} Q(s)\, ds\right), \tag{3.3.2}$$

for $t \in I$, where

$$Q(t) = \sum_{i=1}^{n} \left[k_i(t, \alpha_i(t)) \alpha_i'(t) + \int_{\alpha_i(t_0)}^{\alpha_i(t)} \frac{\partial}{\partial t} k_i(t, \sigma) \, d\sigma \right], \tag{3.3.3}$$

for $t \in I$.

(a_2) If $a(t)$ is nondecreasing for $t \in I$ and

$$u(t) \le a(t) + \sum_{i=1}^{n} \int_{\alpha_i(t_0)}^{\alpha_i(t)} k_i(t, s) u(s) \, ds, \tag{3.3.4}$$

for $t \in I$, then

$$u(t) \le a(t) \exp\left(\int_{t_0}^{t} Q(s) \, ds \right), \tag{3.3.5}$$

for $t \in I$, where $Q(t)$ is given by (3.3.3).

Proof. From the hypotheses on $\alpha_i(t)$ we observe that $\alpha_i'(t) \ge 0$ for $t \in I$ and $i = 1, ..., n$.

(a_1) Define a function $z(t)$ by the right hand side of (3.3.1). Then $z(t_0) = c, u(t) \le z(t)$ and

$$\begin{aligned} z'(t) &= \sum_{i=1}^{n} \left[k_i(t, \alpha_i(t)) u(\alpha_i(t)) \alpha_i'(t) + \int_{\alpha_i(t_0)}^{\alpha_i(t)} \frac{\partial}{\partial t} k_i(t, s) u(s) \, ds \right] \\ &\le \sum_{i=1}^{n} \left[k_i(t, \alpha_i(t)) z(\alpha_i(t)) \alpha_i'(t) + \int_{\alpha_i(t_0)}^{\alpha_i(t)} \frac{\partial}{\partial t} k_i(t, s) z(s) \, ds \right] \\ &\le \sum_{i=1}^{n} \left[k_i(t, \alpha_i(t)) \alpha_i'(t) + \int_{\alpha_i(t_0)}^{\alpha_i(t)} \frac{\partial}{\partial t} k_i(t, s) \, ds \right] z(\alpha_i(t)) \\ &\le Q(t) z(t), \end{aligned}$$

which implies

$$z(t) \le c \exp\left(\int_{t_0}^{t} Q(s) \, ds \right). \tag{3.3.6}$$

Using (3.3.6) in $u(t) \le z(t)$ we get the desired inequality in (3.3.2).

(a_2) First we assume that $a(t) > 0$ for all $t \in I$. It is easy to observe that for $s \leq \alpha_i(t) \leq t$ we have $a(s) \leq a(\alpha_i(t)) \leq a(t)$. In view of this, from (3.3.4) we observe that

$$\frac{u(t)}{a(t)} \leq 1 + \sum_{i=1}^{n} \int_{\alpha_i(t_0)}^{\alpha_i(t)} k_i(t,s) \frac{u(s)}{a(s)} ds. \tag{3.3.7}$$

Now an application of the inequality in part (a_1) to (3.3.7) yields (3.3.5). If $a(t) = 0$, then from (3.3.4) we observe that

$$u(t) \leq \varepsilon + \sum_{i=1}^{n} \int_{\alpha_i(t_0)}^{\alpha_i(t)} k_i(t,s) u(s) ds, \tag{3.3.8}$$

where $\varepsilon > 0$ is an arbitrary small constant. An application of the inequality in part (a_1) to (3.3.8) yields

$$u(t) \leq \varepsilon \exp\left(\int_{t_0}^{t} Q(s) ds\right). \tag{3.3.9}$$

Now by letting $\varepsilon \to 0$ in (3.3.9) we have $u(t) = 0$ and hence (3.3.5) holds.

Theorem 3.3.2 . Let $u(t), a(t), k_i(t,s), \frac{\partial}{\partial t}k_i(t,s), \alpha_i(t)$ be as in Theorem 3.3.1.

(b_1) Let $c \geq 0$ be a real constant, $g \in C(R_+, R_+)$ be a nondecreasing function with $g(u) > 0$ for $u > 0$. If for $t \in I$,

$$u(t) \leq c + \sum_{i=1}^{n} \int_{\alpha_i(t_0)}^{\alpha_i(t)} k_i(t,s) g(u(s)) ds, \tag{3.3.10}$$

then for $t_0 \leq t \leq t_1; t, t_1 \in I$,

$$u(t) \leq G^{-1}\left[G(c) + \int_{t_0}^{t} Q(s) ds\right], \tag{3.3.11}$$

where $Q(t)$ is given by (3.3.3) and G^{-1} is the inverse function of

$$G(r) = \int_{r_0}^{r} \frac{ds}{g(s)}, r > 0, \tag{3.3.12}$$

$r_0 > 0$ is arbitrary and $t_1 \in I$ is chosen so that

$$G(c) + \int_{t_0}^{t} Q(s)\,ds \in Dom\left(G^{-1}\right),$$

for all t lying in the interval $t_0 \leq t \leq t_1$.

(b_2) Let $g(u)$ be as in (b_1) and suppose in addition it is subadditive. If for $t \in I$,

$$u(t) \leq a(t) + \sum_{i=1}^{n} \int_{\alpha_i(t_0)}^{\alpha_i(t)} k_i(t,s)\, g(u(s))\,ds, \tag{3.3.13}$$

then for $t_0 \leq t \leq t_2; t, t_2 \in I$,

$$u(t) \leq a(t) + G^{-1}\left[G(A(t)) + \int_{t_0}^{t} Q(s)\,ds\right], \tag{3.3.14}$$

where G, G^{-1}, $Q(t)$ be as in (b_1),

$$A(t) = \sum_{i=1}^{n} \int_{\alpha_i(t_0)}^{\alpha_i(t)} k_i(t,s)\, g(a(s))\,ds, \tag{3.3.15}$$

for $t \in I$ and $t_2 \in I$ is chosen so that

$$G(A(t)) + \int_{t_0}^{t} Q(s)\,ds \in Dom\left(G^{-1}\right),$$

for all t lying in the interval $t_0 \leq t \leq t_2$.

Proof. From the hypotheses on $\alpha_i(t)$ we observe that $\alpha_i'(t) \geq 0$ for $t \in I$ and $i = 1, ..., n$.

(b_1) We first assume that $c > 0$ and define a function $z(t)$ by the right hand side of (3.3.10). Then $z(t_0) = c, u(t) \leq z(t)$, $z(t)$ is positive and nondecreasing for $t \in I$ and by following the proof of Theorem 3.3.1, part (a_1) with suitable changes we have

$$z'(t) \leq Q(t)\, g(z(t)). \tag{3.3.16}$$

The rest of the proof can be completed by following the proof of Theorem 3.2.5, part (d_1).

(b_2) Define a function $z(t)$ by

$$z(t) = \sum_{i=1}^{n} \int_{\alpha_i(t_0)}^{\alpha_i(t)} k_i(t,s)\, g(u(s))\, ds. \tag{3.3.17}$$

Then $z(t_0) = 0$ and from (3.3.13) we have

$$u(t) \leq a(t) + z(t). \tag{3.3.18}$$

Using (3.3.18) in (3.3.17) we have

$$z(t) \leq \sum_{i=1}^{n} \int_{\alpha_i(t_0)}^{\alpha_i(t)} k_i(t,s)\, g(a(s) + z(s))\, ds$$

$$\leq A(t) + \sum_{i=1}^{n} \int_{\alpha_i(t_0)}^{\alpha_i(t)} k_i(t,s)\, g(z(s))\, ds,$$

where $A(t)$ is given by (3.3.15). It is easy to observe that $A(t)$ is nonnegative and nondecreasing for $t \in I$. Now by following the similar arguments as in the proof of Theorem 2.4.2 given in [34] and in view of the proof of Theorem 3.2.5, part (d_1) we get

$$z(t) \leq G^{-1}\left[G(A(t)) + \int_{t_0}^{t} Q(s)\, ds\right]. \tag{3.3.19}$$

Using (3.3.19) in (3.3.18) we get the required inequality in (3.3.14). The subinterval $t_0 \leq t \leq t_2$ is obvious.

The next theorem contains the inequalities established in [77] involving Lipschitizian type kernel function.

Theorem 3.3.3. Let $u(t), a(t), b(t) \in C(I, R_+)$ and $\alpha(t) \in C^1(I, I)$ be nondecreasing with $\alpha(t) \leq t$ on $I = [t_0, T)$.

(c_1) Let $L \in C(I \times R_+, R_+)$ and

$$0 \leq L(t,x) - L(t,y) \leq M(t,y)(x-y), \tag{3.3.20}$$

for $t \in I$ and $x \geq y \geq 0$,where $M \in C(I \times R_+, R_+)$. If

$$u(t) \leq a(t) + b(t) \int_{\alpha(t_0)}^{\alpha(t)} L(s, u(s))\, ds, \tag{3.3.21}$$

for $t \in I$, then

$$u(t) \le a(t) + b(t) \int_{\alpha(t_0)}^{\alpha(t)} L(\sigma, a(\sigma)) \exp\left(\int_{\sigma}^{\alpha(t)} M(\tau, a(\tau))\, b(\tau)\, d\tau \right) d\sigma, \quad (3.3.22)$$

for $t \in I$

(c_2) Let $L \in C(I \times R_+, R_+)$ and $\psi \in C(R_+, R_+)$ be strictly increasing function with $\psi(0) = 0$ and

$$0 \le L(t, x) - L(t, y) \le M(t, y)\, \psi^{-1}(x - y), \quad (3.3.23)$$

for $t \in I$ and $x \ge y \ge 0$,, where $M \in C(I \times R_+, R_+)$ and ψ^{-1} is the inverse of ψ. If

$$u(t) \le a(t) + \psi\left(b(t) \int_{\alpha(t_0)}^{\alpha(t)} L(s, u(s))\, ds \right), \quad (3.3.24)$$

for $t \in I$, then

$$u(t) \le a(t) + \psi\left(b(t) \int_{\alpha(t_0)}^{\alpha(t)} L(\sigma, a(\sigma)) \right.$$

$$\left. \times \exp\left(\int_{\sigma}^{\alpha(t)} M(\tau, a(\tau))\, b(\tau) d\tau \right) d\sigma \right), \quad (3.3.25)$$

for $t \in I$.

(c_3) Let L, ψ, M be as in (c_2) and the condition (3.3.23) holds. Suppose in addition $\psi^{-1}(xy) \le \psi^{-1}(x)\, \psi^{-1}(y)$ for $x, y \in R_+$.If

$$u(t) \le a(t) + b(t)\, \psi\left(\int_{\alpha(t_0)}^{\alpha(t)} L(s, u(s))\, ds \right), \quad (3.3.26)$$

for $t \in I$, then

$$u(t) \le a(t) + b(t)\, \psi\left(\int_{\alpha(t_0)}^{\alpha(t)} L(\sigma, a(\sigma)) \right.$$

$$\left. \times \exp\left(\int_{\sigma}^{\alpha(t)} M(\tau, a(\tau))\, \psi^{-1}(b(\tau)) d\tau \right) d\sigma \right), \quad (3.3.27)$$

for $t \in I$.

(c_4) Let L, M be as in (c_1) and the condition (3.3.20) holds. Let $g \in C(R_+, R_+)$ be nondecreasing function with $g(u) > 0$ for $u > 0$. If for $t \in I$,

$$u(t) \leq a(t) + b(t) g\left(\int_{\alpha(t_0)}^{\alpha(t)} L(s, u(s))\, ds \right), \tag{3.3.28}$$

for $t_0 \leq t \leq t_1; t, t_1 \in I$,

$$u(t) \leq a(t) + b(t) g\left(G^{-1}\left[G\left(\int_{\alpha(t_0)}^{\alpha(t)} L(\sigma, a(\sigma)) \right) \right.\right.$$

$$\left.\left. + \int_{\alpha(t_0)}^{\alpha(t)} M(\sigma, a(\sigma))\, b(\sigma)\, d\sigma \right] \right), \tag{3.3.29}$$

where G, G^{-1} be as in Theorem 3.2.5, part (d_1) and $t_1 \in I$ is chosen so that

$$G\left(\int_{\alpha_i(t_0)}^{\alpha_i(t)} L(\sigma, a(\sigma)) d\sigma \right) + \int_{\alpha_i(t_0)}^{\alpha_i(t)} M(\sigma, a(\sigma))\, b(\sigma) d\sigma \in Dom\left(G^{-1}\right),$$

for all t lying in the interval $t_0 \leq t \leq t_1$.

Proof. From the hypotheses on $\alpha(t)$ we observe that $\alpha'(t) \geq 0$ fot $t \in I$.

(c_1) Define a function $z(t)$ by

$$z(t) = \int_{\alpha_i(t_0)}^{\alpha_i(t)} L(s, u(s)) ds. \tag{3.3.30}$$

Then $z(t_0) = 0$ and from (3.3.21) we have

$$u(t) \leq a(t) + b(t) z(t), \tag{3.3.31}$$

for $t \in I$. From (3.3.30), (3.3.31) and the condition (3.3.20) it follows that

$$\begin{aligned}
&z'(t) = L(\alpha(t), u(\alpha(t)))\, \alpha'(t) \\
&\leq L(\alpha(t), a(\alpha(t)) + b(\alpha(t))\, z(\alpha(t)))\, \alpha'(t) - L(\alpha(t), a(\alpha(t)))\, \alpha'(t) \\
&+ L(\alpha(t), a(\alpha(t)))\, \alpha'(t) \\
&\leq M(\alpha(t), a(\alpha(t)))\, b(\alpha(t))\, z(\alpha(t))\, \alpha'(t) + L(\alpha(t), a(\alpha(t)))\, \alpha'(t),
\end{aligned}$$

which implies

$$z(t) \leq \int_{t_0}^{t} L(\alpha(s), a(\alpha(s)))\, \alpha'(s)$$

$$\times \exp\left(\int_{s}^{t} M(\alpha(\sigma), a(\alpha(\sigma)))\, b(\alpha(\sigma))\, \alpha'(\sigma)\, d\sigma\right) ds. \tag{3.3.32}$$

By making the change of variable on the right hand side of (3.3.32) we get

$$z(t) \leq \int_{\alpha(t_0)}^{\alpha(t)} L(\sigma, a(\sigma)) \exp\left(\int_{\sigma}^{\alpha(t)} M(\tau, a(\tau))\, a(\tau)\, b(\tau)\, d\tau\right) d\sigma. \tag{3.3.33}$$

Using (3.3.33) in (3.3.31) we get the required inequality in (3.3.22).

The proofs of $(c_2)-(c_4)$ can be completed by following the proof of (c_1) given above and closely looking at the proof of Theorem 1.4.4 parts $(d_2)-(d_4)$. We omit the details.

Another useful inequality proved in [74] is embodied in the following theorem.

Theorem 3.3.4. Let $u(t), a(t), b(t) \in C(I, R_+)$, $\alpha(t) \in C^1(I, I)$ be nondecreasing with $\alpha(t) \leq t$ on $I = [t_0, T)$ and $k \geq 0$ be a real constant. If

$$u(t) \leq k + \int_{\alpha(t_0)}^{\alpha(t)} a(s)\left[u(s) + \int_{\alpha(t_0)}^{s} b(\sigma)\, u(\sigma)\, d\sigma\right] ds, \tag{3.3.34}$$

for $t \in I$, then

$$u(t) \leq k\left[1 + \int_{\alpha(t_0)}^{\alpha(t)} a(s) \exp\left(\int_{\alpha(t_0)}^{s} [a(\sigma) + b(\sigma)]\, d\sigma\right) ds\right], \tag{3.3.35}$$

for $t \in I$.

Proof. From the hypotheses on $\alpha(t)$ we have $\alpha'(t) \geq 0$ for $t \in I$. Define a function $z(t)$ by the right hand side of (3.3.34). Then $z(t_0) = k,$, $u(t) \leq z(t)$ and

$$z'(t) = a(\alpha(t))\left[u(\alpha(t)) + \int_{\alpha(t_0)}^{\alpha(t)} b(\sigma)\, u(\sigma)\, d\sigma\right] \alpha'(t)$$

$$\leq a(\alpha(t))\left[z(\alpha(t))+\int_{\alpha(t_0)}^{\alpha(t)} b(\sigma)z(\sigma)\,d\sigma\right]\alpha'(t)$$

$$\leq a(\alpha(t))\left[z(t)+\int_{\alpha(t_0)}^{\alpha(t)} b(\sigma)z(\sigma)\,d\sigma\right]\alpha'(t). \tag{3.3.36}$$

Let

$$v(t)=z(t)+\int_{\alpha(t_0)}^{\alpha(t)} b(\sigma)z(\sigma)\,d\sigma, \tag{3.3.37}$$

then $v(t_0)=z(t_0)=k$, $z(t)\leq v(t)$ and from (3.3.36) we get

$$z'(t)\leq a(\alpha(t))\,v(t)\,\alpha'(t). \tag{3.3.38}$$

From (3.3.37), (3.3.38) and the fact that $z(t)\leq v(t)$ we have

$$v'(t)=z'(t)+b(\alpha(t))\,z(\alpha(t))\,\alpha'(t)$$

$$\leq a(\alpha(t))\,v(t)\,\alpha'(t)+b(\alpha(t))\,z(\alpha(t))\,\alpha'(t)$$

$$\leq [a(\alpha(t))+b(\alpha(t))]\,\alpha'(t)\,v(t),$$

which implies

$$v(t)\leq k\exp\left(\int_{t_0}^{t}[a(\alpha(s))+b(\alpha(s))]\,\alpha'(s)\,ds\right). \tag{3.3.39}$$

By making the change of variable on the right hand side of (3.3.39) we get

$$v(t)\leq k\exp\left(\int_{\alpha(t_0)}^{\alpha(t)}[a(\sigma)+b(\sigma)]\,d\sigma\right). \tag{3.3.40}$$

Using (3.3.40) in (3.3.38) and integrating it from t_0 to t; $t\in I$ and then making the change of variable and using the fact that $u(t)\leq z(t)$ we get the desired inequality in (3.3.35).

Remark 3.3.1. In the special case when $\alpha(t)=t$, the inequality given in Theorem 3.3.4 reduces to the inequality established earlier by Pachpatte, see[34, Theorem 1.7.1].

We shall now give the following theorem which deals with the inequalities proved in [58].

Theorem 3.3.5. Let $u(t), a(t) \in C(I, R_+)$, $b(t,s) \in C(I^2, R_+)$ for $t_0 \leq s \leq t < T$, $\alpha(t) \in C^1(I, I)$ be nondecreasing with $\alpha(t) < t$ on $I = [t_0, T)$ and $k \geq 0$ be a real constant.

(d_1) If

$$u(t) \leq k + \int_{\alpha(t_0)}^{\alpha(t)} \left[a(s) u(s) + \int_{\alpha(t_0)}^{s} b(s,\sigma) u(\sigma) d\sigma \right] ds, \tag{3.3.41}$$

for $t \in I$, then

$$u(t) \leq k \exp(B(t)), \tag{3.3.42}$$

for $t \in I$, where

$$B(t) = \int_{\alpha(t_0)}^{\alpha(t)} \left[a(s) + \int_{\alpha(t_0)}^{s} b(s,\sigma) d\sigma \right] ds, \tag{3.3.43}$$

for $t \in I$.

(d_2) Let $g \in C(R_+, R_+)$ be a nondecreasing function with $g(u) > 0$ for $u > 0$. If for $t \in I$,

$$u(t) \leq k + \int_{\alpha(t_0)}^{\alpha(t)} \left[a(s) g(u(s)) + \int_{\alpha(t_0)}^{s} b(s,\sigma) g(u(\sigma)) d\sigma \right] ds, \tag{3.3.44}$$

then for $t_0 \leq t \leq t_1; t, t_1 \in I$,

$$u(t) \leq G^{-1}[G(k) + B(t)], \tag{3.3.45}$$

where $B(t)$ is given by (3.3.43), G, G^{-1} be as in Theorem 3.3.2, part (b_1) and $t_1 \in I$ is chosen so that

$$G(k) + B(t) \in Dom\left(G^{-1}\right),$$

for all t lying in the interval $t_0 \leq t \leq t_1$.

Proof. From the hypotheses on $\alpha(t)$ we have $\alpha'(t) \geq 0$ for $t \in I$.

(d_1) Define a function $z(t)$ by the right hand side of (3.3.41). Then $z(t_0) = k, u(t) \leq z(t)$, $z(t)$ is positive and nondecreasing for $t \in I$ and

$$z'(t) = \left[a(\alpha(t)) u(\alpha(t)) + \int_{\alpha(t_0)}^{\alpha(t)} b(\alpha(t), \sigma) u(\sigma) d\sigma \right] \alpha'(t)$$

$$\leq \left[a(\alpha(t))\, z(\alpha(t)) + \int_{\alpha(t_0)}^{\alpha(t)} b(\alpha(t), \sigma)\, z(\sigma)\, d\sigma \right] \alpha'(t)$$

$$\leq \left[a(\alpha(t)) + \int_{\alpha(t_0)}^{\alpha(t)} b(\alpha(t), \sigma)\, d\sigma \right] \alpha'(t)\, z(t),$$

which implies

$$z(t) \leq k \exp\left(\int_{t_0}^{t} \left[a(\alpha(s)) + \int_{\alpha(t_0)}^{\alpha(s)} b(\alpha(s), \sigma)\, d\sigma \right] \alpha'(s)\, ds \right). \tag{3.3.46}$$

By making the change of variable on the right hand side in (3.3.46) and using the fact that $u(t) \leq z(t)$ we get the inequality in (3.3.42).

(d_2) The proof follows by the similar arguments as in the proof of (d_1) and the proof of Theorem 3.2.5, part (d_1). Here we omit the details.

Remark 3.3.2. We note that the inequalities given in Theorem 3.3.5 contains in the special case when $b(t,s) = 0$, the well known inequalities due to Gronwall, Bellman and Bihari (see [34, Theorems 1.2.2 and 2.3.1]).

To the end of this section we present the inequality established in [64].

Theorem 3.3.6. Let $I = [\alpha, \beta]$, $D = \{(t,s) \in I^2 : \alpha \leq s \leq t \leq \beta\}$ and $u(t)$, $f(t) \in C(I, R_+)$ $a(t,s), b(t,s), c(t,s) \in C(D, R_+)$. Suppose that $a(t,s)$, $b(t,s)$ be nondecreasing in t for each $s \in I$, $h(t) \in C^1(I, I)$ be nondecreasing with $h(t) \leq t$ on I, $k \geq 0$ be a real constant and

$$u(t) \leq k + \int_{h(\alpha)}^{h(t)} a(t,s) \left[f(s)\, u(s) + \int_{h(\alpha)}^{s} c(s,\sigma)\, u(\sigma)\, d\sigma \right] ds$$

$$+ \int_{h(\alpha)}^{h(\beta)} b(t,s)\, u(s)\, ds, \tag{3.3.47}$$

for $t \in I$. If

$$p(t) = \int_{h(\alpha)}^{h(\beta)} b(t,s) \exp(E(s))\, ds < 1, \tag{3.3.48}$$

for $t \in I$, where

$$E(t) = \int_{h(\alpha)}^{h(t)} a(t,\xi) \left[f(\xi) + \int_{h(\alpha)}^{\xi} c(\xi,\sigma)\, d\sigma \right] d\xi, \tag{3.3.49}$$

for $t \in I$, then

$$u(t) \leq \frac{k}{1 - p(t)} \exp(E(t)), \tag{3.3.50}$$

for $t \in I$.

Proof. From the hypotheses on $h(t)$ we have $h'(t) \geq 0$ for $t \in I$. Let $k > 0$ and fix $T \in I$, then for $\alpha \leq t \leq T$, from (3.3.47) we have

$$u(t) \leq k + \int_{h(\alpha)}^{h(t)} a(T,s) \left[f(s)\, u(s) + \int_{h(\alpha)}^{s} c(s,\sigma)\, u(\sigma)\, d\sigma \right] ds$$

$$+ \int_{h(\alpha)}^{h(\beta)} b(T,s)\, u(s)\, ds. \tag{3.3.51}$$

Define a function $z(t,T)$, $t \in [\alpha, T]$ by the right hand side of (3.3.51). Then for $t \in [\alpha, T]$, $u(t) \leq z(t,T)$, $z(t,T)$ is positive and nondecreasing in t,

$$z(\alpha,T) = k + \int_{h(\alpha)}^{h(\beta)} b(T,s)\, u(s)\, ds, \tag{3.3.52}$$

and

$$D_1 z(t,T) = a(T,h(t)) \left[f(h(t))\, u(h(t)) + \int_{h(\alpha)}^{h(t)} c(h(t),\sigma)\, u(\sigma)\, d\sigma \right] h'(t)$$

$$\leq a(T,h(t)) \left[f(h(t))\, z(h(t)) + \int_{h(\alpha)}^{h(t)} c(h(t),\sigma)\, z(\sigma,T)\, d\sigma \right] h'(t)$$

$$\leq a(T,h(t)) \left[f(h(t)) + \int_{h(\alpha)}^{h(t)} c(h(t),\sigma)\, d\sigma \right] h'(t)\, z(t,T)$$

i.e.,

$$\frac{D_1 z(t,T)}{z(t,T)} \leq a(T,h(t)) \left[f(h(t)) + \int_{h(\alpha)}^{h(t)} c(h(t),\sigma)\, d\sigma \right] h'(t). \tag{3.3.53}$$

By setting $t = s$ in (3.3.53) and integrating it with respect to s from α to T we get

$$z(T) \leq z(\alpha) \exp\left(\int_{\alpha}^{T} a(T,h(s)) [f(h(s)) + \int_{h(\alpha)}^{h(t)} c(h(s),\sigma)\, d\sigma \right] h'(s)\, ds \right). \tag{3.3.54}$$

Since T is arbitrary, from (3.3.54), (3.3.52) with T replaced by t we have

$$z(t) \leq z(\alpha) \exp\left(\int_{\alpha}^{t} a(t,h(s)) [f(h(s)) + \int_{h(\alpha)}^{h(t)} c(h(s),\sigma)\, d\sigma \right] h'(s)\, ds \right), \tag{3.3.55}$$

$$z(\alpha,t) = k + \int_{h(\alpha)}^{h(\beta)} b(t,s)\, u(s)\, ds. \tag{3.3.56}$$

By making the change of variable on the right hand side of (3.3.55) and using $u(t) \leq z(t,t)$, $t \in I$ we get

$$u(t) \leq z(\alpha,t) \exp(E(t)), \tag{3.3.57}$$

for $t \in I$. Using (3.3.57) in (3.3.56) and in view of (3.3.48), it is easy to observe that

$$z(\alpha,t) \leq \frac{k}{1-p(t)}. \tag{3.3.58}$$

The required inequality in (3.3.50) follows by using (3.3.58) in (3.3.57). The case $k \geq 0$ can be completed as mentioned in the proof of Theorem 3.2.4, part (c_1).

Remark 3.3.3. If we take in Theorem 3.3.6, (i) $c(t,s) = 0, b(t,s) = 0, a(t,s) = a(s)$, then we get the inequality given by Lipovan in [21, Corollary on p. 391] for $t \in I$, (ii) $c(t,s) = 0, h(t) = t$, then we get the inequality given by Pachpatte in [52, Theorem 1]

3.4 Retarded integral inequalities in two variables

The study of various classes of partial differential and integral equations has led to the investigation of a number of new integral inequalities which provide explicit bounds on the unknown functions. In this section we present some fundamental retarded integral inequalities in two independent variables, recently investigated by Pachpatte in [43,58,69,77], which can be used more conveniently in certain applications.

In what follows R denote the set of real numbers; $R_+ = [0, \infty), R_1 = [1, \infty), I_1 = [x_0, X), I_2 = [y_0, Y)$ are the given subsets of R, $\Delta = I_1 \times I_2$ and $'$ denotes the derivative. The partial derivatives of a function $z(x, y)$ for $x, y \in R$ with respect to x, y and xy are denoted by $D_1 z(x, y)$ (or $\frac{\partial}{\partial x} z(x, y)$), $D_2 z(x, y)$ (or $\frac{\partial}{\partial y} z(x, y)$) and $D_1 D_2 z(x, y) = D_2 D_1 z(x, y)$ (or $\frac{\partial^2}{\partial y \partial x} z(x, y)$ or $z_{xy}(x, y)$) respectively.

We begin with the following theorems which contains the inequalities proved in [43].

Theorem 3.4.1. Let $a(x, y), b(x, y) \in C(\Delta, R_+)$ and $\alpha(x) \in C^1(I_1, I_1), \beta(y) \in C^1(I_2, I_2)$ be nondecreasing with $\alpha(x) \leq x$ on I_1, $\beta(y) \leq y$ on I_2. Let $k \geq 0, c \geq 1$ and $p > 1$ be real constants.

(a_1) If $u(x, y) \in C(\Delta, R_+)$ and

$$u(x, y) \leq k + \int_{x_0}^{x} \int_{y_0}^{y} a(s, t) u(s, t) \, dt ds + \int_{\alpha(x_0)}^{\alpha(x)} \int_{\beta(y_0)}^{\beta(y)} b(s, t) u(s, t) \, dt ds, \tag{3.4.1}$$

for $(x, y) \in \Delta$, then

$$u(x, y) \leq k \exp(A(x, y) + B(x, y)), \tag{3.4.2}$$

for $(x, y) \in \Delta$, where

$$A(x, y) = \int_{x_0}^{x} \int_{y_0}^{y} a(s, t) \, dt ds, \tag{3.4.3}$$

$$B(x, y) = \int_{\alpha(x_0)}^{\alpha(x)} \int_{\beta(y_0)}^{\beta(y)} b(s, t) \, dt ds, \tag{3.4.4}$$

for $(x, y) \in \Delta$.

(a_2) If $u(x,y) \in C(\Delta, R_1)$ and

$$u(x,y) \leq c + \int_{x_0}^{x}\int_{y_0}^{y} a(s,t)\, u(s,t) \log u(s,t)\, dtds$$

$$+ \int_{\alpha(x_0)}^{\alpha(x)}\int_{\beta(y_0)}^{\beta(y)} b(s,t)\, u(s,t) \log u(s,t)\, dtds, \tag{3.4.5}$$

for $(x,y) \in \Delta$, then

$$u(x,y) \leq c^{\exp(A(x,y)+B(x,y))}, \tag{3.4.6}$$

for $(x,y) \in \Delta$,where $A(x,y)$ and $B(x,y)$ are given by (3.4.3) and (3.4.4).

(a_3) If $u(x,y) \in C(\Delta, R_+)$ and

$$u^p(x,y) \leq c + \int_{x_0}^{x}\int_{y_0}^{y} a(s,t)\, u(s,t)\, dtds + \int_{\alpha(x_0)}^{\alpha(x)}\int_{\beta(y_0)}^{\beta(y)} b(s,t)\, u(s,t)\, dtds, \tag{3.4.7}$$

for $(x,y) \in \Delta$, then

$$u(x,y) \leq \left[k^{\frac{p-1}{p}} + \left(\frac{p-1}{p}\right)(A(x,y) + B(x,y))\right]^{\frac{1}{p-1}}, \tag{3.4.8}$$

for $(x,y) \in \Delta$, where $A(x,y)$ and $B(x,y)$ are given by (3.4.3) and (3.4.4).

Theorem 3.4.2. Let $a(x,y), b(x,y),\ \alpha(x), \beta(y),\ k, c, p$ be as in Theorem 3.4.1. For $i = 1, 2$, let $g_i \in C(R_+, R_+)$ be nondecreasing with $g_i(u) > 0$ for $u > 0$.

(b_1) If $u(x,y) \in C(\Delta, R_+)$ and for $(x,y) \in \Delta$,

$$u(x,y) \leq k + \int_{x_0}^{x}\int_{y_0}^{y} a(s,t)\, g_1(u(s,t))\, dtds$$

$$+ \int_{\alpha(x_0)}^{\alpha(x)}\int_{\beta(y_0)}^{\beta(y)} b(s,t)\, g_2(u(s,t))\, dtds, \tag{3.4.9}$$

then for $x_0 \leq x \leq x_1, y_0 \leq y \leq y_1; x, x_1 \in I_1, y, y_1 \in I_2$,

(i) in case $g_2(u) \le g_1(u)$,

$$u(x,y) \le G_1^{-1}\left[G_1(k) + A(x,y) + B(x,y)\right], \tag{3.4.10}$$

(ii) in case $g_1(u) \le g_2(u)$,

$$u(x,y) \le G_2^{-1}\left[G_2(k) + A(x,y) + B(x,y)\right], \tag{3.4.11}$$

where G_i, G_i^{-1} are as in Theorem 3.2.2, part (b_1) and $A(x,y), B(x,y)$ are given by (3.4.3), (3.4.4) and $x_1 \in I_1$, $y_1 \in I_2$ are chosen so that for $i = 1, 2$,

$$G_i(k) + A(x,y) + B(x,y) \in Dom\left(G_i^{-1}\right),$$

for all x and y lying in $[x_0, x_1]$ and $[y_0, y_1]$ respectively.

(b_2) If $u(x,y) \in C(\Delta, R_1)$ and for $(x,y) \in \Delta$,

$$u(x,y) \le c + \int_{x_0}^{x}\int_{y_0}^{y} a(s,t)\, u(s,t)\, g_1(\log u(s,t))\, dtds$$

$$+ \int_{\alpha(x_0)}^{\alpha(x)}\int_{\beta(y_0)}^{\beta(y)} b(s,t)\, u(s,t)\, g_2(\log u(s,t))\, dtds, \tag{3.4.12}$$

then for $x_0 \le x \le x_2, y_0 \le y \le y_2; x, x_2 \in I_1, y, y_2 \in I_2$,

(i) in case $g_2(u) \le g_1(u)$,

$$u(x,y) \le \exp\left(G_1^{-1}\left[G_1(\log c) + A(x,y) + B(x,y)\right]\right), \tag{3.4.13}$$

(ii) in case $g_1(u) \le g_2(u)$,

$$u(x,y) \le \exp\left(G_2^{-1}\left[G_2(\log c) + A(x,y) + B(x,y)\right]\right), \tag{3.4.14}$$

where G_i, G_i^{-1}, $A(x,y), B(x,y)$ are as in (b_1) and $x_2 \in I_1$, $y_2 \in I_2$ are chosen so that for $i = 1, 2$,

$$G_i(\log c) + A(x,y) + B(x,y) \in Dom\left(G_i^{-1}\right),$$

for all x and y lying in $[x_0, x_2]$ and $[y_0, y_2]$ respectively.

(b_3) If $u(x,y) \in C(\Delta, R_+)$ and for $(x,y) \in \Delta$,

$$u^p(x,y) \le k + \int_{x_0}^{x}\int_{y_0}^{y} a(s,t)\, g_1(u(s,t))\, dtds$$

$$+ \int_{\alpha(x_0)}^{\alpha(t)}\int_{\beta(y_0)}^{\beta(t)} b(s,t)\, g_2(u(s,t))\, dtds, \tag{3.4.15}$$

then for $x_0 \le x \le x_3, y_0 \le y \le y_3; x, x_3 \in I_1, y, y_3 \in I_2$, (*i*) in case $g_2(u) \leq g_1(u)$,

$$u(x,y) \leq \left\{H_1^{-1}\left[H_1(k) + A(x,y) + B(x,y)\right]\right\}^{\frac{1}{p}}, \tag{3.4.16}$$

(*ii*) in case $g_1(u) \leq g_2(u)$,

$$u(x,y) \leq \left\{H_2^{-1}\left[H_2(k) + A(x,y) + B(x,y)\right]\right\}^{\frac{1}{p}}, \tag{3.4.17}$$

where H_i, H_i^{-1} are as in Theorem 3.2.2, part (b_3) and $A(x,y), B(x,y)$ are given by (3.4.3), (3.4.4) and $x_3 \in I_1$, $y_3 \in I_2$ are chosen so that for $i = 1, 2$,

$$H_i(k) + A(x,y) + B(x,y) \in Dom\left(H_i^{-1}\right),$$

for all x and y lying in $[x_0, x_3]$ and $[y_0, y_3]$ respectively.

Proofs of Theorems 3.4.1 and 3.4.2. Since the proofs resemble one another, we give the details for (a_1) and (b_3) only; the proofs of the remaining inequalities can be completed by following the proofs of the above mentioned inequalities and closely looking at the proofs of Theorems 3.2.1 and 3.2.2.

From the hypotheses we observe that $\alpha'(x) \geq 0$ for $x_1 \in I_1$, $\beta'(y) \geq 0$ for $y \in I_2$.

(a_1) Let $k > 0$ and define a function $z(x,y)$ by the right hand side of (3.4.1).Then $z(x_0,y) = z(x,y_0) = k$, $u(x,y) \leq z(x,y)$, $z(x,y)$ is positive and nondecreasing for $(x,y) \in \Delta$ and

$$D_1 z(x,y) = \int_{y_0}^{y} a(x,t)\, u(x,t)\, dt + \left(\int_{\beta(y_0)}^{\beta(y)} b(\alpha(x),t)\, u(\alpha(x),t)\, dt\right)\alpha'(x)$$

$$\leq \int_{y_0}^{y} a(x,t)\, z(x,t)\, dt + \left(\int_{\beta(y_0)}^{\beta(y)} b(\alpha(x),t) z(\alpha(x),t)\, dt\right)\alpha'(x)$$

$$\leq z(x,t) \int_{y_0}^{y} a(x,t)\, dt + z(\alpha(x),\beta(y)) \left(\int_{\beta(y_0)}^{\beta(y)} b(\alpha(x),t)\, dt\right)\alpha'(x)$$

$$\leq z(x,y)\left[\int_{y_0}^{y} a(x,t)\, dt + \left(\int_{\beta(y_0)}^{\beta(y)} b(\alpha(x),t)\, dt\right)\alpha'(x)\right];$$

i.e.,

$$\frac{D_1 z(x,y)}{z(x,y)} \leq \int_{y_0}^{y} a(x,t)\, dt + \left(\int_{\beta(y_0)}^{\beta(y)} b(\alpha(x),t)\, dt\right)\alpha'(x). \tag{3.4.18}$$

Keeping y fixed in (3.4.18), setting $x = \sigma$, and integrating it with respect to σ from x_0 to x, $x \in I_1$, and by making the change of variable we get

$$z(x,y) \leq k \exp(A(x,y) + B(x,y)). \tag{3.4.19}$$

Using (3.4.19) in $u(x,y) \leq z(x,y)$ we get the required inequality in (3.4.2). The case $k \geq 0$ follows as mentioned in the proof of Theorem 3.2.1, part (a_1).

(b_3) Let $k > 0$ and define a function $z(x,y)$ by the right hand side of (3.4.15). Then $z(x_0,y) = z(x,y_0) = k, u(x,y) \leq \{z(x,y)\}^{\frac{1}{p}}$, $z(x,y)$ is positive and nondecreasing for $(x,y) \in \Delta$ and

$$D_1 z(x,y) = \int_{y_0}^{y} a(x,t)\, g_1(u(x,t))\, dt$$

$$+ \left(\int_{\beta(y_0)}^{\beta(y)} b(\alpha(x),t)\, g_2(u(\alpha(x),t))\, dt \right) \alpha'(x)$$

$$\leq \int_{y_0}^{y} a(x,t)\, g_1\left(\{z(x,t)\}^{\frac{1}{p}}\right) dt$$

$$+ \left(\int_{\beta(y_0)}^{\beta(y)} b(\alpha(x),t)\, g_2\left(\{z(\alpha(x),t)\}^{\frac{1}{p}}\right) dt \right) \alpha'(x)$$

$$\leq g_1\left(\{z(x,t)\}^{\frac{1}{p}}\right) \int_{y_0}^{y} a(x,t)\, dt$$

$$+ g_2\left(\{z(\alpha(x),\beta(y))\}^{\frac{1}{p}}\right) \left(\int_{\beta(y_0)}^{\beta(y)} b(\alpha(x),t) \right) \alpha'(x). \tag{3.4.20}$$

(i) When $g_2(u) \leq g_1(u)$, then from (3.4.20) we observe that

$$\frac{D_1 z(x,y)}{g_1\left(\{z(x,y)\}^{\frac{1}{p}}\right)} \leq \int_{y_0}^{y} a(x,t)\, dt + \left(\int_{\beta(y_0)}^{\beta(y)} b(\alpha(x),t)\, dt \right) \alpha'(x). \tag{3.4.21}$$

From (3.2.27) and (3.4.21) we have

$$D_1 H_1(z(x,y)) = \frac{D_1 z(x,y)}{g_1\left(\{z(x,y)\}^{\frac{1}{p}}\right)}$$

$$\leq \int_{y_0}^{y} a(x,t)\,dt + \left(\int_{\beta(y_0)}^{\beta(y)} b(\alpha(x),t)\,dt \right) \alpha'(x). \tag{3.4.22}$$

Keeping y fixed in (3.4.22), setting $x = \sigma$,then integrating with respect to σ from x_0 to x; $x \in I_1$, and making the change of variable we have

$$H_1(z(x,y)) \leq H_1(k) + A(x,y) + B(x,y). \tag{3.4.23}$$

Using the bound on $z(x,y)$ from (3.4.23) in $u(x,y) \leq \{z(x,y)\}^{\frac{1}{p}}$ we get (3.4.16). The case $k \geq 0$ follows as mentioned in the proof of Theorem 3.2.1, part (a_1). The subdomain for x, y is obvious. The proof of the case when $g_1(u) \leq g_2(u)$ can be completed similarly.

Remark 3.4.1. We note that the above proofs can be carried out by differentiation of $z(x,y)$ defined therein, with respect to y. Similar remarks apply to the proofs of other inequalities in Theorems 3.4.1 and 3.4.2.

A more general version of the inequality (3.4.9) in Theorem 3.4.2, recently established in [69] is embodied in the following theorem.

For some suitable functions defined on the respective domains of their definitions we set

$$F[x,y,m;\phi_1,\psi_1,a_1,p_1;\phi_2,\psi_2,a_2,p_2]$$

$$= \int_{\phi_1(x_0)}^{\phi_1(x)} \int_{\psi_1(y_0)}^{\psi_1(y)} a_1(s,t)\,p_1(m(s,t))\,dtds$$

$$+ \int_{\phi_2(x_0)}^{\phi_2(x)} \int_{\psi_2(y_0)}^{\psi_2(y)} a_2(s,t)\,p_2(m(s,t))\,dtds,$$

to simplify the details of presentation.

Theorem 3.4.3. Let $u(x,y), f(x,y), b(x,y), a_1(x,y), a_2(x,y) \in C(\Delta, R_+)$ and for $i = 1,2$ $\phi_i(x) \in C^1(I_1, I_1), \psi_i(y) \in C^1(I_2, I_2)$ be nondecreasing with $\phi_i(x) \leq x$ on I_1, $\psi_i(y) \leq y$ on I_2. Let $g_i(u), i = 1,2$ be as in Theorem 3.2.3 and for $(x,y) \in \Delta$,

$$u(x,y) \leq f(x,y) + b(x,y)\,F[x,y,u;\phi_1,\psi_1,a_1,g_1;\phi_2,\psi_2,a_2,g_2], \tag{3.4.24}$$

then for $x_0 \leq x \leq x_1, y_0 \leq y \leq y_1; x, x_1 \in I_1, y, y_1 \in I_2$,

(i) in case $g_2(u) \leq g_1(u)$,

$$u(x,y) \leq f(x,y) + b(x,y) G_1^{-1}[G_1(E(x,y))$$
$$+F[x,y,b;\phi_1,\psi_1,a_1,g_1;\phi_2,\psi_2,a_2,g_2]], \tag{3.4.25}$$

(ii) in case $g_1(u) \leq g_2(u)$,

$$u(x,y) \leq f(x,y) + b(x,y) G_2^{-1}[G_2(E(x,y))$$
$$+F[x,y,b;\phi_1,\psi_1,a_1,g_1;\phi_2,\psi_2,a_2,g_2]], \tag{3.4.26}$$

where $G_i, G_i^{-1}, i = 1,2$ are as in Theorem 3.3.2, part (b_1),

$$E(x,y) = F[x,y,f;\phi_1,\psi_1,a_1,g_1;\phi_2,\psi_2,a_2,g_2], \tag{3.4.27}$$

and $x_1 \in I_1$, $y_1 \in I_2$ are chosen so that for $i = 1,2$,

$$G_i(E(x,y)) + F[x,y,b;\phi_1,\psi_1,a_1,g_1;\phi_2,\psi_2,a_2,g_2] \in Dom(G_i^{-1}),$$

for all x and y lying in $[x_0,x_1]$ and $[y_0,y_1]$ respectively.

Proof. From the hypotheses we observe that for $i = 1,2$, $\phi_i'(x) \geq 0$ for $x \in I_1$, $\psi_i'(y) \geq 0$ for $y \in I_2$. Define a function $z(x,y)$ by

$$z(x,y) = F[x,y,u;\phi_1,\psi_1,a_1,g_1;\phi_2,\psi_2,a_2,g_2]. \tag{3.4.28}$$

Then $z(x_0,y) = z(x,y_0) = 0$ and (3.4.24) can be restated as

$$u(x,y) \leq f(x,y) + b(x,y) z(x,y). \tag{3.4.29}$$

Using (3.4.29) in (3.4.28) and making use of the hypotheses on g_1, g_2 we get

$$z(x,y) \leq E(x,y) + \int_{\phi_1(x_0)}^{\phi_1(x)} \int_{\psi_1(y_0)}^{\psi_1(y)} a_1(s,t) g_1(b(s,t)) g_1(z(s,t)) dtds$$
$$+ \int_{\phi_2(x_0)}^{\phi_2(x)} \int_{\psi_2(y_0)}^{\psi_2(y)} a_2(s,t) g_2(b(s,t)) g_2(u(s,t)) dtds. \tag{3.4.30}$$

The rest of the proof can be completed by closely looking at the proof of Theorem 3.2.3 given in section 3.2 and following the proofs of similar results given in [43] and [34] with suitable changes. Here we omit the further details.

Remark 3.4.2. We note that the inequalities given in Theorem 3.4.2, parts (b_2) and (b_3) can be extended very easily in the framework of Theorem 3.4.3, when $b(x,y) = 1$ and $f(x,y)$ is equal to the respective constants given therein. Since these translations are quite straightforward in view of Theorem 3.4.3, we leave it for the readers to fill in where needed.

The following theorem involving Lipschitzian type kernel functions, deals with the inequalities proved in [77].

Theorem 3.4.4. Let $u(x,y), a(x,y), b(x,y) \in C(\Delta, R_+)$ and $\alpha(x) \in C^1(I_1, I_1)$, $\beta(y) \in C^1(I_2, I_2)$ be nondecreasing with $\alpha(x) \le x$,for $x \in I_1$ $\beta(y) \le y$ for $y \in I_2$.

(c_1) Let $L \in C(\Delta \times R_+, R_+)$ and

$$0 \le L(x,y,u) - L(x,y,v) \le M(x,y,v)(u-v), \tag{3.4.31}$$

for $(x,y) \in \Delta$ and $u \ge v \ge 0$, where $M \in C(\Delta \times R_+, R_+)$. If

$$u(x,y) \le a(x,y) + b(x,y) \int_{\alpha(x_0)}^{\alpha(x)} \int_{\beta(y_0)}^{\beta(y)} L(s,t,u(s,t))\,dtds, \tag{3.4.32}$$

for $(x,y) \in \Delta$, then

$$u(x,y) \le a(x,y) + b(x,y)e(x,y) \times \exp\left(\int_{\alpha(x_0)}^{\alpha(x)} \int_{\beta(y_0)}^{\beta(y)} M(\sigma,\tau,a(\sigma,\tau))\,b(\sigma,\tau)\,d\tau d\sigma\right), \tag{3.4.33}$$

for $(x,y) \in \Delta$, where

$$e(x,y) = \int_{\alpha(x_0)}^{\alpha(x)} \int_{\beta(y_0)}^{\beta(y)} L(s,t,a(\sigma,\tau))\,d\tau d\sigma, \tag{3.4.34}$$

for $(x,y) \in \Delta$.

(c_2) Let $L \in C(\Delta \times R_+, R_+)$ and $\psi \in C(R_+, R_+)$ be strictly increasing function with $\psi(0) = 0$ and

$$0 \le L(x,y,u) - L(x,y,v) \le M(x,y,v)\,\psi^{-1}(u-v), \tag{3.4.35}$$

for $(x,y) \in \Delta$ and $u \ge v \ge 0$, where $M \in C(\Delta \times R_+, R_+)$ and ψ^{-1} is the inverse of ψ. If

$$u(x,y) \le a(x,y) + \psi\left(b(x,y) \int_{\alpha(x_0)}^{\alpha(x)} \int_{\beta(y_0)}^{\beta(y)} L(s,t,u(s,t))\,dtds\right), \tag{3.4.36}$$

for $(x,y) \in \Delta$,, then

$$u(x,y) \le a(x,y) + \psi\Bigg(b(x,y)e(x,y) \times \exp\left(\int_{\alpha(x_0)}^{\alpha(x)} \int_{\beta(y_0)}^{\beta(y)} M(\sigma,\tau,a(\sigma,\tau))\,b(\sigma,\tau)\,d\tau d\sigma\right)\Bigg), \tag{3.4.37}$$

for $(x,y) \in \Delta$, where $e(x,y)$ is given by (3.4.34).

(c_3) Let L, ψ and M be as in (c_2) and the condition (3.4.35) holds. Suppose in addition $\psi^{-1}(xy) \leq \psi^{-1}(x)\psi^{-1}(y)$ for $x, y \in R_+$. If

$$u(x,y) \leq a(x,y) + b(x,y)\psi\left(\int_{\alpha(x_0)}^{\alpha(x)}\int_{\beta(y_0)}^{\beta(y)} L(s,t,u(s,t))\,dtds\right), \qquad (3.4.38)$$

for $(x,y) \in \Delta$, then

$$u(x,y) \leq a(x,y) + b(x,y)\psi\Bigg(e(x,y)$$
$$\times \exp\left(\int_{\alpha(x_0)}^{\alpha(x)}\int_{\beta(y_0)}^{\beta(y)} M(\sigma,\tau,a(\sigma,\tau))\psi^{-1}(b(\sigma,\tau))\,d\tau d\sigma\right)\Bigg), \qquad (3.4.39)$$

for $(x,y) \in \Delta$, where $e(x,y)$ is given by (3.4.34).

(c_4) Let L, M be as in (c_1) and the condition (3.4.31) holds. Let g, G, G^{-1} be as in Theorem 3.2.5, part (d_1). If for $(x,y) \in \Delta$,

$$u(x,y) \leq a(x,y) + b(x,y)\,g\left(\int_{\alpha(x_0)}^{\alpha(x)}\int_{\beta(y_0)}^{\beta(y)} L(s,t,u(s,t))\,dtds\right), \qquad (3.4.40)$$

then for $x_0 \leq x \leq x_1, y_0 \leq y \leq y_1; x, x_1 \in I_1, y, y_1 \in I_2$,

$$u(x,y) \leq a(x,y) + b(x,y)\,g\Bigg(G^{-1}\Bigg[G(e(x,y))$$
$$+ \int_{\alpha(x_0)}^{\alpha(x)}\int_{\beta(y_0)}^{\beta(y)} M(\sigma,\tau,a(\sigma,\tau))\,b(\sigma,\tau)\,d\tau d\sigma\Bigg]\Bigg), \qquad (3.4.41)$$

where $e(x,y)$ is given by (3.4.34) and $x_1 \in I_1$, $y_1 \in I_2$ are chosen so that

$$G(e(x,y)) + \int_{\alpha(x_0)}^{\alpha(x)}\int_{\beta(y_0)}^{\beta(y)} M(\sigma,\tau,a(\sigma,\tau))\,b(\sigma,\tau)\,d\tau d\sigma \in Dom\left(G^{-1}\right),$$

for all x and y lying in $[x_0, x_1]$ and $[y_0, y_1]$ respectively.

Proof. (c_1) Define a function $z(x,y)$ by

$$z(x,y) = \int_{\alpha(x_0)}^{\alpha(x)}\int_{\beta(y_0)}^{\beta(y)} L(s,t,u(s,t))\,dtds. \qquad (3.4.42)$$

Then (3.4.32) can be restated as

$$u(x,y) \leq a(x,y) + b(x,y)\, z(x,y), \tag{3.4.43}$$

for $(x,y) \in \Delta$. From (3.4.42) it is easy to see that

$$D_2 D_1 z(x,y) = L(\alpha(x), \beta(y), u(\alpha(x), \beta(y)))\, \alpha'(x)\, \beta'(y). \tag{3.4.44}$$

From (3.4.44), (3.4.31) and following the idea of the proof of Theorem 3.3.3 part (c_1) it follows that

$$D_2 D_1 z(x,y) \leq L(\alpha(x), \beta(y), a(\alpha(x), \beta(y)))\, \alpha'(x)\, \beta'(y)$$

$$+ M(\alpha(x), \beta(y), a(\alpha(x), \beta(y)))\, b(\alpha(x), \beta(y))\, z(\alpha(x), \beta(y))\, \alpha'(x)\, \beta'(y),$$

which implies

$$z(x,y) \leq \int_{x_0}^{x} \int_{y_0}^{y} L(\alpha(s), \beta(t), a(\alpha(s), \beta(t)))\, \alpha'(s)\, \beta'(t)\, dt ds$$

$$+ \int_{x_0}^{x} \int_{y_0}^{y} M(\alpha(s), \beta(t), a(\alpha(s), \beta(t)))$$

$$\times b(\alpha(s), \beta(t))\, z(s,t)\, \alpha'(s)\, \beta'(t)\, dt ds. \tag{3.4.45}$$

Clearly the first integral on the right hand side in (3.4.45) is nonnegative and nondecreasing in both the variables x and y. Now a suitable application of Theorem 4.2.2 given in [34, p. 325] yields

$$z(x,y) \leq \left(\int_{x_0}^{x} \int_{y_0}^{y} L(\alpha(s), \beta(t), a(\alpha(s), \beta(t)))\, \alpha'(s)\, \beta'(t)\, dt ds \right)$$

$$\times \exp\left(\int_{x_0}^{x} \int_{y_0}^{y} M(\alpha(s), \beta(t), a(\alpha(s), \beta(t))) \right.$$

$$\left. \times b(\alpha(s), \beta(t))\, b(\alpha(s), \beta(t))\, \alpha'(s)\, \beta'(t)\, dt ds \right). \tag{3.4.46}$$

Now by making the change of variables on the right hand side of (3.4.46) and substituting the resulting estimate on $z(x,y)$ in (3.4.43) we get (3.4.33).

The proofs of $(c_2) - (c_4)$ can be completed by following the proof of (c_1) given above and closely looking at the proof of Theorem 1.4.4, parts $(d_2) - (d_4)$. Here we leave the details to the reader.

Remark 3.4.3. We note that from Theorem 3.4.4, one can very easily obtain the corollaries similar to those of given in [12] (see [34]) with suitable changes.

The following theorem offer another useful inequality established in [58].

Theorem 3.4.5. Let $I_1 = [x_0, M]$ $I_2 = [y_0, N]$, $\Delta = I_1 \times I_2$ and $D = \{(x,y,s,t) \in \Delta^2 : x_0 \le s \le x \le M, y_0 \le t \le y \le N\}$. Let $u(x,y) \in C(\Delta, R_+)$ and $a(x,y,s,t), b(x,y,s,t) \in C(D, R_+)$ be nondecreasing in x and y for $(s,t) \in \Delta$. Let $\alpha(x) \in C^1(I_1, I_1)$, $\beta(y) \in C^1(I_2, I_2)$ be nondecreasing with $\alpha(x) \le x$ on I_1, $\beta(y) \le y$ on I_2 and suppose that

$$u(x,y) \le c + \int_{\alpha(x_0)}^{\alpha(x)} \int_{\beta(y_0)}^{\beta(y)} a(x,y,s,t)\, u(s,t)\, dtds$$

$$+ \int_{\alpha(x_0)}^{\alpha(M)} \int_{\beta(y_0)}^{\beta(N)} b(x,y,s,t)\, u(s,t)\, dtds, \tag{3.4.47}$$

for $(x,y) \in \Delta$, where $c \ge 0$ is a real constant.If

$$p(x,y) = \int_{\alpha(x_0)}^{\alpha(M)} \int_{\beta(y_0)}^{\beta(N)} b(x,y,s,t)$$

$$\times \exp\left(\int_{\alpha(x_0)}^{\alpha(x)} \int_{\beta(y_0)}^{\beta(y)} a(x,y,\sigma,\tau)\, d\tau d\sigma \right) dtds < 1, \tag{3.4.48}$$

for $(x,y) \in \Delta$, then

$$u(x,y) \le \frac{c}{1 - p(x,y)} \exp\left(\int_{\alpha(x_0)}^{\alpha(x)} \int_{\beta(y_0)}^{\beta(y)} a(x,y,s,t)\, dtds \right), \tag{3.4.49}$$

for $(x,y) \in \Delta$.

Proof. Fix any arbitrary element $(X,Y) \in \Delta$. Then for $x_0 \le x \le X, y_0 \le y \le Y$ we have

$$u(x,y) \le c + \int_{\alpha(x_0)}^{\alpha(x)} \int_{\beta(y_0)}^{\beta(y)} a(X,Y,s,t)\, u(s,t)\, dtds$$

$$+ \int_{\alpha(x_0)}^{\alpha(M)} \int_{\beta(y_0)}^{\beta(N)} b(X,Y,s,t)\, u(s,t)\, dtds. \tag{3.4.50}$$

Let

$$k(X,Y) = c + \int_{\alpha(x_0)}^{\alpha(M)} \int_{\beta(y_0)}^{\beta(N)} b(X,Y,s,t)\, u(s,t)\, dtds, \tag{3.4.51}$$

then (3.4.50) can be restated as

$$u(x,y) \le k(X,Y) + \int_{\alpha(x_0)}^{\alpha(x)} \int_{\beta(y_0)}^{\beta(y)} a(X,Y,s,t)\, u(s,t)\, dtds \tag{3.4.52}$$

for $x_0 \le x \le X, y_0 \le y \le Y$. Now a suitable application of the inequality given in Theorem 3.4.1, part (a_1) to (3.4.52) yields

$$u(x,y) \le k(X,Y) \exp\left(\int_{\alpha(x_0)}^{\alpha(x)} \int_{\beta(y_0)}^{\beta(y)} a(X,Y,\sigma,\tau)\, d\sigma d\tau \right), \tag{3.4.53}$$

for $x_0 \le x \le X, y_0 \le y \le Y$. Since $(X,Y) \in \Delta$ is arbitrary, from (3.4.53) and (3.4.51) with X and Y replaced by x and y we have

$$u(x,y) \le k(x,y) \exp\left(\int_{\alpha(x_0)}^{\alpha(x)} \int_{\beta(y_0)}^{\beta(y)} a(x,y,\sigma,\tau)\, d\sigma d\tau \right), \tag{3.4.54}$$

where

$$k(x,y) = c + \int_{\alpha(x_0)}^{\alpha(M)} \int_{\beta(y_0)}^{\beta(N)} b(x,y,s,t)\, u(s,t)\, dtds, \tag{3.4.55}$$

for all $(x,y) \in \Delta$. Using (3.4.54) on the right hand side of (3.4.55) and in view of (3.4.48) we have

$$k(x,y) \le \frac{c}{1 - p(x,y)}, \tag{3.4.56}$$

for $(x,y) \in \Delta$. Using (3.4.56) in (3.4.54) we get the desired inequality in (3.4.49).

Remark 3.4.4. If we take in Theorem 3.4.5, *(i)* $b(x,y,s,t) = 0$, *(ii)* $\alpha(x) = x$, $\beta(y) = y$, then we get new inequalities which can also be used as tools in certain applications.

3.5 More retarded integral inequalities in two variables

In [47,59,60,74] Pachpatte has investigated a number of integral inequalities in two independent variables, which play a vital role in the study of various classes of retarded partial differential and integral equations. This section is devoted to some retarded integral inequalities established in the above cited references, which can be used as basic tools in variety of applications. In what follows, we shall use the notations and definitions as given in section 3.4.

First we give the following theorem which deals with the integral inequality proved in [74].

Theorem 3.5.1. Let $u(x,y), a(x,y), b(x,y) \in C(\Delta, R_+)$ and $\alpha(x) \in C^1(I_1, I_1), \beta(y) \in C^1(I_2, I_2)$ be nondecreasing with $\alpha(x) \leq x$ on I_1, $\beta(y) \leq y$ on I_2. If

$$u(x,y) \leq k + \int_{\alpha(x_0)}^{\alpha(x)} \int_{\beta(y_0)}^{\beta(y)} a(s,t) \left[u(s,t) + \int_{\alpha(x_0)}^{s} \int_{\beta(y_0)}^{t} b(\sigma,\eta) u(\sigma,\eta) d\eta d\sigma \right] dtds, \tag{3.5.1}$$

for $(x,y) \in \Delta$,where $k \geq 0$ is a real constant, then

$$u(x,y) \leq k \left[1 + \int_{\alpha(x_0)}^{\alpha(x)} \int_{\beta(y_0)}^{\beta(y)} a(m,n) \times \exp\left(\int_{\alpha(x_0)}^{m} \int_{\beta(y_0)}^{n} [a(\sigma,\eta) + b(\sigma,\eta)] d\eta d\sigma \right) dndm \right], \tag{3.5.2}$$

for $(x,y) \in \Delta$.

Proof. From the hypotheses we observe that $\alpha'(x) \geq 0$ for $x \in I_1$, $\beta'(y) \geq 0$ for $y \in I_2$. Let $k > 0$ and define a function $z(x,y)$ by the right hand side of (3.5.1). Then $z(x_0,y) = z(x,y_0) = k, u(x,y) \leq z(x,y)$, $z(x,y)$ is positive and nondecreasing for $(x,y) \in \Delta$ and

$$D_2D_1z(x,y) = a(\alpha(x),\beta(y))\left[u(\alpha(x),\beta(y))\right.$$

$$\left.+\int_{\alpha(x_0)}^{\alpha(x)}\int_{\beta(y_0)}^{\beta(y)} b(\sigma,\eta)u(\sigma,\eta)\,d\eta d\sigma\right]\beta'(y)\alpha'(x)$$

$$\leq a(\alpha(x),\beta(y))\left[z(\alpha(x),\beta(y)) + \int_{\alpha(x_0)}^{\alpha(x)}\int_{\beta(y_0)}^{\beta(y)} b(\sigma,\eta)z(\sigma,\eta)\,d\eta d\sigma\right]\beta'(y)\alpha'(x)$$

$$\leq a(\alpha(x),\beta(y))\left[z(x,y) + \int_{\alpha(x_0)}^{\alpha(x)}\int_{\beta(y_0)}^{\beta(y)} b(\sigma,\eta)z(\sigma,\eta)\,d\eta d\sigma\right]\beta'(y)\alpha'(x).$$

Let

$$v(x,y) = z(x,y) + \int_{\alpha(x_0)}^{\alpha(x)}\int_{\beta(y_0)}^{\beta(y)} b(\sigma,\eta)z(\sigma,\eta)\,d\eta d\sigma. \tag{3.5.3}$$

Then $v(x_0,y) = z(x_0,y) = k$, $v(x,y_0) = z(x,y_0) = k$, $z(x,y) \leq v(x,y)$, $v(x,y)$ is positive and nondecreasing for $(x,y) \in \Delta$,

$$D_2D_1z(x,y) \leq a(\alpha(x),\beta(y))v(x,y)\beta'(y)\alpha'(x), \tag{3.5.4}$$

and

$$D_2D_1v(x,y) = D_2D_1z(x,y) + b(\alpha(x),\beta(y))z(\alpha(x),\beta(y))\beta'(y)\alpha'(x)$$

$$\leq a(\alpha(x),\beta(y))v(x,y)\beta'(y)\alpha'(x) + b(\alpha(x),\beta(y))v(\alpha(x),\beta(y))\beta'(y)\alpha'(x)$$

$$\leq [a(\alpha(x),\beta(y)) + b(\alpha(x),\beta(y))]v(x,y)\beta'(y)\alpha'(x). \tag{3.5.5}$$

Now by following the proof of Theorem 4.2.1 given in [34] with suitable changes, from (3.5.5) we obtain

$$v(x,y) \leq k\exp\left(\int_{x_0}^{x}\int_{y_0}^{y}[a(\alpha(s),\beta(t)) + b(\alpha(s),\beta(t))]\right.$$

$$\left.\times\beta'(t)\alpha'(s)\,dtds\right). \tag{3.5.6}$$

Making the change of variables on the right hand side of (3.5.6) yields

$$v(x,y) \le k \exp\left(\int_{\alpha(x_0)}^{\alpha(x)} \int_{\beta(y_0)}^{\beta(y)} [a(\sigma,\eta) + b(\sigma,\eta)]\, d\eta d\sigma\right). \tag{3.5.7}$$

Using (3.5.7) in (3.5.4) we have

$$D_2 D_1 z(x,y) \le ka(\alpha(x), \beta(y))$$

$$\times \exp\left(\int_{\alpha(x_0)}^{\alpha(x)} \int_{\beta(y_0)}^{\beta(y)} [a(\sigma,\eta) + b(\sigma,\eta)] d\eta d\sigma\right) \beta'(y)\, \alpha'(x)). \tag{3.5.8}$$

Keeping x fixed in (3.5.8), set $y = t$ and integrate with respect to t from y_0 to y for $y \in I_2$, then keeping y fixed in the resulting inequality, set $x = s$ and integrate with respect to s from x_0 to x for $x \in I_1$ to obtain the estimate

$$z(x,y) \le k \left[1 + \int_{\alpha(x_0)}^{x} \int_{\beta(y_0)}^{y} a(\alpha(s), \beta(t))\right.$$

$$\times \exp\left(\int_{\alpha(x_0)}^{\alpha(x)} \int_{\beta(y_0)}^{\beta(y)} [a(\sigma,\eta) + b(\sigma,\eta)] d\eta d\sigma\right)$$

$$\left.\times\, \beta'(t)\, \alpha'(s))\, dt ds\right]. \tag{3.5.9}$$

By making the change of variables on the right hand side of (3.5.9) and using the fact that $u(x,y) \le z(x,y)$ we obtain the desired inequality in (3.5.2). The case $k \ge 0$ follows as mentioned in the proof of Theorem 3.2.1, part (a_1).

Remark 3.5.1. In the special case when $\alpha(x) = x$, $\beta(y) = y$ the inequality given in Theorem 3.5.1 reduces to the inequality given in [34, p. 336].

Next we shall give the following theorem which contains the inequalities established in [47].

Theorem 3.5.2. Let $u(x,y), a(x,y) \in C(\Delta, R_+)$ and $b(x,y,s,t) \in C(\Delta^2, R_+)$ for $x_0 \le s \le x < X, y_0 \le t \le y < Y$. Let $\alpha(x) \in C^1(I_1, I_1)$, $\beta(y) \in C^1(I_2, I_2)$ be nondecreasing with $\alpha(x) \le x$ on I_1, $\beta(y) \le y$ on I_2 and $k \ge 0$ be a real constant.

(a_1) If

$$u(x,y) \leq k + \int_{\alpha(x_0)}^{\alpha(x)} \int_{\beta(y_0)}^{\beta(y)} \left[a(s,t)\,u(s,t) \right.$$

$$\left. + \int_{\alpha(x_0)}^{s} \int_{\beta(y_0)}^{t} b(s,t,\sigma,\eta)\,u(\sigma,\eta) d\eta d\sigma \right] dtds, \tag{3.5.10}$$

for $(x,y) \in \Delta$, then

$$u(x,y) \leq k \exp(A(x,y)), \tag{3.5.11}$$

for $(x,y) \in \Delta$, where

$$A(x,y) = \int_{\alpha(x_0)}^{\alpha(x)} \int_{\beta(y_0)}^{\beta(y)} \left[a(s,t) + \int_{\alpha(x_0)}^{s} \int_{\beta(y_0)}^{t} b(s,t,\sigma,\eta)\, d\eta d\sigma \right] dtds, \tag{3.5.12}$$

for $(x,y) \in \Delta$.

(a_2) Let $g(u)$ be as in Theorem 3.3.2, part(b_1). If for $(x,y) \in \Delta$,

$$u(x,y) \leq k + \int_{\alpha(x_0)}^{\alpha(x)} \int_{\beta(y_0)}^{\beta(y)} \left[a(s,t)\,g(u(s,t)) \right.$$

$$\left. + \int_{\alpha(x_0)}^{s} \int_{\beta(y_0)}^{t} b(s,t,\sigma,\eta)\,g(u(\sigma,\eta)) d\eta d\sigma \right] dtds, \tag{3.5.13}$$

then for $x_0 \leq x \leq x_1, y_0 \leq y \leq y_1; x, x_1 \in I_1, y, y_1 \in I_2$,

$$u(x,y) \leq G^{-1}\left[G(k) + A(x,y)\right], \tag{3.5.14}$$

where $A(x,y)$ is given by (3.5.12), G, G^{-1} are as given in Theorem 3.3.2, part (b_1) and $x_1 \in I_1$, $y_1 \in I_2$ are chosen so that

$$G(k) + A(x,y) \in Dom\left(G^{-1}\right), \tag{3.5.15}$$

for all x and y lying in $[x_0, x_1]$ and $[y_0, y_1]$ respectively.

Proof. (a_1) From the hypotheses we observe that $\alpha'(x) \geq 0$ for $x \in I_1$, $\beta'(y) \geq 0$ for $y \in I_2$. Let $k > 0$ and define a function $z(x,y)$ by the right hand side of (3.5.10). Then $z(x_0,y) = z(x,y_0) = k, u(x,y) \leq z(x,y)$, $z(x,y)$ is positive and nondecreasing for $(x,y) \in \Delta$ and

$$D_1 z(x,y) = \left[\int_{\beta(y_0)}^{\beta(y)} \left[a(\alpha(x),t)\, u(\alpha(x),t) + \int_{\alpha(x_0)}^{\alpha(x)} \int_{\beta(y_0)}^{t} b(\alpha(x),t,\sigma,\eta)\, u(\sigma,\eta) d\eta d\sigma \right] dt \right] \alpha'(x)$$

$$\leq \left[\int_{\beta(y_0)}^{\beta(y)} \left[a(\alpha(x),t)\, z(\alpha(x),t) + \int_{\alpha(x_0)}^{\alpha(x)} \int_{\beta(y_0)}^{t} b(\alpha(x),t,\sigma,\eta)\, z(\sigma,\eta) d\eta d\sigma \right] dt \right] \alpha'(x). \tag{3.5.16}$$

From (3.5.16) it is easy to observe that

$$\frac{D_1 z(x,y)}{z(x,y)} \leq \left[\int_{\beta(y_0)}^{\beta(y)} \left[a(\alpha(x),t) + \int_{\alpha(x_0)}^{\alpha(x)} \int_{\beta(y_0)}^{t} b(\alpha(x),t,\sigma,\eta) d\eta d\sigma \right] dt \right] \alpha'(x). \tag{3.5.17}$$

Keeping y fixed in (3.5.17), setting $x = \xi$ and integrating it with respect to ξ from x_0 to x for $x \in I_1$ and making the change of variables we get

$$z(x,y) \leq k \exp(A(x,y)). \tag{3.5.18}$$

Using (3.5.18) in $u(x,y) \leq z(x,y)$ we get the required inequality in (3.5.11). The case $k \geq 0$ follows as mentioned in the proof of Theorem 3.4.1, part (a_1).

(a_2) The proof can be completed by following the proof of (a_1) given above and closely looking at the proof of Theorem 3.4.2. Here we omit the details.

In the following theorems we present the inequalities investigated in [59].

Theorem 3.5.3. Let $u(x,y), a(x,y), b_i(x,y) \in C(\Delta, R_+)$ and $\alpha_i(x) \in C^1(I_1, I_1), \beta_i(y) \in C^1(I_2, I_2)$ be nondecreasing with $\alpha_i(x) \leq x$ on I_1, $\beta_i(y) \leq y$ on I_2 for $i = 1, ..., n$ and $k \geq 0$ be a real constant.

(b_1) If

$$u(x,y) \leq k + \int_{x_0}^{x} a(s,y)\,u(s,y)\,ds + \sum_{i=1}^{n} \int_{\alpha_i(x_0)}^{\alpha_i(x)} \int_{\beta_i(y_0)}^{\beta_i(y_0)} b_i(s,t)\,u(s,t)\,dtds, \quad (3.5.19)$$

for $(x,y) \in \Delta$, then

$$u(x,y) \leq kq(x,y) \exp\left(\sum_{i=1}^{n} \int_{\alpha_i(x_0)}^{\alpha_i(x)} \int_{\beta_i(y_0)}^{\beta_i(y_0)} b_i(s,t)\,q(s,t)\,dtds \right), \quad (3.5.20)$$

for $(x,y) \in \Delta$, where

$$q(x,y) = \exp\left(\int_{x_0}^{x} a(\xi,y)\,d\xi \right), \quad (3.5.21)$$

for $(x,y) \in \Delta$.

(b_2) Let $g \in C(R_+, R_+)$ be nondecreasing and submultiplicative function with $g(u) > 0$ for $u > 0$. If for $(x,y) \in \Delta$,

$$u(x,y) \leq k + \int_{x_0}^{x} a(s,y)\,u(s,y)\,ds$$

$$+ \sum_{i=1}^{n} \int_{\alpha_i(x_0)}^{\alpha_i(x)} \int_{\beta_i(y_0)}^{\beta_i(y)} b_i(s,t)\,g(u(s,t))\,dtds, \quad (3.5.22)$$

then for $x_0 \leq x \leq x_2, y_0 \leq y \leq y_2; x, x_2 \in I_1, y, y_2 \in I_2$,

$$u(x,y) \leq q(x,y)\,G^{-1}\Big[G(k)$$

$$+ \sum_{i=1}^{n} \int_{\alpha_i(x_0)}^{\alpha_i(x)} \int_{\beta_i(y_0)}^{\beta_i(y)} b_i(s,t)\,g(q(s,t))\,dtds\Big], \quad (3.5.23)$$

where $q(x,y)$ is given by (3.5.21) and G^{-1} is the inverse function of

$$G(r) = \int_{r_0}^{r} \frac{ds}{g(s)}, r > 0, \quad (3.5.24)$$

$r_0 > 0$ is arbitrary and $x_2 \in I_1$, $y_2 \in I_2$ are chosen so that

$$G(k) + \sum_{i=1}^{n} \int_{\alpha_i(x_0)}^{\alpha_i(x)} \int_{\beta_i(y_0)}^{\beta_i(y)} b_i(s,t)\,g(q(s,t))\,dtds \in Dom\left(G^{-1}\right),$$

for all x and y lying in $[x_0, x_2]$ and $[y_0, y_2]$ respectively.

Proof. From the hypotheses we observe that $\alpha'(x) \geq 0$ for $x \in I_1$, $\beta'(y) \geq 0$ for $y \in I_2$.

(b_1) Let $k > 0$ and define a function $z(x,y)$ by

$$z(x,y) = k + \sum_{i=1}^{n} \int_{\alpha_i(x_0)}^{\alpha_i(x)} \int_{\beta_i(y_0)}^{\beta_i(y)} b_i(s,t)\, u(s,t)\, dtds. \tag{3.5.25}$$

Then (3.5.19) can be restated as

$$u(x,y) \leq z(x,y) + \int_{x_0}^{x} a(s,y)\, u(s,y)\, ds. \tag{3.5.26}$$

It is easy to observe that $z(x,y)$ is positive, continuous and nondecreasing function for $(x,y) \in \Delta$. Treating y fixed in (3.5.26) and using Theorem 1.3.1 given in [34] to (3.5.26) we get

$$u(x,y) \leq q(x,y)\, z(x,y), \tag{3.5.27}$$

for $(x,y) \in \Delta$, where $q(x,y)$ is given by (3.5.21). From (3.5.25)and (3.5.27) we have

$$z(x,y) \leq k + \sum_{i=1}^{n} \int_{\alpha_i(x_0)}^{\alpha_i(x)} \int_{\beta_i(y_0)}^{\beta_i(y)} b_i(s,t)\, q(s,t)\, z(s,t)\, dtds. \tag{3.5.28}$$

Define a function $v(x,y)$ by the right hand side of (3.5.28). Then $v(x_0,y) = z(x_0,y) = k$, $z(x,y) \leq v(x,y)$, $v(x,y)$ is positive and nondecreasing for $(x,y) \in \Delta$ and

$$D_1 v(x,y) = \sum_{i=1}^{n} \left(\int_{\beta_i(y_0)}^{\beta_i(y)} b_i(\alpha_i(x),t)\, q(\alpha_i(x),t) z(\alpha_i(x),t)\, dt \right) \alpha_i'(x)$$

$$\leq \sum_{i=1}^{n} \left(\int_{\beta_i(y_0)}^{\beta_i(y)} b_i(\alpha_i(x),t)\, q(\alpha_i(x),t) v(\alpha_i(x),t)\, dt \right) \alpha_i'(x)$$

$$\leq v(x,y) \sum_{i=1}^{n} \left(\int_{\beta_i(y_0)}^{\beta_i(y)} b_i(\alpha_i(x),t)\, q(\alpha_i(x),t) dt \right) \alpha_i'(x)$$

i.e.,

$$\frac{D_1 v(x,y)}{v(x,y)} \leq \sum_{i=1}^{n} \left(\int_{\beta_i(y_0)}^{\beta_i(y)} b_i(\alpha_i(x),t)\, q(\alpha_i(x),t) dt \right) \alpha_i'(x). \tag{3.5.29}$$

Keeping y fixed in (2.5.29), setting $x = \sigma$ and integrating it with respect to σ from x_0 to x for $x \in I_1$, making the change of variables and using the fact that $z(x,y) \leq v(x,y)$ we get

$$z(x,y) \leq k \exp\left(\sum_{i=1}^{n} \int_{\alpha_i(x_0)}^{\alpha_i(x)} \int_{\beta_i(y_0)}^{\beta(y)} b_i(s,t)\, q(s,t)\, dtds\right), \tag{3.5.30}$$

for $(x,y) \in \Delta$. Using (3.5.30) in (3.5.27) we get the required inequality in (3.5.20). The case $k \geq 0$ follows as noted in the proof of Theorem 3.2.1, part (a_1).

(b_2) Let $k > 0$ and define a function $z(x,y)$ by

$$z(x,y) = k + \sum_{i=1}^{n} \int_{\alpha_i(x_0)}^{\alpha_i(x)} \int_{\beta_i(y_0)}^{\beta(y)} b_i(s,t)\, g(u(s,t))\, dtds. \tag{3.5.31}$$

Then (3.5.22) can be restated as

$$u(x,y) \leq z(x,y) + \int_{x_0}^{x} a(s,y)\, u(s,y)\, ds. \tag{3.5.32}$$

As in the proof of part (b_1), using Theorem 1.3.1 given in [34] to (3.5.32) we have

$$u(x,y) \leq q(x,y)\, z(x,y), \tag{3.5.33}$$

for $(x,y) \in \Delta$, where $q(x,y)$ and $z(x,y)$ are given by (3.5.21) and (3.5.31). From (3.5.31), (3.5.33) and the hypotheses on g we have

$$z(x,y) \leq k + \sum_{i=1}^{n} \int_{\alpha_i(x_0)}^{\alpha_i(x)} \int_{\beta_i(y_0)}^{\beta_i(y)} b_i(s,t)\, g(q(s,t)\, z(s,t))\, dtds$$

$$\leq k + \sum_{i=1}^{n} \int_{\alpha_i(x_0)}^{\alpha_i(x)} \int_{\beta_i(y_0)}^{\beta_i(y)} b_i(s,t)\, g(q(s,t))\, g(z(s,t))\, dtds. \tag{3.5.34}$$

Define a function $v(x,y)$ by the right hand side of (3.5.34). Then $v(x_0,y) = v(x,y_0) = k$, $z(x,y) \leq v(x,y)$, $v(x,y)$ is positive and nondecreasing for $(x,y) \in \Delta$ and

$$D_1 v(x,y) = \sum_{i=1}^{n} \left(\int_{\beta_i(y_0)}^{\beta_i(y)} b_i(\alpha_i(x),t)\, g(q(\alpha_i(x),t))\, g(z(\alpha_i(x),t))\, dt \right) \alpha_i'(x)$$

$$\leq \sum_{i=1}^{n}\left(\int_{\beta_i(y_0)}^{\beta_i(y)} b_i\left(\alpha_i\left(x\right),t\right) g\left(q\left(\alpha_i\left(x\right),t\right)\right) g\left(v\left(\alpha_i\left(x\right),t\right)\right) dt\right)\alpha_i'\left(x\right)$$

$$\leq g\left(v\left(x,y\right)\right)\sum_{i=1}^{n}\left(\int_{\beta_i(y_0)}^{\beta_i(y)} b_i\left(\alpha_i\left(x\right),t\right) g\left(q\left(\alpha_i\left(x\right),t\right)\right) dt\right)\alpha_i'\left(x\right). \quad (3.5.35)$$

From (3.5.24) and (3.5.35) we have

$$D_1G\left(v\left(x,y\right)\right) = \frac{D_1v\left(x,y\right)}{g\left(v\left(x,y\right)\right)}$$

$$\leq \sum_{i=1}^{n}\left(\int_{\beta_i(y_0)}^{\beta_i(y)} b_i\left(\alpha_i\left(x\right),t\right) g\left(q\left(\alpha_i\left(x\right),t\right)\right) dt\right)\alpha_i'\left(x\right). \quad (3.5.36)$$

Keeping y fixed in (3.5.36), setting $x = \sigma$ and integrating it with respect to σ from x_0 to x for $x \in I_1$ and making the change of variables we get

$$G\left(v\left(x,y\right)\right) \leq G\left(k\right) + \sum_{i=1}^{n}\int_{\alpha_i(x_0)}^{\alpha_i(x)}\int_{\beta_i(y_0)}^{\beta_i(y)} b_i\left(s,t\right) g\left(q\left(s,t\right)\right) dtds. \quad (3.5.37)$$

From (3.5.37) and (3.5.33) we get the required inequality in (3.5.23). The case $k \geq 0$ follows as mentioned in the proof of Theorem 3.2.1,part (a_1). The subdomain for x, y is obvious.

Theorem 3.5.4. Let $u\left(x,y\right), a\left(x,y\right), b_i\left(x,y\right), \alpha_i\left(x\right), \beta_i\left(y\right), k$ be as in Theorem 3.5.3 and $c\left(x,y\right) \in C\left(\Delta, R_+\right)$.

(c_1) If

$$u\left(x,y\right) \leq k + \int_{x_0}^{x} a\left(s,y\right)\left(u\left(s,y\right) + \int_{x_0}^{s} c\left(\sigma,y\right) u\left(\sigma,y\right) d\sigma\right) ds$$

$$+\sum_{i=1}^{n}\int_{\alpha_i(x_0)}^{\alpha_i(x)}\int_{\beta_i(y_0)}^{\beta_i(y)} b_i\left(s,t\right) u\left(s,t\right) dtds, \quad (3.5.38)$$

for $(x,y) \in \Delta$,then

$$u\left(x,y\right) \leq kp\left(x,y\right)\exp\left(\sum_{i=1}^{n}\int_{\alpha_i(x_0)}^{\alpha_i(x)}\int_{\beta_i(y_0)}^{\beta_i(y)} b_i\left(s,t\right) p\left(s,t\right) dtds\right), \quad (3.5.39)$$

for $(x, y) \in \Delta$, where

$$p(x,y) = 1 + \int_{x_0}^{x} a(\xi, y) \exp\left(\int_{x_0}^{\xi} [a(\sigma, y) + b(\sigma, y)] d\sigma \right) d\xi, \quad (3.5.40)$$

for $(x, y) \in \Delta$.

(c_2) Let g be as in Theorem 3.5.3, part (b_2). If for $(x, y) \in \Delta$,

$$u(x,y) \le k + \int_{x_0}^{x} a(s,y) \left(u(s,y) + \int_{x_0}^{s} c(\sigma, y) u(\sigma, y) d\sigma \right) ds$$

$$+ \sum_{i=1}^{n} \int_{\alpha_i(x_0)}^{\alpha_i(x)} \int_{\beta_i(y_0)}^{\beta_i(y)} b_i(s,t) g(u(s,t)) dtds, \quad (3.5.41)$$

then for $x_0 \le x \le x_3, y_0 \le y \le y_3; x, x_3 \in I_1, y, y_3 \in I_2$,

$$u(x,y) \le p(x,y) G^{-1} [G(k)$$

$$+ \sum_{i=1}^{n} \int_{\alpha_i(x_0)}^{\alpha_i(x)} \int_{\beta_i(y_0)}^{\beta_i(y)} b_i(s,t) g(p(s,t)) dtds \Bigg], \quad (3.5.42)$$

where $p(x, y)$ is given by (3.5.40), G, G^{-1} are given as in Theorem 3.5.3, part (b_2) and $x_3 \in I_1$, $y_3 \in I_2$ are chosen so that

$$G(k) + \sum_{i=1}^{n} \int_{\alpha_i(x_0)}^{\alpha_i(x)} \int_{\beta_i(y_0)}^{\beta_i(y)} b_i(s,t) g(p(s,t)) dtds \in Dom\left(G^{-1}\right),$$

for all x and y lying in $[x_0, x_3]$ and $[y_0, y_3]$ respectively.

Proof. From the hypotheses we have $\alpha_i'(x) \ge 0$ for $x \in I_1$, $\beta_i'(x) \ge 0$ for $y \in I_2$.

(c_1) Let $k > 0$ and define a function $z(x, y)$ by (3.5.25). Then (3.5.38) can be restated as

$$u(x,y) \le z(x,y) + \int_{x_0}^{x} a(s,y) \left(u(s,y) + \int_{x_0}^{s} c(\sigma, y) u(\sigma, y) d\sigma \right) ds. \quad (3.5.43)$$

Clearly, $z(x, y)$ is positive, continuous and nondecreasing function for $(x, y) \in \Delta$. Treating y for $y \in I_2$ fixed in (3.5.43) and applying Theorem 1.7.4 given in [34] to (3.5.43) yields

$$u(x,y) \le p(x,y) z(x,y), \quad (3.5.44)$$

where $p(x, y)$ and $z(x, y)$ are given by (3.5.40) and (3.5.25). Now by following the proof of Theorem 3.5.3, part (b_1) with suitable changes we get the desired inequality in (3.5.39).

(c_2) The proof can be completed by following the proof of part (c_1) given above and the proof of Theorem 3.5.3, part (b_2). Here we omit the details.

Remark 3.5.2. If we take $a(x, y) = 0$ in Theorems 3.5.3 and 3.5.4, then we recapture the inequalities established in Theorem 3, part (C_1) and Theorem 4, part (D_1) in a recent paper [74]. We also note that, if we take in (3.5.19) $a(x, y) = 0$ and replace the constant k by a function $r(x, y) \in C(\Delta, R_+)$, which is nondecreasing for $(x, y) \in \Delta$, then the bound obtained in (3.5.20) takes the form

$$u(x, y) \le r(x, y) \exp\left(\sum_{i=1}^{n} \int_{\alpha_i(x_0)}^{\alpha_i(x)} \int_{\beta_i(y_0)}^{\beta_i(y)} b_i(s, t)\, dt ds\right),$$

for $(x, y) \in \Delta$, which is the inequality given in Theorem 3, part (C_2) in [74]. These inequalities can be used as basic tools in some applications.

Finally we give the following theorem which deals with the inequalities proved in [60].

Theorem 3.5.5. Let $u(x, y), a_i(x, y), b_i(x, y) \in C(\Delta, R_+)$ and $\alpha_i(x) \in C^1(I_1, I_1)$, $\beta_i(y) \in C^1(I_2, I_2)$ be nondecreasing with $\alpha_i(x) \le x$ on I_1, $\beta_i(y) \le y$ on I_2 for $i = 1, ..., n$. Let $p > 1$ and $c \ge 0$ be real constants.

(d_1) If

$$u^p(x, y) \le c + p \sum_{i=1}^{n} \int_{\alpha_i(x_0)}^{\alpha_i(x)} \int_{\beta_i(y_0)}^{\beta_i(y)} [a_i(s, t) u^p(s, t) + b_i(s, t) u(s, t)]\, dt ds, \quad (3.5.45)$$

for $(x, y) \in \Delta$, then

$$u(x, y) \le \left\{B(x, y) \exp\left((p - 1) \sum_{i=1}^{n} \int_{\alpha_i(x_0)}^{\alpha_i(x)} \int_{\beta_i(y_0)}^{\beta_i(y)} a_i(\sigma, \tau)\, d\tau d\sigma\right)\right\}^{\frac{1}{p-1}}, \quad (3.5.46)$$

for $(x, y) \in \Delta$, where

$$B(x, y) = \{c\}^{\frac{p-1}{p}} + (p - 1) \sum_{i=1}^{n} \int_{\alpha_i(x_0)}^{\alpha_i(x)} \int_{\beta_i(y_0)}^{\beta_i(y)} b_i(\sigma, \tau)\, d\tau d\sigma, \quad (3.5.47)$$

for $(x, y) \in \Delta$.

(d_2) Let $w(u)$ be as in Theorem 3.2.6, part (q_2).If for $(x,y) \in \Delta$,

$$u^p(x,y) \le c + p\sum_{i=1}^{n} \int_{\alpha_i(x_0)}^{\alpha_i(x)} \int_{\beta_i(y_0)}^{\beta_i(y)} [a_i(s,t)\,u(s,t)\,w(u(s,t))$$

$$+ b_i(s,t)\,u(s,t)]\,dtds, \tag{3.5.48}$$

then for $x_0 \le x \le x_4, y_0 \le y \le y_4; x, x_4 \in I_1, y, y_4 \in I_2$,

$$u(x,y) \le \{F^{-1}[F(B(x,y)) + (p-1)$$

$$\times \sum_{i=1}^{n} \int_{\alpha_i(x_0)}^{\alpha_i(x)} \int_{\beta_i(y_0)}^{\beta_i(y)} a_i(\sigma,\tau)d\tau d\sigma \Bigg]\Bigg\}^{\frac{1}{p-1}}, \tag{3.5.49}$$

where $B(x,y)$ is given by (3.5.47), F, F^{-1} are as in Theorem 3.2.6, part (q_2) and $x_4 \in I_1$, $y_4 \in I_2$ are chosen so that

$$F(B(x,y)) + (p-1)\sum_{i=1}^{n} \int_{\alpha_i(x_0)}^{\alpha_i(x)} \int_{\beta_i(y_0)}^{\beta_i(y)} a_i(\sigma,\tau)\,d\tau d\sigma \in Dom\left(F^{-1}\right),$$

for all x and y lying in $[x_0, x_4]$ and $[y_0, y_4]$ respectively.

Proof. From the hypotheses we have $\alpha_i'(x) \ge 0$ for $x \in I_1$, $\beta_i'(y) \ge 0$ for $y \in I_2$. We give the details of the proof of (d_2) only; the proof of (d_1) can be completed by following the proof of (d_2) with suitable modifications.

(d_2) Let $c > 0$ and define a function $z(x,y)$ by the right hand side of (3.5.48).Then $z(x_0,y) = z(x,y_0) = c$, $z(x,y)$ is positive and nondecreasing for $(x,y) \in \Delta$, $u(x,y) \le \{z(x,y)\}^{\frac{1}{p}}$ and

$$D_2D_1z(x,y) = p\sum_{i=1}^{n} [a_i(\alpha_i(x),\beta_i(y))\,u(\alpha_i(x),\beta_i(y))\,w(u(\alpha_i(x),\beta_i(y)))$$

$$+ b_i(\alpha_i(x),\beta_i(y))\,u(\alpha_i(x),\beta_i(y))]\,\beta_i'(y)\,\alpha_i'(x)$$

$$\le p\sum_{i=1}^{n} \left[a_i(\alpha_i(x),\beta_i(y))\{z(\alpha_i(x),\beta_i(y))\}^{\frac{1}{p}} w\left(\{z(\alpha_i(x),\beta_i(y))\}^{\frac{1}{p}}\right)\right.$$

$$\left. + b_i(\alpha_i(x),\beta_i(y))\{z(\alpha_i(x),\beta_i(y))\}^{\frac{1}{p}}\right] \beta_i'(y)\,\alpha_i'(x)$$

$$\le p\sum_{i=1}^{n} \left[a_i(\alpha_i(x),\beta_i(y))\,w\left(\{z(\alpha_i(x),\beta_i(y))\}^{\frac{1}{p}}\right)\right.$$

$$+b_i\left(\alpha_i\left(x\right),\beta_i\left(y\right)\right)\right]\left\{z\left(x,y\right)\right\}^{\frac{1}{p}}\beta_i'\left(y\right)\alpha_i'\left(x\right). \tag{3.5.50}$$

From (3.5.50) and the facts that $D_1 z\left(x,y\right), D_2 z\left(x,y\right)$ are nonnegative, we observe that (see [34])

$$\frac{D_2 D_1 z\left(x,y\right)}{\left\{z\left(x,y\right)\right\}^{\frac{1}{p}}} \leq p\sum_{i=1}^{n}\left[a_i\left(\alpha_i\left(x\right),\beta_i\left(y\right)\right)w\left(\left\{z\left(\alpha_i\left(x\right),\beta_i\left(y\right)\right)\right\}^{\frac{1}{p}}\right)\right.$$

$$\left.+b_i\left(\alpha_i\left(x\right),\beta_i\left(y\right)\right)\right]\beta_i'\left(y\right)\alpha_i'\left(x\right)+\frac{D_1 z\left(x,y\right)\left[D_2\left\{z\left(x,y\right)\right\}^{\frac{1}{p}}\right]}{\left[\left\{z\left(x,y\right)\right\}^{\frac{1}{p}}\right]^2},$$

i.e.,

$$D_2\left(\frac{D_1 z\left(x,y\right)}{\left\{z\left(x,y\right)\right\}^{\frac{1}{p}}}\right) \leq p\sum_{i=1}^{n}\left[a_i\left(\alpha_i\left(x\right),\beta_i\left(y\right)\right)w\left(\left\{z\left(\alpha_i\left(x\right),\beta_i\left(y\right)\right)\right\}^{\frac{1}{p}}\right)\right.$$

$$\left.+b_i\left(\alpha_i\left(x\right),\beta_i\left(y\right)\right)\right]\beta_i'\left(y\right)\alpha_i'\left(x\right), \tag{3.5.51}$$

for $(x,y)\in\Delta$. By keeping x fixed in (3.5.51), we set $y=t$ and then, by integrating with respect to t from y_0 to y and using the fact that $D_1 z\left(x,y_0\right)=0$, we have

$$\frac{D_1 z\left(x,y\right)}{\left\{z\left(x,y\right)\right\}^{\frac{1}{p}}} \leq p\int_{y_0}^{y}\sum_{i=1}^{n}\left[a_i\left(\alpha_i\left(x\right),\beta_i\left(t\right)\right)w\left(\left\{z\left(\alpha_i\left(x\right),\beta_i\left(t\right)\right)\right\}^{\frac{1}{p}}\right)\right.$$

$$\left.+b_i\left(\alpha_i\left(x\right),\beta_i\left(t\right)\right)\right]\beta_i'\left(t\right)\alpha_i'\left(x\right)dt. \tag{3.5.52}$$

Now keeping y fixed in (3.5.52) and setting $x=s$ and integrating with respect to s from x_0 to x we have

$$\left\{z\left(x,y\right)\right\}^{\frac{p-1}{p}} \leq \left\{c\right\}^{\frac{p-1}{p}}+(p-1)\int_{x_0}^{x}\int_{y_0}^{y}\sum_{i=1}^{n}\left[a_i\left(\alpha_i\left(s\right),\beta_i\left(t\right)\right)\right.$$

$$\left.\times w\left(\left\{z\left(\alpha_i\left(s\right),\beta_i\left(t\right)\right)\right\}^{\frac{1}{p}}\right)+b_i\left(\alpha_i\left(s\right),\beta_i\left(t\right)\right)\right]$$

$$\times\beta_i'\left(t\right)\alpha_i'\left(s\right)dtds. \tag{3.5.53}$$

By making the change of variables on the right hand side of (3.5.53) and rewriting we have

$$\left\{z\left(x,y\right)\right\}^{\frac{p-1}{p}} \leq B\left(x,y\right)+(p-1)$$

$$\times\sum_{i=1}^{n}\int_{\alpha_i(x_0)}^{\alpha_i(x)}\int_{\beta_i(y_0)}^{\beta_i(y)}a_i\left(\sigma,\tau\right)w\left(\left\{z\left(\sigma,\tau\right)\right\}^{\frac{1}{p}}\right)d\tau d\sigma. \tag{3.5.54}$$

Now fix $(\lambda, \mu) \in \Delta$ such that $x_0 \leq x \leq \lambda \leq x_4, y_0 \leq y \leq \mu \leq y_4$. Then from (3.5.54) we observe that

$$\{z(x,y)\}^{\frac{p-1}{p}} \leq B(\lambda, \mu) + (p-1)$$

$$\times \sum_{i=1}^{n} \int_{\alpha_i(x_0)}^{\alpha_i(x)} \int_{\beta_i(y_0)}^{\beta_i(y)} a_i(\sigma, \tau)\, w\left(\{z(\sigma, \tau)\}^{\frac{1}{p}}\right) d\tau d\sigma, \tag{3.5.55}$$

for $x_0 \leq x \leq \lambda, y_0 \leq y \leq \mu$. Define a function $v(x,y)$ by the right hand side of (3.5.55). Then $v(x_0, y) = v(x, y_0) = B(\lambda, \mu)$, $v(x,y)$ is positive and nondecreasing for $x_0 \leq x \leq \lambda, y_0 \leq y \leq \mu$, $\{z(x,y)\}^{\frac{p-1}{p}} \leq v(x,y)$ and

$$v(x,y) \leq B(\lambda, \mu) + (p-1) \sum_{i=1}^{n} \int_{\alpha_i(x_0)}^{\alpha_i(x)} \int_{\beta_i(y_0)}^{\beta_i(y)} a_i(\sigma, \tau)\, w\left(\{v(\sigma, \tau)\}^{\frac{1}{p-1}}\right) d\tau d\sigma,$$

for $x_0 \leq x \leq \lambda, y_0 \leq y \leq \mu$. Now by following the proof of Theorem 3.5.3, part (b_2) (see also [34]) we get

$$v(x,y) \leq F^{-1}\left[F(B(\lambda, \mu)) + (p-1) \sum_{i=1}^{n} \int_{\alpha_i(x_0)}^{\alpha_i(x)} \int_{\beta_i(y_0)}^{\beta_i(y)} a_i(\sigma, \tau) d\tau d\sigma\right], \tag{3.5.56}$$

for $x_0 \leq x \leq \lambda \leq x_4, y_0 \leq y \leq \mu \leq y_4$. Since $(\lambda, \mu) \in \Delta$ is arbitrary, we get the desired inequality in (3.5.49) from (3.5.56) and the fact that

$$u(x,y) \leq \{z(x,y)\}^{\frac{1}{p}} \leq \left\{[v(x,y)]^{\frac{p}{p-1}}\right\}^{\frac{1}{p}} = \{v(x,y)\}^{\frac{1}{p-1}}.$$

The proof of the case when $c \geq 0$ can be completed as mentioned in the proof of Theorem 3.2.1, part (a_1). The subdomain $x_0 \leq x \leq x_4, y_0 \leq y \leq y_4$ is obvious.

Remark 3.5.3. If we take $p = 2, n = 1$ $\alpha_1 = \alpha$, $\beta_1 = \beta, a_1 = f, b_1 = g$ in Theorem 3.5.5, then we get the two independent variable generalizations of the inequalities given in [22, see Corollary 2 and Theorem 1].

3.6 Applications

In the literature, a number of new methods and tools are developed by different investigators to study various types of differential and integral equations. In this section we present applications of some of the inequalities given in earlier sections and it is hoped that these inequalities will assure greater importance in the near future. In what follows we shall use the notations and definitions as given in sections 3.2 and 3.4 and explained if necessary at appropriate places.

3.6.1 Differential equations with many retarded arguments

Consider the following differential equations involving several retarded arguments

$$x'(t) = f(t, x(t - h_1(t)), ..., x(t - h_n(t))), \tag{3.6.1}$$

and

$$x^{p-1}(t)\, x'(t) = f(t, x(t - h_1(t)), ..., x(t - h_n(t))), \tag{3.6.2}$$

for $t \in I$, with the given initial condition

$$x(t_0) = x_0, \tag{3.6.3}$$

where $p > 1$ and x_0 are constants, $f \in C(I \times R^n, R)$ and for $i = 1, ..., n$, $h_i(t) \in C(I, R_+)$ be nonincreasing and such that $t - h_i(t) \geq 0$, $t - h_i(t) \in C^1(I, I)$, $h_i'(t) < 1$, $h_i(t_0) = 0$. For the theory and applications of differential equations with deviating arguments, see [7,13,18].

The following theorems deals with the estimates on the solutions of equations (3.6.1), (3.6.2) with the given initial condition (3.6.3), see Pachpatte [60,74].

Theorem 3.6.1. Suppose that

$$|f(t, u_1, ..., u_n)| \leq \sum_{i=1}^{n} b_i(t)\, |u_i|, \tag{3.6.4}$$

where $b_i(t)$ are as in Theorem 3.2.4, and let

$$M_i = \max_{t \in I} \frac{1}{1 - h_i'(t)}, i = 1, ..., n. \tag{3.6.5}$$

If $x(t)$ is any solution of the initial value problem (3.6.1)-(3.6.3), then

$$|x(t)| \leq |x_0| \exp\left(\sum_{i=1}^{n} \int_{t_0}^{t-h_i(t)} \bar{b}_i(\sigma)\, d\sigma\right), \tag{3.6.6}$$

for $t \in I$, where $\bar{b}_i(\sigma) = M_i b_i(\sigma + h_i(s)), \sigma, s \in I$.

Proof. The solution $x(t)$ of the initial value problem (3.6.1)-(3.6.3) can be written as

$$x(t) = x_0 + \int_{t_0}^{t} f(s, x(s - h_1(s)), ..., x(s - h_n(s)))ds. \tag{3.6.7}$$

Using (3.6.4) in (3.6.7) and making the change of variables, then using (3.6.5) we have

$$|x(t)| \leq |x_0| + \sum_{i=1}^{n} \int_{t_0}^{t} b_i(s) |x(s - h_i(s))|\, ds$$

$$\leq |x_0| + \sum_{i=1}^{n} \int_{t_0}^{t-h_i(t)} \bar{b}_i(\sigma) |x(\sigma)|\, d\sigma, \tag{3.6.8}$$

for $t \in I$. Now a suitable application of the inequality given in Theorem 3.2.4, part (c_1) to (3.6.8) yields the required estimate in (3.6.6).

Theorem 3.6.2. Suppose that the function f in (3.6.2) satisfies the condition (3.6.4). Let M_i and $\bar{b}_i(\sigma)$ be as given in Theorem 3.6.1. If $x(t)$ is any solution of the initial value problem (3.6.2)-(3.6.3), then

$$|x(t)| \leq \left\{ |x_0|^{p-1} + (p-1) \sum_{i=1}^{n} \int_{t_0}^{t-h_i(t)} \bar{b}_i(\sigma)\, d\sigma \right\}^{\frac{1}{p-1}}, \tag{3.6.9}$$

for $t \in I$.

Proof. The solution $x(t)$ of the initial value problem (3.6.2)-(3.6.3) can be written as

$$x^p(t) = x_0^p + p \int_{t_0}^{t} f(s, x(s - h_1(s)), ..., x(s - h_n(s)))ds. \tag{3.6.10}$$

From (3.6.10), (3.6.4), (3.6.5) and making the change of variables we have

$$|x(t)|^p \leq |x_0|^p + p\sum_{i=1}^{n}\int_{t_0}^{t} b_i(s)\,|x(s - h_i(s))|\,ds$$

$$\leq |x_0|^p + p\sum_{i=1}^{n}\int_{t_0}^{t} \bar{b}_i(\sigma)\,|x(\sigma)|\,d\sigma, \tag{3.6.11}$$

for $t \in I$. Now a suitable application of the inequality given in Theorem 3.2.6,part (q_1) (when $a_i(t) = 0$) to (3.6.11) yields the required estimate in (3.6.9).

3.6.2 Retarded differential and integrodifferential equations

First we consider the initial value problem (IVP for short) for higher order retarded differential equation of the form

$$y^{(n)}(t) = f(t, y(t), y(t - h(t))), \tag{3.6.12}$$

$$y^{(k)}(t_0) = c_k, k = 0, 1, 2, ..., n-1, \tag{3.6.13}$$

for $t \in J = [t_0, T]$, where $f \in C(J \times R^2, R)$ and $h \in C(J, R_+)$ be nonincreasing with $h(t) \leq t$ on J, $t - h(t) \in C^1(J, J)$, $h'(t) < 1, h(t_0) = 0$ and $n \geq 2$ is a natural number and c_k are real constants.

As an application of the inequality given in Theorem 3.2.1, part (a_1) we present the following theorem which deals with certain properties of solutions of IVP (3.6.12)-(3.6.13), see [63].

Theorem 3.6.3. (i) Assume that

$$|f(t, y, z)| \leq a(t)\,|y| + b(t)\,|z|, \tag{3.6.14}$$

where $a(t), b(t) \in C(J, R_+)$ and let

$$L = \max_{t \in J} \frac{1}{1 - h'(t)}. \tag{3.6.15}$$

If $y(t)$ is any solution of IVP (3.6.12)-(3.6.13), then

$$|y(t)| \leq M \exp\left(\int_{t_0}^{t} \bar{a}(s)\,ds + \int_{t_0}^{\phi(t)} \bar{b}(s)\,ds\right), \tag{3.6.16}$$

for $t \in J$, where $\bar{a}(t) = Na(t)$, $\bar{b}(t) = NLb(t + h(s))$, $t, s \in J$, $\phi(t) = t - h(t)$,

$$M = \sum_{i=0}^{n-1} \frac{|c_i| (T - t_0)^i}{i!}, \tag{3.6.17}$$

and

$$N = \frac{(T - t_0)^{n-1}}{(n-1)!}. \tag{3.6.18}$$

(ii) Suppose that

$$|f(t, y, z) - f(t, \bar{y}, \bar{z})| \leq a(t) |y - \bar{y}| + b(t) |z - \bar{z}|, \tag{3.6.19}$$

where $a(t), b(t) \in C(J, R_+)$. Let L, M, N, $\bar{a}(t), \bar{b}(t), \phi(t)$ be as in part (i). Then the IVP (3.6.12)-(3.6.13) has at most one solution on J.

Proof. (i) It is easy to see that the solution $y(t)$ of IVP (3.6.12)-(3.6.13) satisfies the equivalent integral equation

$$y(t) = \sum_{i=1}^{n-1} \frac{c_i (t - t_0)^i}{i!} + \int_{t_0}^{t} \frac{(t-s)^{n-1}}{(n-1)!} f(s, y(s), y(s - h(s))) \, ds. \tag{3.6.20}$$

From (3.6.20), (3.6.14), (3.6.17), (3.6.18) we have

$$|y(t)| \leq \sum_{i=1}^{n-1} \frac{|c_i| (t - t_0)^i}{i!} + \int_{t_0}^{t} \frac{(t-s)^{n-1}}{(n-1)!} |f(s, y(s), y(s - h(s)))| ds$$

$$\leq M + N \left[\int_{t_0}^{t} a(s) |y(s)| \, ds + \int_{t_0}^{t} b(s) |y(s - h(s))| \, ds \right]. \tag{3.6.21}$$

By making the change of variable in the second integral in (3.6.21) and using (3.6.15) we have

$$|y(t)| \leq M + \int_{t_0}^{t} \bar{a}(s) |y(s)| \, ds + \int_{t_0}^{\phi(t)} \bar{b}(\sigma) |y(\sigma)| d\sigma. \tag{3.6.22}$$

Now a suitable application of the inequality in Theorem 3.2.1, part (a_1) to (3.6.22) yields the required estimate in (3.6.16).

(*ii*) Let $y_1(t)$ and $y_2(t)$ be two solutions of IVP (3.6.12)-(3.6.13) on J, then we have

$$y_1(t) - y_2(t) = \int_{t_0}^{t} \frac{(t-s)^n}{(n-1)!} \{f(s, y_1(s), y_1(s-h(s)))$$

$$-f(s, y_2(s), y_2(s-h(s)))\} \, ds. \qquad (3.6.23)$$

From (3.6.23), (3.6.19), (3.6.18) we have

$$|y_1(t) - y_2(t)| \le \int_{t_0}^{t} Na(s) |y_1(s) - y_2(s)| \, ds$$

$$+ \int_{t_0}^{t} Nb(s) |y_1(s-h(s)) - y_2(s-h(s))| \, ds. \qquad (3.6.24)$$

Making the change of variable in the second integral on the right side in (3.6.24) we get

$$|y_1(t) - y_2(t)| \le \int_{t_0}^{t} \bar{a}(s) |y_1(s) - y_2(s)| ds$$

$$+ \int_{t_0}^{\phi(t)} \bar{b}(\sigma) |y_1(\sigma) - y_2(\sigma)| d\sigma. \qquad (3.6.25)$$

A suitable application of the inequality in Theorem 3.2.1, part (a_1) to (3.6.25) yields $|y_1(t) - y_2(t)| \le 0$ .Therefore $y_1(t) = y_2(t)$ i.e., there is at most one solution of IVP (3.6.12)-(3.6.13).

Next, we apply the inequality given in Theorem 3.3.5, part (d_1) to study certain properties of solutions of the retarded integrodifferential equation

$$x'(t) = F\left(t, x(t-h(t)), \int_{t_0}^{t} f(t, \sigma, x(\sigma - h(\sigma)))\right) d\sigma, \qquad (3.6.26)$$

with the given initial condition

$$x(t_0) = x_0, \qquad (3.6.27)$$

for $t \in I$, where $f \in C(I^2 \times R, R), F \in C(I \times R^2, R), x_0$ is a real constant and $h(t) \in C(I, R_+)$ be nonincreasing with $h(t) \le t$ on I, $t - h(t) \in C^1(I, I)$, $h'(t) < 1, h(t_0) = 0$, see [47].

Theorem 3.6.4. (*i*) Suppose that

$$|f(t,s,x)| \leq b(t,s)|x|, \tag{3.6.28}$$

$$|F(t,z,w)| \leq a(t)|z| + |w|, \tag{3.6.29}$$

where $a(t), b(t,s)$ are as given in Theorem 3.3.5 and let

$$M = \max_{t \in I} \frac{1}{1-h'(t)}. \tag{3.6.30}$$

If $x(t)$ is any solution of (3.6.26)-(3.6.27), then

$$|x(t)| \leq |x_0| \exp\left(\int_{t_0}^{t-h(t)} \left[Ma(s+h(\eta)) + \int_{t_0}^{s} M^2 b(s+h(\eta), \sigma+h(\tau))\, d\sigma \right] ds \right), \tag{3.6.31}$$

for $t, \eta, \tau \in I$.

(*ii*) Suppose that the functions f, F in (3.6.26) satisfy the conditions

$$|f(t,s,x) - f(t,s,y)| \leq b(t,s)|x-y|, \tag{3.6.32}$$

$$|F(t,x,\bar{x}) - F(t,y,\bar{y})| \leq a(t)|x-y| + |\bar{x}-\bar{y}|, \tag{3.6.33}$$

where $a(t), b(t,s)$ are as given in Theorem 3.3.5 and let M be given by (3.6.30). Then the problem (3.6.26)-(3.6.27) has at most one solution on I.

Proof. (*i*) The solution $x(t)$ of (3.6.26)-(3.6.27) can be written as

$$x(t) = x_0 + \int_{t_0}^{t} F\left(s, x(s-h(s)), \int_{t_0}^{s} f(s,\sigma,x(\sigma-h(\sigma)))\, d\sigma \right) ds. \tag{3.6.34}$$

Using (3.6.28)-(3.6.30) in (3.6.34) and making the change of variables we have

$$|x(t)| \leq |x_0| + \int_{t_0}^{t-h(t)} \left[Ma(s+h(\eta))|x(s)| + \int_{t_0}^{s} M^2 b(s+h(\eta), \sigma+h(\tau))|x(\sigma)| d\sigma \right] ds, \tag{3.6.35}$$

for $t, \eta, \tau \in I$. Now a suitable application of the inequality given in Theorem 3.3.5,part (d_1) to (3.6.35) yields the required estimate in (3.6.31).

(ii) Let $x(t)$ and $\bar{x}(t)$ be two solutions of (3.6.26)-(3.6.27) on I, then we have

$$x(t) - \bar{x}(t) = \int_{t_0}^{t} \left\{ F\left(s, x(s-h(s)), \int_{t_0}^{s} f(s,\sigma,x(\sigma - h(\sigma)))\,d\sigma\right)\right.$$

$$\left. -F\left(s, \bar{x}(s-h(s)), \int_{t_0}^{s} f(s,\sigma,\bar{x}(\sigma - h(\sigma)))\,d\sigma\right)\right\} ds. \qquad (3.6.36)$$

Using (3.6.32), (3.6.33) in (3.6.36) and making the change of variables, we have

$$|x(t) - \bar{x}(t)| \le \int_{t_0}^{t-h(t)} \left[Ma(s+h(\eta))\,|x(s) - \bar{x}(s)| \right.$$

$$\left. + \int_{t_0}^{s} M^2 b(s+h(\eta), \sigma + h(\tau))\,|x(\sigma) - \bar{x}(\sigma)| d\sigma \right] ds, \qquad (3.6.37)$$

for $t, \eta, \tau \in I$. A suitable application of the inequality given in Theorem 3.3.5, part (d_1) to (3.6.37) yields $|x(t) - \bar{x}(t)| \le 0$. Therefore $x(t) = \bar{x}(t)$ i.e., there is at most one solution of (3.6.26)-(3.6.27).

3.6.3 Retarded partial differential equations in two variables

Consider the following retarded non-self-adjoint hyperbolic partial differential equation

$$z_{xy}(x,y) = \frac{\partial}{\partial y}(a(x,y)\,z(x,y))$$

$$+f(x,y,z(x-h_1(x), y-g_1(y)), \ldots, z(x-h_n(x), y-g_n(y))), \qquad (3.6.38)$$

with the given initial boundary conditions

$$z(x,y_0) = a_1(x), z(x_0,y) = a_2(y), a_1(x_0) = a_2(y_0) = 0, \qquad (3.6.39)$$

where $a \in C(\Delta, R)$ is differentiable with respect to y , $f \in C(\Delta \times R^n, R)$, $a_1 \in C^1(I_1, R)$, $a_2 \in C^1(I_2, R)$,; $h_i \in C(I_1, R_+)$, $g_i \in C(I_2, R_+)$ are nonincreasing and such that $x - h_i(x) \ge 0, x - h_i(x) \in C^1(I_1, I_1)$, $y - g_i(y) \ge 0, y - h_i(y) \in C^1(I_2, I_2)$, $h_i'(x) < 1, g_i'(y) < 1, h_i(x_0) = g_i(y_0) = 0$ for $i = 1, \ldots, n$; $x \in I_1, y \in I_2$. Let

$$M_i = \max_{x \in I_1} \frac{1}{1 - h_i'(x)}, N_i = \max_{y \in I_2} \frac{1}{1 - g_i'(y)}, \qquad (3.6.40)$$

for $i = 1, \ldots, n$.

The following theorem deals with the estimate and uniqueness of solutions of (3.6.38)-(3.6.39), see [59].

Theorem 3.6.5. (i) Suppose that

$$|f(x, y, u_1, ..., u_n)| \leq \sum_{i=1}^{n} b_i(x, y) |u_i|, \tag{3.6.41}$$

$$|e(x, y)| \leq k, \tag{3.6.42}$$

where $b_i(x, y)$, k are as in Theorem 3.5.3 and

$$e(x, y) = a_1(x) + a_2(y) - \int_{x_0}^{x} a(s, y_o) a_1(s) \, ds. \tag{3.6.43}$$

If $z(x, y)$ is any solution of (3.6.38)-(3.6.39), then

$$|z(x, y)| \leq k\bar{q}(x, y) \exp\left(\sum_{i=1}^{n} \int_{\phi_i(x_0)}^{\phi_i(x)} \int_{\psi_i(y_0)}^{\psi_i(y)} \bar{b}_i(\sigma, \tau) \bar{q}_i(\sigma, \tau) \, d\tau d\sigma \right), \tag{3.6.44}$$

for $(x, y) \in \Delta$,where $\phi_i(x) = x - h_i(x), x \in I_1$, $\psi_i(y) = y - g_i(y), y \in I_2$, $\bar{b}_i(\sigma, \tau) = M_i N_i b_i(\sigma + h_i(s), \tau + g_i(t))$ for $\sigma, s \in I_1$, $\tau, t \in I_2$ and

$$\bar{q}(x, y) = \exp\left(\int_{x_0}^{x} |a(\xi, y)| \, d\xi \right), \tag{3.6.45}$$

for $(x, y) \in \Delta$ and M_i, N_i are given by (3.6.40).

(ii) Suppose that the function f in (3.6.38) satisfies the condition

$$|f(x, y, u_1, ..., u_n) - f(x, y, v_1, ..., v_n)| \leq \sum_{i=1}^{n} b_i(x, y) |u_i - v_i|, \tag{3.6.46}$$

where $b_i(x, y)$ are as in Theorem 3.5.3. Let $\phi_i, \psi_i, \bar{b}_i, M_i, N_i$ be as in part (i). Then the problem (3.6.38)-(3.6.39) has at most one solution on Δ.

Proof. (i) It is easy to see that the solution $z(x, y)$ of the problem (3.6.38)-(3.6.39) satisfies the equivalent integral equation

$$z(x, y) = e(x, y) + \int_{x_0}^{x} a(s, y) z(s, y) \, ds$$

$$+\int_{x_0}^{x}\int_{y_0}^{y} f\left(s,t,z\left(s-h_1\left(s\right),t-g_1\left(t\right)\right),...,z\left(s-h_n\left(s\right),t-g_n\left(t\right)\right)\right)dtds, \tag{3.6.47}$$

where $e(x,y)$ is given by (3.6.43). From (3.6.47), (3.6.41), (3.6.42), (3.6.40) and making the change of variables we have

$$|z\left(x,y\right)| \leq k+\int_{x_0}^{x}|a\left(s,y\right)|\,|z\left(s,y\right)|\,ds$$

$$+\int_{x_0}^{x}\int_{y_0}^{y}\sum_{i=1}^{n} b_i\left(s,t\right)|z\left(s-h_i\left(s\right),t-g_i\left(t\right)\right)|\,dtds$$

$$\leq k+\int_{x_0}^{x}|a\left(s,y\right)|\,|z\left(s,y\right)|\,ds$$

$$+\sum_{i=1}^{n}\int_{\phi_i(x_0)}^{\phi_i(x)}\int_{\psi(y_0)}^{\psi(y)}\bar{b}_i\left(\sigma,\tau\right)|z\left(\sigma,\tau\right)|\,d\tau d\sigma. \tag{3.6.48}$$

Now a suitable application of the inequality given in Theorem 3.5.3, part (b_1) to (3.6.48) yields (3.6.44).

(ii) Let $u(x,y)$ and $v(x,y)$ be two solutions of the problem (3.6.38)-(3.6.39) on Δ, then

$$u\left(x,y\right)-v\left(x,y\right)=\int_{x_0}^{x}a\left(s,y\right)\left\{u\left(s,y\right)-v\left(s,y\right)\right\}ds$$

$$+\int_{x_0}^{x}\int_{y_0}^{y}\left\{f\left(s,t,u\left(s-h_1\left(s\right),t-g_1\left(t\right)\right),...,u\left(s-h_n\left(s\right),t-g_n\left(t\right)\right)\right)\right.$$

$$\left.-f\left(s,t,v\left(s-h_1\left(s\right),t-g_1\left(t\right)\right),...,v\left(s-h_n\left(s\right),t-g_n\left(t\right)\right)\right)\right\}dtds. \tag{3.6.49}$$

From (3.6.49), (3.6.46), making the change of variables and in view of (3.6.40) we have

$$|u\left(x,y\right)-v\left(x,y\right)| \leq \int_{x_0}^{x}|a\left(s,y\right)|\,|u\left(s,y\right)-v\left(s,y\right)|\,ds$$

$$+\int_{x_0}^{x}\int_{y_0}^{y}\sum_{i=1}^{n} b_i\left(s,t\right)|u\left(s-h_i\left(s\right),t-g_i\left(t\right)\right)-v\left(s-h_i\left(s\right),t-g_i\left(t\right)\right)|dtds$$

$$\leq \int_{x_0}^{x} |a(s,y)|\,|u(s,y) - v(s,y)|\,ds$$

$$+\sum_{i=1}^{n} \int_{\phi_i(x_0)}^{\phi_i(x)} \int_{\psi_i(y_0)}^{\psi_i(y)} \bar{b}_i(\sigma,\tau)\,|u(\sigma,\tau) - v(\sigma,\tau)|d\tau d\sigma. \quad (3.6.50)$$

A suitable application of the inequality given in Theorem 3.5.3, part (b_1) to (3.6.50) yields $|u(x,y) - v(x,y)| \leq 0$. Therefore $u(x,y) = v(x,y)$ i.e., there is at most one solution of the problem (3.6.38)-(3.6.39) on Δ.

Next, as an application of Theorem 3.5.5, part (d_1) we obtain the explict bound on the solution of retarded partial differential equation of the form

$$\frac{\partial}{\partial y}\left(z^{p-1}(x,y)\frac{\partial}{\partial x}z(x,y)\right)$$

$$= f(x,y,z(x-h_1(x),y-g_1(y)),...,z(x-h_n(x),y-g_n(y))), \quad (3.5.51)$$

with the given initial boundary conditions (3.6.39), where $p > 1$ is a constant and the functions involved in the problem (3.6.51)-(3.6.39) are as given in the problem (3.6.38)-(3.6.39), see [60].

Theorem 3.6.6. Suppose that

$$|f(x,y,u_1,...,u_n)| \leq \sum_{i=1}^{n} b_i(x,y)\,|u_i|, \quad (3.6.52)$$

$$|a_1^p(x)| + |a_2^p(y)| \leq c, \quad (3.6.53)$$

where $b_i(x,y)$ and c are as in Theorem 3.5.5. Let M_i, N_i for $i = 1,...,n$ be as in (3.6.40). If $z(x,y)$ is any solution of the problem (3.6.51)-(3.6.39), then

$$|z(x,y)| \leq \left\{\{c\}^{\frac{p-1}{p}} + (p-1)\sum_{i=1}^{n} \int_{\phi_i(x_0)}^{\phi_i(x)} \int_{\psi_i(y_0)}^{\psi_i(y)} \bar{b}_i(\sigma,\tau)\,d\tau d\sigma\right\}^{\frac{1}{p-1}}, \quad (3.6.54)$$

for $(x,y) \in \Delta$,where $\phi_i(x), \psi_i(y), \bar{b}_i(\sigma,\tau)$ be as in Theorem 3.6.5, part (i).

Proof. It is easy to see that the solution $z(x,y)$ of the problem (3.6.51)-(3.6.39) satisfies the equivalent integral equation

$$z^p(x,y) = a_1^p(x) + a_2^p(y)$$

$$z^p(x,y) = a_1^p(x) + a_2^p(y) + p\int_{x_0}^{x}\int_{y_0}^{y}$$

$$\times f\left(s,t,z\left(xs-h_1\left(s\right),t-g_1\left(t\right)\right),...,z\left(s-h_n\left(s\right),t-g_n\left(t\right)\right)\right)dtds. \quad (3.6.55)$$

From (3.6.55), (3.6.52), (3.6.53), (3.6.40) and making the change of variables we have

$$|z\left(x,y\right)|^p \le c+p\int_{x_0}^{x}\int_{y_0}^{y}\sum_{i=1}^{n} b_i\left(s,t\right)|z\left(s-h_i\left(s\right),t-g_i\left(t\right)\right)|\,dtds$$

$$\le c+p\sum_{i=1}^{n}\int_{\phi_i(x_0)}^{\phi_i(x)}\int_{\psi_i(y_0)}^{\psi_i(y)}\bar{b}_i\left(\sigma,\tau\right)|z\left(\sigma,\tau\right)|d\tau ds. \quad (3.6.56)$$

Now a suitable application of the inequality given in Theorem 3.3.5, part (d_1) to (3.6.56) yields (3.6.54).

3.6.4 Retarded Volterra-Fredholm integral equation in two variables

In this section, we present applications of Theorem 3.4.5 given in [58] to study certain properties of solutions of the retarded Volterra-Fredholm integral equation in two independent variables of the form

$$z\left(x,y\right)=f\left(x,y\right)+\int_{x_0}^{x}\int_{y_0}^{y}A\left(x,y,s,t,z\left(s-h_1\left(s\right),t-h_2\left(t\right)\right)\right)dtds$$

$$+\int_{x_0}^{M}\int_{y_0}^{N}B\left(x,y,s,t,z\left(s-h_1\left(s\right),t-h_2\left(t\right)\right)\right)dtds, \quad (3.6.57)$$

for $(x,y)\in\Delta$, where $f\in C\left(\Delta,R\right), A,B\in C\left(D\times R,R\right)$ and $h_1\in C\left(I_1,R_+\right), h_2 \in C\left(I_2,R_+\right)$ are nonincreasing, $x-h_1\left(x\right)\ge 0, x\in I_1; y-h_2\left(y\right)\ge 0, y\in I_2$; $x-h_1\left(x\right)\in C^1\left(I_1,I_1\right), y-h_2\left(y\right)\in C^1\left(I_2,I_2\right), h_1'\left(x\right)<1, h_2'\left(y\right)<1, h_1\left(x_0\right)= h_2\left(y_0\right)=0$, in which $I_1=\left[x_0,M\right], I_2=\left[y_0,N\right]$ are the given subsets of R, $\Delta=I_1\times I_2$ and $D=\left\{(x,y,s,t)\in\Delta^2: x_0\le s\le x\le M, y_0\le t\le y\le N\right\}$.

Theorem 3.6.7. (i) Suppose that the functions f,A,B in equation (3.6.57) satisfy the conditions

$$|f\left(x,y\right)|\le c, \quad (3.6.58)$$

$$|A\left(x,y,s,t,z\right)|\le a\left(x,y,s,t\right)|z|, \quad (3.6.59)$$

$$|B\left(x,y,s,t,z\right)|\le b\left(x,y,s,t\right)|z|, \quad (3.6.60)$$

where $c, a(x,y,s,t), b(x,y,s,t)$ are as in Theorem 3.4.5. Let

$$M_1 = \max_{x \in I_1} \frac{1}{1-h_1'(x)}, M_2 = \max_{y \in I_2} \frac{1}{1-h_1'(y)}, \tag{3.6.61}$$

and

$$\bar{p}(x,y) = \int_{\phi(x_0)}^{\phi(M)} \int_{\psi(y_0)}^{\psi(N)} \bar{b}(x,y,s,t) \exp\left(\int_{\phi(x_0)}^{\phi(s)} \int_{\psi(y_0)}^{\psi(t)} \bar{a}(s,t,\sigma,\tau)\, d\tau d\sigma \right)$$

$$\times dtds < 1, \tag{3.6.62}$$

where $\phi(x) = x - h_1(x), x \in I_1, \psi(y) = y - h_2(y), y \in I_2$ and

$$\bar{a}(x,y,\sigma,\tau) = M_1 M_2 a(x,y,\sigma + h_1(s), \tau + h_2(t)),$$

$$\bar{b}(x,y,\sigma,\tau) = M_1 M_2 b(x,y,\sigma + h_1(s), \tau + h_2(t)).$$

If $z(x,y)$ is a solution of equation (3.6.57) on Δ, then

$$|z(x,y)| \le \frac{c}{1-\bar{p}(x,y)} \exp\left(\int_{\phi(x_0)}^{\phi(x)} \int_{\psi(y_0)}^{\psi(y)} \bar{a}(x,y,\sigma,\tau)\, d\tau d\sigma \right), \tag{3.6.63}$$

for $(x,y) \in \Delta$.

(ii) Suppose that the functions A, B in equation (3.6.57) satisfy the conditions

$$|A(x,y,s,t,z) - A(x,y,s,t,\bar{z})| \le a(x,y,s,t)|z - \bar{z}|, \tag{3.6.64}$$

$$|B(x,y,s,t,z) - B(x,y,s,t,\bar{z})| \le b(x,y,s,t)|z - \bar{z}|, \tag{3.6.65}$$

where $a(x,y,s,t), b(x,y,s,t)$ are as in Theorem 3.4.5. Let $M_1, M_2, \phi, \psi, \bar{a}, \bar{b}, \bar{p}$ be as in part (i). Then the equation (3.6.47) has at most one solution on Δ.

Proof. Since $z(x,y)$ is a solution of equation (3.6.57), from (3.6.57)-(3.6.60) we have

$$|z(x,y)| \le c + \int_{x_0}^{x} \int_{y_0}^{y} a(x,y,s,t)|z(s - h_1(s), t - h_2(t))| dtds$$

$$+ \int_{x_0}^{M} \int_{y_0}^{N} b(x,y,s,t)|z(s - h_1(s), t - h_2(t))|\, dtds. \tag{3.6.66}$$

Now by making the change of variables on the right hand side of (3.6.66) and using (3.6.61) we have

$$|z(x,y)| \leq c + \int_{\phi(x_0)}^{\phi(x)} \int_{\psi(y_0)}^{\psi(y)} \bar{a}(x,y,s,t)\,|z(\sigma,\tau)|d\tau d\sigma$$

$$+ \int_{\phi(x_0)}^{\phi(M)} \int_{\psi(y_0)}^{\psi(N)} \bar{b}(x,y,s,t)\,|z(\sigma,\tau)|\,d\tau d\sigma. \tag{3.6.67}$$

A suitable application of Theorem 3.4.5 to (3.6.67) yields (3.6.63).

(*ii*) Let $z(x,y)$ and $\bar{z}(x,y)$ be two solutions of equation (3.6.57) on Δ. From (3.6.57), (3.6.64), (3.6.65) we have

$$|z(x,y) - \bar{z}(x,y)|$$

$$\leq \int_{x_0}^{x} \int_{y_0}^{y} a(x,y,s,t)\,|z(s-h_1(s), t-h_2(t))$$

$$-\bar{z}(s-h_1(s), t-h_2(t))|\,dtds$$

$$+ \int_{x_0}^{M} \int_{y_0}^{N} b(x,y,s,t)\,|z(s-h_1(s), t-h_2(t))$$

$$-\bar{z}(s-h_1(s), t-h_2(t))|\,dtds. \tag{3.6.68}$$

By making the change of variables on the right hand side of (3.6.68) and using (3.6.61) we have

$$|z(x,y) - \bar{z}(x,y)| \leq \int_{\phi(x_0)}^{\phi(x)} \int_{\psi(y_0)}^{\psi(y)} \bar{a}(x,y,s,t)\,|z(\sigma,\tau) - \bar{z}(\sigma,\tau)|d\tau d\sigma$$

$$+ \int_{\phi(x_0)}^{\phi(M)} \int_{\psi(y_0)}^{\psi(N)} (x,y,s,t)\,|z(\sigma,\tau) - \bar{z}(\sigma,\tau)|d\tau d\sigma. \tag{3.6.69}$$

Now a suitable application of Theorem 3.4.5 to (3.6.69) yields $|z(x,y) - \bar{z}(x,y)| \leq 0$.Therefore $z(x,y) = \bar{z}(x,y)$ i.e., there is at most one solution to the equation (3.6.57).

We next consider the following retarded Volterra-Fredholm integral equations

$$z(x,y) = f(x,y) + \int_{x_0}^{x}\int_{y_0}^{y} A(x,y,s,t,z(s-h_1(s),t-h_2(t)),\mu)dtds$$

$$+\int_{x_0}^{M}\int_{y_0}^{N} B(x,y,s,t,z(s-h_1(s),t-h_2(t)),\mu)dtds, \tag{3.6.70}$$

$$z(x,y) = f(x,y) + \int_{x_0}^{x}\int_{y_0}^{y} A(x,y,s,t,z(s-h_1(s),t-h_2(t)),\mu_0)dtds$$

$$+\int_{x_0}^{M}\int_{y_0}^{N} B(x,y,s,t,z(s-h_1(s),t-h_2(t)),\mu_0)dtds, \tag{3.6.71}$$

for $(x,y) \in \Delta$, where $f \in C(\Delta, R)$, $A, B \in C(D \times R \times R, R)$ and μ, μ_0 are real parameters.

The following theorem shows the dependency of solutions of equations (3.6.70) and (3.6.71) on parameters.

Theorem 3.6.8 . Suppose that

$$|A(x,y,s,t,z,\mu) - A(x,y,s,t,\bar{z},\mu)| \le a(x,y,s,t)|z-\bar{z}|, \tag{3.6.72}$$

$$|A(x,y,s,t,z,\mu) - A(x,y,s,t,z,\mu_0)| \le r(x,y,s,t)|\mu-\mu_0|, \tag{3.6.73}$$

$$|B(x,y,s,t,z,\mu) - B(x,y,s,t,\bar{z},\mu)| \le b(x,y,s,t)|z-\bar{z}|, \tag{3.6.74}$$

$$|B(x,y,s,t,z,\mu) - B(x,y,s,t,z,\mu_0)| \le e(x,y,s,t)|\mu-\mu_0|, \tag{3.6.75}$$

where $a(x,y,s,t), b(x,y,s,t)$ are as in Theorem 3.4.5 and $r, e \in C(D, R_+)$ are such that

$$\int_{x_0}^{x}\int_{y_0}^{y} r(x,y,s,t)\,dtds \le k_1, \tag{3.6.76}$$

$$\int_{x_0}^{M}\int_{y_0}^{N} e(x,y,s,t)\,dtds \le k_2, \tag{3.6.77}$$

for $(x,y) \in \Delta$, where k_1, k_2 are positive constants. Let $M_1, M_2, \phi, \psi, \bar{a}, \bar{b}, \bar{p}$ be as in Theorem 3.6.7, part (i). Let $z_1(x,y)$ and $z_2(x,y)$ for $(x,y) \in \Delta$ be the solutions of (3.6.70) and (3.6.71) respectively. Then

$$|z_1(x,y) - z_2(x,y)| \le \frac{(k_1+k_2)|\mu-\mu_0|}{1-\bar{p}(x,y)}$$

$$\times \exp\left(\int_{\phi(x_0)}^{\phi(x)}\int_{\psi(y_0)}^{\psi(y)} \bar{a}\left(x,y,s,t\right)dtds\right), \tag{3.6.78}$$

for $(x, y) \in \Delta$.

Proof. Let $z(x,y) = z_1(x,y) - z_2(x,y), (x,y) \in \Delta$. Then

$$z(x,y) = \int_{x_0}^{x}\int_{y_0}^{y} \{A(x,y,s,t,z_1(s-h_1(s),t-h_2(t)),\mu)$$

$$-A(x,y,s,t,z_2(s-h_1(s),t-h_2(t)),\mu)\}dtds$$

$$+\int_{x_0}^{x}\int_{y_0}^{y} \{A(x,y,s,t,z_2(s-h_1(s),t-h_2(t)),\mu)$$

$$-A(x,y,s,t,z_2(s-h_1(s),t-h_2(t)),\mu_0)\}dtds$$

$$+\int_{x_0}^{M}\int_{y_0}^{N} \{B(x,y,s,t,z_1(s-h_2(s),t-h_2(t)),\mu)$$

$$-B(x,y,s,t,z_2(s-h_1(s),t-h_2(t)),\mu)\}dtds$$

$$+\int_{x_0}^{M}\int_{y_0}^{N} \{B(x,y,s,t,z_2(s-h_2(s),t-h_2(t)),\mu)$$

$$-B(x,y,s,t,z_2(s-h_1(s),t-h_2(t)),\mu_0)\}dtds. \tag{3.6.79}$$

Using (3.6.72)-(3.6.77) in (3.6.79) we get

$$|z(x,y)| \le |\mu-\mu_0|k_1 + |\mu-\mu_0|k_2$$

$$+\int_{x_0}^{x}\int_{y_0}^{y} a(x,y,s,t)|z(s-h_1(s),t-h_2(t))|dtds$$

$$+\int_{x_0}^{M}\int_{y_0}^{N} b(x,y,s,t)|z(s-h_1(s),t-h_2(t))|dtds. \tag{3.6.80}$$

By making the change of variables on the right hand side of (3.6.80) and using (3.6.61) we get

$$|z(x,y)| \le (k_1+k_2)|\mu-\mu_0| + \int_{\phi(x_0)}^{\phi(x)}\int_{\psi(y_0)}^{\psi(y)} \bar{a}(x,y,\sigma,\tau)|z(\sigma,\tau)|d\tau d\sigma$$

$$+\int_{\phi(x_0)}^{\phi(M)}\int_{\psi(y_0)}^{\psi(N)} \bar{b}(x,y,\sigma,\tau)\,|z(\sigma,\tau)|d\tau d\sigma. \tag{3.6.81}$$

Now a suitable application of Theorem 3.4.5 to (3.6.81) yields (3.6.78) which shows the dependency of solutions of (3.6.70) and (3.6.71) on parameters.

In conclusion, we note that the inequalities given in earlier sections are recently established and still admit various generalizations and extensions in different directions. Here we have given some basic and immediate applications of few of the inequalities which will encourage to widen the scope of their applications.

3.7 Notes

The search for more efficient methods to study certain retarded differential and integral equations has recently led to the discoveries of some basic retarded integral inequalities with explicit estimates. The main advantage of obtaining such results lies in the fact that they can serve as effective tools when the earlier results do not apply directly and certainly a good source to further development.Sections 3.2 and 3.3 are devoted to some basic retarded integral inequalities in one independent variable, recently investigated and used by Pachpatte in [43,58,60,61,64,69,77]. For some earlier results on such inequalities, we refer the reader to the book by Bainov and Simeonov [3, pp. 142-145]. The results given in sections 3.4 and 3.5 deals with a number of new retarded integral inequalities involving functions of two independent variables, recently investigated by Pachpatte in [43,47,58,59, 60,69,74,77]. Section 3.6 contains applications of some of the inequalities given in earlier sections.

Chapter 4

Finite difference inequalities in one variable

4.1 Introduction

The theory of finite difference equations has gained increasing significance in the last decades as is apparent from the large number of publications on the subject. A great variety of methods and tools are available for handling such equations. In the study of many finite difference and sum-difference equations, one often needs some new and specific type of finite difference inequalities for proving various theorems or approximating functions. The desire to widen the scope of applications of such inequalities resulted in the necessity of discovering new finite difference inequalities which are directly applicable in the new situations. In this chapter, we offer various basic finite difference inequalities recently investigated in [35,37,39,44,45,53,55,57,67,68,70,73,75] which can be used as powerful tools in certain applications. Some fundamental applications are given to illustrate the usefulness of certain inequalities.

4.2 Fundamental finite difference inequalities

In this section, we focus our attention on some basic inequalities established by Pachpatte in [57] (see also [42]) which provide explicit bounds on unknown functions and can be used as an effective tool in the development of the theory of finite difference equations and numerical analysis.

We start with the following theorems which deals with the finite difference inequalities proved in [57].

Theorem 4.2.1. Let $u(n), a(n), b(n), c(n), p(n) \in D(N_0, R_+)$ and $\Delta c(n) \geq 0$ for $n \in N_0$., If

$$u(n) \leq a(n) + b(n)\left(c(n) + \sum_{s=0}^{n-1} p(s)\,u(s)\right), \tag{4.2.1}$$

for $n \in N_0$, then

$$u(n) \leq a(n) + b(n)\left(c(0)\prod_{s=0}^{n-1}[1 + b(s)\,p(s)]\right.$$

$$\left. + \sum_{s=0}^{n-1}[\Delta c(s) + a(s)\,p(s)]\prod_{\sigma=s+1}^{n-1}[1 + b(\sigma)\,p(\sigma)]\right), \tag{4.2.2}$$

for $n \in N_0$.

Proof. Define a function $z(n)$ by

$$z(n) = c(n) + \sum_{s=0}^{n-1} p(s)\,u(s). \tag{4.2.3}$$

Then $z(0) = c(0)$ and (4.2.1) can be restated as

$$u(n) \leq a(n) + b(n)\,z(n). \tag{4.2.4}$$

From (4.2.3) and (4.2.4) we observe that

$$\Delta z(n) = \Delta c(n) + p(n)\,u(n)$$

$$\leq b(n)\,p(n)\,z(n) + [\Delta c(n) + a(n)\,p(n)]. \tag{4.2.5}$$

Now by applying Theorem 1.2.1 given in [42, p. 11] to (4.2.5) we get

$$z(n) \leq c(0)\prod_{s=0}^{n-1}[1 + b(\sigma)\,p(\sigma)]$$

$$+ \sum_{s=0}^{n-1}[\Delta c(s) + a(s)\,p(s)]\prod_{\sigma=s+1}^{n-1}[1 + b(\sigma)\,p(\sigma)]. \tag{4.2.6}$$

Using (4.2.6) in (4.2.4) we get the required inequality in (4.2.2).

Remark 4.2.1. We note that in the special case when $c(n) = 0$, the inequality given in Theorem 4.2.1 reduces to the inequality given by Pachpatte, see [42, Theorem 1.2.3, p. 13].

Theorem 4.2.2. Let $u(n), a(n), b(n), c(n), \Delta c(n)$ be as in Theorem 4.2.1.

(a_1) Let $L : N_0 \times R_+ \to R_+$ be a function such that

$$0 \leq L(n,x) - L(n,y) \leq M(n,y)(x-y), \tag{4.2.7}$$

for $n \in N_0$, $x \geq y \geq 0$, where $M(n,y)$ is a real-valued nonnegative function defined for $n \in N_0$, $y \in R_+$. If

$$u(n) \leq a(n) + b(n)\left(c(n) + \sum_{s=0}^{n-1} L(s,u(s))\right), \tag{4.2.8}$$

for $n \in N_0$, then

$$u(n) \leq a(n) + b(n)\left(c(0)\prod_{s=0}^{n-1}[1 + M(s,a(s))b(s)]\right.$$

$$\left. + \sum_{s=0}^{n-1}[\Delta c(s) + L(s,a(s))]\prod_{\sigma=s+1}^{n-1}[1 + M(\sigma,a(\sigma))b(\sigma)]\right), \tag{4.2.9}$$

for $n \in N_0$.

(a_2) Let $L : N_0 \times R_+ \to R_+$ be a function which satisfies the condition

$$0 \leq L(n,x) - L(n,y) \leq M(n,y)\phi^{-1}(x-y), \tag{4.2.10}$$

for $n \in N_0, x \geq y \geq 0$, where $M(n,y)$ is as in (a_1), $\phi : R_+ \to R_+$ is a continuous and strictly increasing function with $\phi(0) = 0$, ϕ^{-1} is the inverse function of ϕ and

$$\phi^{-1}(xy) \leq \phi^{-1}(x)\phi^{-1}(y), \tag{4.2.11}$$

for $x, y \in R_+$. If

$$u(n) \leq a(n) + b(n)\phi\left(c(n) + \sum_{s=0}^{n-1} L(s,u(s))\right), \tag{4.2.12}$$

for $n \in N_0$, then

$$u(n) \leq a(n) + b(n)\phi\left(c(0)\prod_{s=0}^{n-1}\left[1 + M(s,a(s))\phi^{-1}(b(s))\right]\right.$$

$$\left. + \sum_{s=0}^{n-1}[\Delta c(s) + L(s,a(s))]\prod_{\sigma=s+1}^{n-1}\left[1 + M(\sigma,a(\sigma))\phi^{-1}(b(\sigma))\right]\right), \tag{4.2.13}$$

for $n \in N_0$.

Proof. (a_1) Define a function $z(n)$ by

$$z(n) = c(n) + \sum_{s=0}^{n-1} L(s, u(s)). \tag{4.2.14}$$

Then $z(0) = c(0)$ and (4.2.8) can be restated as

$$u(n) \leq a(n) + b(n) z(n). \tag{4.2.15}$$

From (4.2.14), (4.2.15) and (4.2.7) we have

$$\begin{aligned} \Delta z(n) &= \Delta c(n) + L(n, u(n)) \\ &\leq \Delta c(n) + L(n, a(n) + b(n) z(n)) - L(n, a(n)) + L(n, a(n)) \\ &\leq M(n, a(n)) b(n) z(n) + [\Delta c(n) + L(n, a(n))]. \end{aligned} \tag{4.3.16}$$

Now by applying Theorem 1.2.1 given in [42] to (4.2.16) we get

$$z(n) \leq c(0) \prod_{s=0}^{n-1} [1 + M(s, a(s)) b(s)]$$

$$+ \sum_{s=0}^{n-1} [\Delta c(s) + L(s, a(s))] \prod_{\sigma=s+1}^{n-1} [1 + M(\sigma, a(\sigma)) b(\sigma)]. \tag{4.2.17}$$

Using (4.2.17) in (4.2.15) we get the desired inequality in (4.2.9).

(a_2) Define a function $z(n)$ by (4.2.14). Then $z(0) = c(0)$ and (4.2.12) can be restated as

$$u(n) \leq a(n) + b(n) \phi(z(n)). \tag{4.2.18}$$

From (4.2.14), (4.2.18), (4.2.10) and (4.2.11) we have

$$\begin{aligned} \Delta z(n) &= \Delta c(n) + L(n, u(n)) \\ &\leq \Delta c(n) + L(n, a(n) + b(n) \phi(z(n))) - L(n, a(n)) + L(n, a(n)) \\ &\leq M(n, a(n)) \phi^{-1}(b(n) \phi(z(n))) + [\Delta c(n) + L(n, a(n))] \\ &\leq M(n, a(n)) \phi^{-1}(b(n)) z(n) + [\Delta c(n) + L(n, a(n))]. \end{aligned} \tag{4.2.19}$$

Now an application of Theorem 1.2.1 given in [42] to (4.2.19) yields

$$z(n) \leq c(0) \prod_{s=0}^{n-1} [1 + M(s, a(s)) \phi^{-1}(b(s))]$$

$$+ \sum_{s=0}^{n-1} [\Delta c(s) + L(s, a(s))] \prod_{\sigma=s+1}^{n-1} [1 + M(\sigma, a(\sigma)) \phi^{-1}(b(\sigma))]. \tag{4.2.20}$$

Using (4.2.20) in (4.2.18) we get (4.2.13).

Remark 4.2.2. If we take $c(n) = 0$ in Theorem 4.2.2, part (a_1), then we recapture the inequality given by Dragomir in [11] (see also [10]). We note that from Theorem 4.2.2, part (a_1), one can easily obtain the corollaries similar to that of given in [10] (see also [11]) which can also be used in certain applications.

In the following theorems we present some useful generalizations of the inequalities given in [42, Theorem 2.3.1, Corollary 3.3.1].

Theorem 4.2.3. Let $u(n), a(n), b(n), c(n), p(n) \in D(N_0, R_+)$.

(b_1) Let $\Delta a(n) \geq 0$ for $n \in N_0; g \in C(R_+, R_+)$ be a nondecreasing function with $g(u) > 0$ for $u > 0$. If

$$u(n) \leq a(n) + \sum_{s=0}^{n-1} p(s) g(u(s)), \tag{4.2.21}$$

for $n \in N_0$, then for $0 \leq n \leq n_1; n, n_1 \in N_0$,

$$u(n) \leq G^{-1}\left[G(a(0)) + \sum_{s=0}^{n-1}\left(\frac{\Delta a(s)}{g(a(s))} + p(s)\right)\right], \tag{4.2.22}$$

where

$$G(r) = \int_{r_0}^{r} \frac{ds}{g(s)}, r > 0, \tag{4.2.23}$$

$r_0 > 0$ is arbitrary and G^{-1} is the inverse of G and $n_1 \in N_0$ is chosen so that

$$G(a(0)) + \sum_{s=0}^{n-1}\left(\frac{\Delta a(s)}{g(a(s))} + p(s)\right) \in Dom\left(G^{-1}\right)$$

for all $n \in N_0$ lying in $0 \leq n \leq n_1$..

(b_2) Let $\Delta c(n) \geq 0$ for $n \in N_0$; g, G, G^{-1} be as in part (b_1) and suppose in addition, $g(u)$ is subadditive and submultiplicative. If

$$u(n) \leq a(n) + b(n)\left(c(n) + \sum_{s=0}^{n-1} p(s) g(u(s))\right), \tag{4.2.24}$$

for $n \in N_0$, then for $0 \leq n \leq n_2; n, n_2 \in N_0$,

$$u(n) \leq a(n) + b(n) G^{-1}\left[G(c(0))\right.$$

$$\left. + \sum_{s=0}^{n-1}\left(\frac{\Delta c(s) + p(s) g(a(s))}{g(c(s))}\right) + p(s) g(b(s))\right], \tag{4.2.25}$$

and $n_2 \in N_0$ is chosen so that

$$G(c(0)) + \sum_{s=0}^{n-1} \left(\frac{\Delta c(s) + p(s) g(a(s))}{g(c(s))} + p(s) g(b(s)) \right) \in Dom\left(G^{-1}\right),$$

for all $n \in N_0$ lying in $0 \leq n \leq n_2$.

Proof. (b_1) Let $a(n) > 0$ for $n \in N_0$ and define a function $z(n)$ by the right hand side of (4.2.21). Then $z(0) = a(0)$, $u(n) \leq z(n)$, $z(n) > 0$ and

$$\Delta z(n) = \Delta a(n) + p(n) g(u(n))$$

$$\leq \Delta a(n) + p(n) g(z(n)). \tag{4.2.26}$$

From (4.2.23), (4.2.26) and the fact that $a(n) \leq z(n)$ we observe that

$$G(z(n+1)) - G(z(n)) = \int_{z(n)}^{z(n+1)} \frac{ds}{g(s)}$$

$$\leq \frac{\Delta z(n)}{g(z(n))}$$

$$\leq \frac{\Delta a(n) + p(n) g(z(n))}{g(z(n))}$$

$$\leq \frac{\Delta a(n)}{g(a(n))} + p(n). \tag{4.2.27}$$

By taking $n = s$ in (4.2.27) and summing up over s from 0 to $n-1$ we get

$$G(z(n)) \leq G(z(0)) + \sum_{s=0}^{n-1} \left(\frac{\Delta a(s)}{g(a(s))} + p(s) \right),$$

which implies

$$z(n) \leq G^{-1} \left[G(a(0)) + \sum_{s=0}^{n-1} \left(\frac{\Delta a(s)}{g(a(s))} + p(s) \right) \right]. \tag{4.2.28}$$

Using (4.2.28) in $u(n) \leq z(n)$ we get the required inequality in (4.2.22). If $a(n) \geq 0$ for $n \in N_0$, we carry out the above procedure with $a(n) + \varepsilon$ instead of $a(n)$, where $\varepsilon > 0$ is an arbitrary small constant, and subsequently pass to the limit as $\varepsilon \to 0$ to obtain (4.2.22). The subdomain $0 \leq n \leq n_1$ is obvious.

(b_2) Let $c(n) > 0$ for $n \in N_0$ and define a function $z(n)$ by

$$z(n) = c(n) + \sum_{s=0}^{n-1} p(s) g(u(s)). \tag{4.2.29}$$

Then $z(0) = c(0)$, $z(n) > 0$ for $n \in N_0$ and (4.2.24) can be restated as

$$u(n) \leq a(n) + b(n) z(n). \tag{4.2.30}$$

From (4.2.29) and (4.2.30) we have

$$\begin{aligned} \Delta z(n) &= \Delta c(n) + p(n) g(u(n)) \\ &\leq \Delta c(n) + p(n) g(a(n) + b(n) z(n)) \\ &\leq (\Delta c(n) + p(n) g(a(n))) + p(n) g(b(n)) g(z(n)). \end{aligned} \tag{4.2.31}$$

From (4.2.23), (4.2.31) the fact that $c(n) \leq z(n)$ and following the proof of (b_1) we obtain

$$z(n) \leq G^{-1}\Bigg[G(c(0)) + \sum_{s=0}^{n-1} \left(\frac{\Delta c(s) + p(s) g(a(s))}{g(c(s))}\right) + p(s) g(b(s))\Bigg], \tag{4.2.32}$$

Using (4.2.32) in (4.2.30) we get (4.2.25). The proof of the case when $c(n) \geq 0$ can be completed as mentioned in the proof of (b_1).The subdomain $0 \leq n \leq n_2$ is obvious.

Remark 4.2.3. If we take $a(n) = k$, a nonnegative constant, then the inequality given in Theorem 4.2.3, part (b_1) reduces to the discrete version of the well known Bihar's inequality, see [42, p. 103]. For a detailed account on such inequalities, see [42] and also [85].

Theorem 4.2.4. Let $u(n), a(n), b(n) \in D(N_0, R_+)$.

(c_1) Let $\Delta a(n) \geq 0$ for $n \in N_0$. If

$$u^2(n) \leq a(n) + 2\sum_{s=0}^{n-1} b(s) u(s), \tag{4.2.33}$$

for $n \in N_0$, then

$$u(n) \leq \sqrt{a(0)} + \sum_{s=0}^{n-1} \left(\frac{\Delta a(s)}{2\sqrt{a(s)}} + b(s)\right), \tag{4.2.34}$$

for $n \in N_0$.

(c_2) Let $h \in C(R_+, R_+)$ be a nondecreasing function with $h(u) > 0$ for $u > 0$. If

$$u^2(n) \le a(n) + 2\sum_{s=0}^{n-1} b(s)\, h(u(s)), \tag{4.2.35}$$

for $n \in N_0$, then for $0 \le n \le n_3; n, n_3 \in N_0$,

$$u(n) \le \left\{ H^{-1}\left[H(a(0)) + \sum_{s=0}^{n-1} \left(\frac{\Delta a(s)}{h\left(\sqrt{a(s)}\right)} + 2b(s) \right) \right] \right\}^{\frac{1}{2}}, \tag{4.2.36}$$

where

$$H(r) = \int_{r_0}^{r} \frac{ds}{h(\sqrt{s})}, r > 0, \tag{4.2.37}$$

$r_0 > 0$ is arbitrary and H^{-1} is the inverse of H and $n_3 \in N_0$ is chosen so that

$$H(a(0)) + \sum_{s=0}^{n-1} \left(\frac{\Delta a(s)}{h\left(\sqrt{a(s)}\right)} + 2b(s) \right) \in Dom\left(H^{-1}\right),$$

for all $n \in N_0$ lying in $0 \le n \le n_3$.

Proof. (c_1) Let $a(n) > 0$ for $n \in N_0$ and define a function $z(n)$ by the right hand side of (4.2.33). Then $z(0) = a(0), u(n) \le \sqrt{z(n)}$ and

$$\Delta z(n) = \Delta a(n) + 2b(n)\, u(n)$$

$$\le \Delta a(n) + 2b(n)\sqrt{z(n)}. \tag{4.2.38}$$

Using the facts that $\sqrt{z(n)} > 0$, $\Delta z(n) \ge 0$, $\sqrt{z(n)} \le \sqrt{z(n+1)}$, $a(n) \le z(n)$ for $n \in N_0$ and (4.2.40) we observe that (see [42, p. 212])

$$\Delta\left(\sqrt{z(n)}\right) = \frac{z(n+1) - z(n)}{\sqrt{z(n+1)} + \sqrt{z(n)}}$$

$$\le \frac{\Delta z(n)}{2\sqrt{z(n)}}$$

$$\le \frac{\Delta a(n) + 2b(n)\sqrt{z(n)}}{2\sqrt{z(n)}}$$

$$\le \frac{\Delta a(n)}{2\sqrt{a(n)}} + b(n),$$

which implies

$$\sqrt{z(n)} \leq \sqrt{a(0)} + \sum_{s=0}^{n-1} \left(\frac{\Delta a(s)}{2\sqrt{a(s)}} + b(s) \right). \tag{4.2.39}$$

Using (4.2.39) in $u(n) \leq \sqrt{z(n)}$ we get the required inequality in (4.2.34). The proof of the case when $a(n) \geq 0$ for $n \in N_0$ can be completed as in the proof of Theorem 4.2.3, part (b_1).

(c_2) Let $a(n) > 0$ for $n \in N_0$ and define a function $z(n)$ by the right hand side of (4.2.35). Then $z(0) = a(0)$, $z(n) > 0$, $u(n) \leq \sqrt{z(n)}$ and

$$\Delta z(n) \leq \Delta a(n) + 2b(n)\, h\left(\sqrt{z(n)}\right). \tag{4.2.40}$$

As in the proof of Theorem 4.2.3, part (b_1), from (4.2.37), (4.2.40) and the fact that $a(n) \leq z(n)$ we observe that

$$\Delta H(z(n)) \leq \frac{\Delta z(n)}{h\left(\sqrt{z(n)}\right)}$$

$$\leq \frac{\Delta a(n)}{h\left(\sqrt{a(n)}\right)} + 2b(n).$$

The rest of the proof can be completed by following the proof of Theorem 4.2.3, part (b_1).We omit the details.

Remark 4.2.4. We note that the inequality given in Theorem 4.2.4, part (c_1) can be considered as a generalization of the inequality in Corollary 3.3.1 given in [42], while the inequality in part (c_2) is a slight variant of the special version of the inequality in Theorem 3.3.5 given in [42].

4.3 Some more finite difference inequalities

Due to various motivations, several new finite difference inequalities which yield explicit estimates on unknown functions have been investigated and used extensively in the literature, see [42]. In this section, we offer some more finite difference inequalities recently established by Pachpatte in [35,45,55,68] which can be used as tool in certain new applications.

Our first theorem deals with the finite difference inequalities proved in [68].

Theorem 4.3.1. Let $u(n), a(n) \in D(N_0, R_+)$; $k(n,\sigma), \Delta_1 k(n,\sigma) \in D(E, R_+)$, where $E = \left\{(m,n) \in N_0^2 : 0 \leq n \leq m < \infty\right\}$.

(a_1) Let $g \in C(R_+, R_+)$ be a nondecreasing function with $g(u) > 0$ for $u > 0$. If

$$u(n) \leq c + \sum_{\sigma=0}^{n-1} k(n,\sigma) g(u(\sigma)), \tag{4.3.1}$$

for $n \in N_0$, where $c \geq 0$ is a real constant, then for $0 \leq n \leq n_1; n, n_1 \in N_0$,

$$u(n) \leq G^{-1}\left[G(c) + \sum_{s=0}^{n-1} H(s)\right], \tag{4.3.2}$$

where

$$H(n) = k(n+1,n) + \sum_{\sigma=0}^{n-1} \Delta_1 k(n,\sigma), \tag{4.3.3}$$

$$G(r) = \int_{r_0}^{r} \frac{ds}{g(s)}, r > 0, \tag{4.3.4}$$

$r_0 > 0$ is arbitrary, G^{-1} is the inverse of G and $n_1 \in N_0$ is chosen so that

$$G(c) + \sum_{s=0}^{n-1} H(s) \in Dom\left(G^{-1}\right)$$

for all $n \in N_0$ lying in $0 \leq n \leq n_1$.

(a_2) Let g, G, G^{-1} be as in (a_1) and suppose in addition $g(u)$ is subadditive. If

$$u(n) \leq a(n) + \sum_{\sigma=0}^{n-1} k(n,\sigma) g(u(\sigma)), \tag{4.3.5}$$

for $n \in N_0$, then for $0 \leq n \leq n_2; n, n_2 \in N_0$,

$$u(n) \leq a(n) + G^{-1}\left[G(B(n)) + \sum_{s=0}^{n-1} H(s)\right], \tag{4.3.6}$$

where $H(n)$ is given by (4.3.3),

$$B(n) = \sum_{\sigma=0}^{n-1} k(n,\sigma) g(a(\sigma)), \tag{4.3.7}$$

for $n \in N_0$ and $n_2 \in N_0$ is chosen so that

$$G(B(n)) + \sum_{s=0}^{n-1} H(s) \in Dom\left(G^{-1}\right),$$

for all $n \in N_0$ lying in $0 \leq n \leq n_2$.

Proof. (a_1) Let $c > 0$ and define a function $z(n)$ by the right hand side of (4.3.1). Then $z(0) = c$, $u(n) \le z(n)$, $z(n) > 0$ and

$$\begin{aligned}\Delta z(n) &= k(n+1,n)\,g(u(n)) + \sum_{\sigma=0}^{n-1} \Delta_1 k(n,\sigma)\,g(u(\sigma)) \\ &\le H(n)\,g(z(n)),\end{aligned}$$

where $H(n)$ is given by (4.3.3), see [42, p. 22]. The rest of the proof can be completed by following the similar arguments as in the proof of Theorem 4.2.3, part (b_1). We omit the details.

(a_2) The proof follows by closely looking at the proof of (a_1) and the proof of Theorem 4.2.3, part (b_2). Here we leave the details to the reader.

The next theorem contains the inequalities established in [55].

Theorem 4.3.2. Let $u(n), k(n,\sigma), \Delta_1 k(n,\sigma)$ and c be as in Theorem 4.3.1.

(b_1) If

$$u^2(n) \le c + \sum_{\sigma=0}^{n-1} k(n,\sigma) u(\sigma), \tag{4.3.8}$$

for $n \in N_0$, then

$$u(n) \le \sqrt{c} + \frac{1}{2}\sum_{s=0}^{n-1} H(s), \tag{4.3.9}$$

for $n \in N_0$, where $H(n)$ is given by (4.3.3).

(b_2) Let $g(u)$ be as in Theorem 4.3.1, part (a_1).If

$$u^2(n) \le c + \sum_{\sigma=0}^{n-1} k(n,\sigma) u(\sigma)\, g(u(\sigma)), \tag{4.3.10}$$

for $n \in N_0$, then for $0 \le n \le n_3; n, n_3 \in N_0$,

$$u(n) \le G^{-1}\left[G(\sqrt{c}) + \frac{1}{2}\sum_{s=0}^{n-1} H(s)\right], \tag{4.3.11}$$

where $H(n)$ is given by (4.3.3), G, G^{-1} are as defined in Theorem 4.3.1, part (a_1) and $n_3 \in N_0$ is chosen so that

$$G(\sqrt{c}) + \frac{1}{2}\sum_{s=0}^{n-1} H(s) \in Dom\,(G^{-1}),$$

for all $n \in N_0$ lying in $0 \le n \le n_3$.

Proof. (b_1) Let $c > 0$ and define a function $z(n)$ by the right hand side of (4.3.8). Then $z(0) = c$, $u(n) \le \sqrt{z(n)}$, $z(n)$ is positive and nondecreasing for $n \in N_0$ and

$$\Delta z(n) = \sum_{\sigma=0}^{n} k(n+1,\sigma)\, u(\sigma) - \sum_{\sigma=0}^{n-1} k(n+1,\sigma)\, u(\sigma)$$

$$+\sum_{\sigma=0}^{n-1} k(n+1,\sigma)\, u(\sigma) - \sum_{\sigma=0}^{n-1} k(n,\sigma)\, u(\sigma)$$

$$= k(n+1,n)\, u(n) + \sum_{\sigma=0}^{n-1} \Delta_1 k(n,\sigma) u(\sigma)$$

$$\le k(n+1,n)\sqrt{z(n)} + \sum_{\sigma=0}^{n-1} \Delta_1 k(n,\sigma)\sqrt{z(\sigma)}$$

$$\le H(n)\sqrt{z(n)}. \tag{4.3.12}$$

The rest of the proof follows by using the similar arguments as in the proof of Theorem 4.2.4, part (c_1) below (4.2.40) with suitable changes. We omit the details.

(b_2) The proof follows by closely looking at the proof of part (b_1) given above and the proof of Theorem 3.3.5 given in [42]. We omit it here to avoid repetition.

The discrete inequalities established in [35,45] are embodied in the following theorems.

Theorem 4.3.3. Let $u(n), a(n), b(n), g(n), h(n) \in D(N_0, R_+)$ and $p > 1$ be a real constant.

(c_1) If

$$u^p(n) \le a(n) + b(n) \sum_{s=0}^{n-1} [g(s)\, u^p(s) + h(s)\, u(s)], \tag{4.3.13}$$

for $n \in N_0$, then

$$u(n) \le \Bigg\{ a(n) + b(n) \sum_{s=0}^{n-1} \left(g(s)\, a(s) + h(s) \left(\frac{p-1}{p} + \frac{a(s)}{p} \right) \right)$$

$$\times \prod_{\sigma=s+1}^{n-1} \left[1 + b(\sigma) \left(g(\sigma) + \frac{h(\sigma)}{p} \right) \right] \Bigg\}^{\frac{1}{p}}, \tag{4.3.14}$$

for $n \in N_0$.

(c_2) Let $c(n)$ be a real-valued positive and nondecreasing function defined on N_0. If

$$u^p(n) \leq c^p(n) + b(n) \sum_{s=0}^{n-1} [g(s) u^p(s) + h(s) u(s)], \tag{4.3.15}$$

for $n \in N_0$, then

$$u(n) \leq c(n) \left\{ 1 + b(n) \sum_{s=0}^{n-1} \left[g(s) + h(s) c^{1-p}(s) \right] \right.$$

$$\left. \times \prod_{\sigma=s+1}^{n-1} \left[1 + b(\sigma) \left(g(\sigma) + \frac{h(\sigma)}{p} c^{1-p}(\sigma) \right) \right] \right\}^{\frac{1}{p}}, \tag{4.3.16}$$

for $n \in N_0$.

(c_3) Let $k(n, \sigma), \Delta_1 k(n, \sigma)$ be as in Theorem 4.3.1. If

$$u^p(n) \leq a(n) + b(n) \sum_{s=0}^{n-1} k(n, s) [g(s) u^p(s) + h(s) u(s)], \tag{4.3.17}$$

for $n \in N_0$, then

$$u(n) \leq \left\{ a(n) + b(n) \sum_{\sigma=0}^{n-1} \bar{B}(\sigma) \prod_{\tau=\sigma+1}^{n-1} \left[1 + \bar{A}(\tau) \right] \right\}^{\frac{1}{p}}, \tag{4.3.18}$$

for $n \in N_0$, where

$$\bar{A}(n) = k(n+1, n) b(n) \left(g(n) + \frac{h(n)}{p} \right)$$

$$+ \sum_{s=0}^{n-1} \Delta_1 k(n, s) b(s) \left(g(s) + \frac{h(s)}{p} \right), \tag{4.3.19}$$

$$\bar{B}(n) = k(n+1, n) \left(g(n) a(n) + h(n) \left(\frac{p-1}{p} + \frac{a(n)}{p} \right) \right)$$

$$+ \sum_{s=0}^{n-1} \Delta_1 k(n, s) \left(g(s) a(s) + h(s) \left(\frac{p-1}{p} + \frac{a(s)}{p} \right) \right), \tag{4.3.20}$$

for $n \in N_0$.

Proof. (c_1) Define a function $z(n)$ by

$$z(n) = \sum_{s=0}^{n-1} [g(s) u^p(s) + h(s) u(s)]. \tag{4.3.21}$$

Then $z(0) = 0$ and (4.3.13) can be written as

$$u^p(n) \leq a(n) + b(n) z(n). \tag{4.3.22}$$

From (4.3.22), as in the proof of Theorem 1.3.2, part (a_1) we obtain

$$u(n) \leq \left(\tfrac{p-1}{p} + \frac{a(n)}{p}\right) + \frac{b(n)}{p} z(n). \tag{4.3.23}$$

From (4.3.21) and using (4.3.22), (4.3.23) we get (see [42, p. 13])

$$\Delta z(n) \leq b(n)\left(g(n) + \frac{h(n)}{p}\right) z(n)$$

$$+ \left[g(n) a(n) + h(n)\left(\frac{p-1}{p} + \frac{a(n)}{p}\right)\right]. \tag{4.3.24}$$

Now a suitable application of Theorem 1.2.1 given in [42, p.11] to (4.3.24) yields

$$z(n) \leq \sum_{s=0}^{n-1} \left[g(s) a(s) + h(s)\left(\frac{p-1}{p} + \frac{a(s)}{p}\right)\right]$$

$$\times \prod_{\sigma=s+1}^{n-1} \left[1 + b(\sigma)\left(g(\sigma) + \frac{h(\sigma)}{p}\right)\right]. \tag{4.3.25}$$

Using (4.3.25) in (4.3.22) we get the required inequality in (4.3.14).

(c_2) Since $c(n)$ is positive and nondecreasing function for $n \in N_0$, from (4.3.15) we observe that

$$\left(\frac{u(n)}{c(n)}\right)^p \leq 1 + b(n) \sum_{s=0}^{n-1} \left[g(s)\left(\frac{u(s)}{c(s)}\right)^p + h(s) c^{1-p}(s)\left(\frac{u(s)}{c(s)}\right)\right]. \tag{4.3.26}$$

Now an application of the inequality given in (c_1) to (4.3.26) yields the desired inequality in (4.3.16).

(c_3) Define a function $z(n)$ by

$$z(n) = \sum_{s=0}^{n-1} k(n,s) [g(s) u^p(s) + h(s) u(s)]. \tag{4.3.27}$$

Then $z(0) = 0$ and as in the proof of part (c_1), from (4.3.17) we see that the inequalities (4.3.22) and (4.3.23) hold. From (4.3.27) and using (4.3.22), (4.3.23) and the fact that the function $z(n)$ is nondecreasing in n, we observe that

$$\Delta z(n) = k(n+1,n))\left[g(n)u^p(n) + h(n)u(n)\right]$$
$$+\sum_{s=0}^{n-1}\Delta_1 k(n,s)\left[g(s)u^p(s) + h(s)u(s)\right]$$
$$\leq k(n+1,n)\left[g(n)(a(n)+b(n)z(n))\right.$$
$$\left.+h(n)\left(\frac{p-1}{p}+\frac{a(n)}{p}+\frac{b(n)}{p}z(n)\right)\right]$$
$$+\sum_{s=0}^{n-1}\Delta_1 k(n,s)\left[g(s)(a(s)+b(s)z(s))\right.$$
$$\left.+h(s)\left(\frac{p-1}{p}+\frac{a(s)}{p}+\frac{b(s)}{p}z(s)\right)\right]$$
$$\leq \bar{A}(n)z(n) + \bar{B}(n). \tag{4.3.28}$$

Now a suitable application of Theorem 1.2.1 given in [42, p. 11] to (4.3.28) yields

$$z(n) \leq \sum_{\sigma=0}^{n-1}\bar{B}(\sigma)\prod_{\tau=\sigma+1}^{n-1}\left[1+\bar{A}(\tau)\right]. \tag{4.3.29}$$

From (4.3.29) and (4.3.22) the desired inequality in (4.3.18) follows.

Theorem 4.3.4. Let $u(n), a(n), b(n), g(n) \in D(N_0, R_+)$ and $p > 1$ be a real constant.

(d_1) Let $L : N_0 \times R_+ \to R_+$ be a function such that

$$0 \leq L(n,x) - L(n,y) \leq M(n,y)(x-y), \tag{4.3.30}$$

for $n \in N_0$, $x \geq y \geq 0$, where $M : N_0 \times R_+ \to R_+$.If

$$u^p(n) \leq a(n) + b(n)\sum_{s=0}^{n-1}L(s,u(s)), \tag{4.3.31}$$

for $n \in N_0$, then

$$u(n) \leq \left\{a(n) + b(n)\sum_{s=0}^{n-1}L\left(s,\frac{p-1}{p}+\frac{a(s)}{p}\right)\right.$$
$$\left.\times\prod_{\sigma=s+1}^{n-1}\left[1+M\left(\sigma,\frac{p-1}{p}+\frac{a(\sigma)}{p}\right)\frac{b(\sigma)}{p}\right]\right\}^{\frac{1}{p}}, \tag{4.3.32}$$

for $n \in N_0$.

(d_2) Let $L : N_0 \times R_+ \to R_+$ be a function which satisfies the condition

$$0 \le L(n,x) - L(n,y) \le M(n,y)\,\psi^{-1}(x-y), \tag{4.3.33}$$

for $n \in N_0$, $x \ge y \ge 0$,where $M : N_0 \times R_+ \to R_+$, $\psi : R_+ \to R_+$ is a continuous and strictly increasing function with $\psi(0) = 0$, ψ^{-1} is the inverse function of ψ and $\psi^{-1}(xy) \le \psi^{-1}(x)\,\psi^{-1}(y)$ for $x, y \in R_+$. If

$$u^p(n) \le a(n) + b(n)\,\psi\left(\sum_{s=0}^{n-1} L(s, u(s))\right), \tag{4.3.34}$$

for $n \in N_0$, then

$$u(n) \le \left\{a(n) + b(n)\,\psi\left(\sum_{s=0}^{n-1} L\left(s, \frac{p-1}{p} + \frac{a(s)}{p}\right)\right.\right.$$

$$\left.\left.\times \prod_{\sigma=s+1}^{n-1}\left[1 + M\left(\sigma, \frac{p-1}{p} + \frac{a(\sigma)}{p}\right)\psi^{-1}\left(\frac{b(\sigma)}{p}\right)\right]\right)\right\}^{\frac{1}{p}}, \tag{4.3.35}$$

for $n \in N_0$.

(d_3) Let $W(r), G, G^{-1}$ be as in Theorem 1.3.2, part (b_3). If

$$u^p(n) \le a(n) + b(n)\sum_{s=0}^{n-1} g(s)\,W(u(s)), \tag{4.3.36}$$

for $n \in N_0$, then for $0 \le n \le n_4$; $n, n_4 \in N_0$,

$$u(n) \le \left\{a(n) + b(n)\,G^{-1}\left[G\left(\bar{D}(n)\right) + \sum_{s=0}^{n-1} g(s)\,W\left(\frac{b(s)}{p}\right)\right]\right\}^{\frac{1}{p}}, \tag{4.3.37}$$

where

$$\bar{D}(n) = \sum_{s=0}^{n-1} g(s)\,W\left(\frac{p-1}{p} + \frac{a(s)}{p}\right),$$

and $n_4 \in N_0$ is chosen so that

$$G\left(\bar{D}(n)\right) + \sum_{s=0}^{n-1} g(s)\,W\left(\frac{b(s)}{p}\right) \in Dom\left(G^{-1}\right),$$

for all $n \in N_0$ lying in $0 \le n \le n_4$.

The proof follows by closely looking at the proofs of Theorem 1.3.2 and 4.3.3, see also [42]. Here we omit the details.

Theorem 4.3.5. Let $u(n), f(n) \in D(N_0, R_+)$, $h(n,\sigma) \in D(E, R_+)$ and $c \geq 0$, $p > 1$ are real constants and E is defined as in Theorem 4.3.1.

(e_1) Let g, H, H^{-1} be as in Theorem 1.3.3. If

$$u^p(n) \leq c + \sum_{s=0}^{n-1} \left[f(s) g(u(s)) + \sum_{\sigma=0}^{s-1} h(s,\sigma) g(u(\sigma)) \right], \tag{4.3.38}$$

for $n \in N_0$, then for $0 \leq n \leq n_5; n, n_5 \in N_0$,

$$u(n) \leq \left\{ H^{-1} \left[H(c) + F(n) \right] \right\}^{\frac{1}{p}}, \tag{4.3.39}$$

where

$$F(n) = \sum_{s=0}^{n-1} \left[f(s) + \sum_{\sigma=0}^{s-1} h(s,\sigma) \right], \tag{4.3.40}$$

and $n_5 \in N_0$ is chosen so that

$$H(c) + F(n) \in Dom\left(H^{-1}\right)$$

for all $n \in N_0$ lying in $0 \leq n \leq n_5$.

(e_2) If

$$u^p(n) \leq c + \sum_{s=0}^{n-1} \left[f(s) u(s) + \sum_{\sigma=0}^{s-1} h(s,\sigma) u(s) \right], \tag{4.3.41}$$

for $n \in N_0$, then

$$u(n) \leq \left[c^{\frac{p-1}{p}} + \left(\frac{p-1}{p} \right) F(n) \right]^{\frac{1}{p-1}}, \tag{4.3.42}$$

for $n \in N_0$, where $F(n)$ is given by (4.3.40).

Proof. (e_1) Let $c > 0$ and define a function $z(n)$ by the right hand side of (4.3.38). Then $z(0) = c$, $u(n) \leq \{z(n)\}^{\frac{1}{p}}$, $z(n)$ is positive and nondecreasing for $n \in N_0$ and

$$\Delta z(n) = f(n) g(u(n)) + \sum_{\sigma=0}^{n-1} h(n,\sigma) g(u(\sigma))$$

$$\leq g\left(\{z(n)\}^{\frac{1}{p}} \right) \left[f(n) + \sum_{\sigma=0}^{n-1} h(n,\sigma) \right]. \tag{4.3.43}$$

From (1.3.41) and (4.3.43) we observe that

$$H(z(n+1)) - H(z(n)) = \int_{z(n)}^{z(n+1)} \frac{ds}{g\left(s^{\frac{1}{p}}\right)}$$

$$\leq \frac{\Delta z(n)}{g\left(\{z(n)\}^{\frac{1}{p}}\right)}$$

$$\leq f(n) + \sum_{\sigma=0}^{s-1} h(n,\sigma). \tag{4.3.44}$$

The rest of the proof follows as in the proof of Theorem 4.3.2, part (b_1) with suitable changes. We omit the details.

(e_2) The proof is similar to that of Theorem 1.3.4. We omit it here to avoid repetition.

4.4 Finite difference inequalities with iterated sums

The main concern of this section is to present some finite difference inequalities involving iterated sums, investigated by Pachpatte in [53, 67,73] which can be used as tools in the study of general classes of finite difference and sum-difference equations.

Our first theorem deals with the inequalities proved in [53].

Theorem 4.4.1. Let $u(n), f(n), a(n) \in D(N_0, R_+)$, $k(n,\sigma), \Delta_1 k(n,\sigma) \in D(E, R_+)$ and $c \geq 0$ be a real constant, where $E = \left\{(m,n) \in N_0^2 : 0 \leq n \leq m < \infty\right\}$.

(a_1) If

$$u(n) \leq c + \sum_{s=0}^{n-1} f(s)\left[u(s) + \sum_{\sigma=0}^{s-1} k(s,\sigma)\, u(\sigma)\right], \tag{4.4.1}$$

for $n \in N_0$, then

$$u(n) \leq c\left[1 + \sum_{s=0}^{n-1} f(s) \prod_{\sigma=0}^{s-1} [1 + f(\sigma) + A(\sigma)]\right], \tag{4.4.2}$$

for $n \in N_0$, where

$$A(n) = k(n+1, n) + \sum_{\tau=0}^{n-1} \Delta_1 k(n, \tau), \tag{4.4.3}$$

for $n \in N_0$.

(a_2) If

$$u(n) \leq a(n) + \sum_{s=0}^{n-1} f(s) \left[u(s) + \sum_{\sigma=0}^{s-1} k(s, \sigma) u(\sigma) \right], \tag{4.4.4}$$

for $n \in N_0$, then

$$u(n) \leq a(n) + B(n) \left[1 + \sum_{s=0}^{n-1} f(s) \prod_{\sigma=0}^{s-1} [1 + f(\sigma) + A(\sigma)] \right], \tag{4.4.5}$$

for $n \in N_0$, where

$$B(n) = \sum_{s=0}^{n-1} f(s) \left[a(s) + \sum_{\sigma=0}^{s-1} k(s, \sigma) a(\sigma) \right], \tag{4.4.6}$$

for $n \in N_0$ and $A(n)$ is given by (4.4.3).

Proof. (a_1) Define a function $z(n)$ by the right hand side of (4.4.1). Then $z(0) = c$, $u(n) \leq z(n)$ and

$$\Delta z(n) = f(n) \left[u(n) + \sum_{\sigma=0}^{n-1} k(s, \sigma) u(\sigma) \right]$$

$$\leq f(n) \left[z(n) + \sum_{\sigma=0}^{n-1} k(s, \sigma) z(\sigma) \right]. \tag{4.4.7}$$

Define a function $v(n)$ by

$$v(n) = z(n) + \sum_{\sigma=0}^{n-1} k(s, \sigma) z(\sigma). \tag{4.4.8}$$

Then $v(0) = z(0) = c$, $z(n) \leq v(n)$, $\Delta z(n) \leq f(n) v(n)$, $v(n)$ is nondecreasing for $n \in N_0$, and

$$\Delta v(n) = \Delta z(n) + \sum_{\sigma=0}^{n} k(n+1, \sigma) z(\sigma) - \sum_{\sigma=0}^{n-1} k(n, \sigma) z(\sigma)$$

$$= \Delta z(n) + k(n+1,n)\, z(n) + \sum_{\sigma=0}^{n-1} \Delta_1 k(n,\sigma)\, z(\sigma)$$

$$\leq [f(n) + A(n)]\, v(n), \tag{4.4.9}$$

where $A(n)$ is given by (4.4.3). The inequality (4.4.9) implies (see [42, p. 12])

$$v(n) \leq c \prod_{\sigma=0}^{n-1} [1 + f(\sigma) + A(\sigma)]. \tag{4.4.10}$$

Using (4.4.10) in $\Delta z(n) \leq f(n)\, v(n)$ we get

$$\Delta z(n) \leq c f(n) \prod_{\sigma=0}^{n-1} [1 + f(\sigma) + A(\sigma)]. \tag{4.4.11}$$

The inequality (4.4.11) implies the estimate

$$z(n) \leq c \left[1 + \sum_{s=0}^{n-1} f(s) \prod_{\sigma=0}^{s-1} [1 + f(\sigma) + A(\sigma)] \right]. \tag{4.4.12}$$

Using (4.4.12) in $u(n) \leq z(n)$ we get the desired inequality in (4.4.2).

(a_2) The proof can be completed by closely looking at the proofs of (a_1) given above and Theorem 1.4.1, part (a_2). We omit the details.

Remark 4.4.1. We note that the inequalities given in Theorem 4.4.1 are the discrete analogues of the inequalities given in Theorem 1.4.1. In the special case when $k(n,\sigma) = g(\sigma)$, the inequality in (a_1) reduces to the inequality established earlier by Pachpatte, see [42, Theorem 1.4.1, p. 26]. For slight variants of the inequalities given in Theorem 4.4.1, see [42].

In the following theorems we present the inequalities established in [67].

Theorem 4.4.2. Let $u(n) \in D(N_0, R_+)$, $k(n,\sigma), \Delta_1 k(n,\sigma) \in D(E, R_+)$, $h(n,s,\sigma), \Delta_1 h(n,s,\sigma) \in D(F, R_+)$ and $c \geq 0$ be a real constant, where $E = \{(n,s) \in N_0^2 : 0 \leq s \leq n < \infty\}$, $F = \{(n,s,\sigma) \in N_0^3 : 0 \leq \sigma \leq s \leq n < \infty\}$.

(b_1) If

$$u(n) \leq c + \sum_{s=0}^{n-1} k(n,s)\, u(s) + \sum_{s=0}^{n-1} \left(\sum_{\sigma=0}^{s-1} h(n,s,\sigma)\, u(\sigma) \right), \tag{4.4.13}$$

for $n \in N_0$, then

$$u(n) \leq c \prod_{s=0}^{n-1} [1 + P(s) + Q(s)], \tag{4.4.14}$$

for $n \in N_0$, where

$$P(n) = k(n+1,n) + \sum_{\sigma=0}^{n-1} h(n+1,n,\sigma), \tag{4.4.15}$$

$$Q(n) = \sum_{\tau=0}^{n-1} \Delta_1 k(n,\tau) + \sum_{\tau=0}^{n-1} \left(\sum_{\sigma=0}^{\tau-1} \Delta_1 h(n,\tau,\sigma) \right), \tag{4.4.16}$$

for $n \in N_0$

(b_2) Let $g \in C(R_+, R_+)$ be a nondecreasing function with $g(u) > 0$ for $u > 0$. If

$$u(n) \leq c + \sum_{s=0}^{n-1} k(n,s)\, g(u(s)) + \sum_{s=0}^{n-1} \left(\sum_{\sigma=0}^{s-1} h(n,s,\sigma)\, g(u(\sigma)) \right), \tag{4.4.17}$$

for $n \in N_0$, then for $0 \leq n \leq n_1; n, n_1 \in N_0$,

$$u(n) \leq G^{-1} \left[G(c) + \sum_{s=0}^{n-1} [P(s) + Q(s)] \right], \tag{4.4.18}$$

where $P(n), Q(n)$ are given by (4.4.15), (4.4.16),

$$G(r) = \int_{r_0}^{r} \frac{dt}{g(t)}, r > 0, \tag{4.4.19}$$

$r_0 > 0$ is arbitrary, G^{-1} is the inverse of G and $n_1 \in N_0$ be chosen so that

$$G(c) + \sum_{s=0}^{n-1} [P(s) + Q(s)] \in Dom\left(G^{-1}\right),$$

for all $n \in N_0$ lying in $0 \leq n \leq n_1$.

Proof. (b_1) Define a function $z(n)$ by the right hand side of (4.4.13), then $z(0) = c$ and $u(n) \leq z(n)$. From the hypotheses, we observe that $z(n)$ is nondecreasing for $n \in N_0$ and

$$\Delta z(n) = \sum_{s=0}^{n} k(n+1,s)\, u(s) + \sum_{s=0}^{n} \left(\sum_{\sigma=0}^{s-1} h(n+1,s,\sigma)\, u(\sigma) \right)$$

$$- \sum_{s=0}^{n-1} k(n,s)\, u(s) - \sum_{s=0}^{n-1} \left(\sum_{\sigma=0}^{s-1} h(n,s,\sigma)\, u(\sigma) \right)$$

$$= k(n+1,n)\, u(n) + \sum_{s=0}^{n-1} k(n+1,s)\, u(s) - \sum_{s=0}^{n-1} k(n,s)\, u(s)$$

$$+\sum_{\sigma=0}^{n-1} h(n+1,n,\sigma)\, u(\sigma) + \sum_{s=0}^{n-1} \left(\sum_{\sigma=0}^{s-1} h(n+1,s,\sigma)\, u(\sigma) \right)$$

$$-\sum_{s=0}^{n-1} \left(\sum_{\sigma=0}^{s-1} h(n,s,\sigma)\, u(\sigma) \right)$$

$$= k(n+1,n)\, u(n) + \sum_{s=0}^{n-1} \Delta_1 k(n,s)\, u(s)$$

$$+\sum_{\sigma=0}^{n-1} h(n+1,n,\sigma) u(\sigma) + \sum_{s=0}^{n-1} \left(\sum_{\sigma=0}^{s-1} \Delta_1 h(n,s,\sigma)\, u(\sigma) \right)$$

$$\leq k(n+1,n)\, z(n) + \sum_{s=0}^{n-1} \Delta_1 k(n,s)\, z(s)$$

$$+\sum_{\sigma=0}^{n-1} h(n+1,n,\sigma) z(\sigma) + \sum_{s=0}^{n-1} \left(\sum_{\sigma=0}^{s-1} \Delta_1 h(n,s,\sigma)\, z(\sigma) \right)$$

$$\leq [P(n) + Q(n)]\, z(n). \tag{4.4.20}$$

Now a suitable application of the Corollary 1.2.2 given in [42, p. 12] to (4.4.20) yields

$$z(n) \leq c \prod_{s=0}^{n-1} [1 + P(s) + Q(s)]. \tag{4.4.21}$$

Using (4.4.21) in $u(n) \leq z(n)$ we get the required inequality in (4.4.13).

(b_2) Let $c > 0$ and define a function $z(n)$ by the right hand side of (4.4.17). Then $z(0) = c$, $u(n) \leq z(n)$, $z(n)$ is positive and nondecreasing for $n \in N_0$ and by following the proof of (b_1) with suitable modifications we get

$$\Delta z(n) \leq [P(n) + Q(n)]\, g(z(n)). \tag{4.4.22}$$

The rest of the proof can be completed by following the proof of Theorem 4.2.3, part (b_1). Here we omit the details.

Theorem 4.4.3. Let $u(n), k(n,s)$, $h(n,s,\sigma)$, c be as in Theorem 4.4.2 and $b(n) \in D(N_0, R_+)$.

(c_1) If

$$u(n) \leq c + \sum_{s=0}^{n-1} b(s) u(s) + \sum_{s=0}^{n-1} \left(\sum_{\tau=0}^{s-1} k(s,\tau) u(\tau) \right)$$

$$+ \sum_{s=0}^{n-1} \left(\sum_{\tau=0}^{s-1} \left(\sum_{\sigma=0}^{\tau-1} h(s,\tau,\sigma) u(\sigma) \right) \right), \tag{4.4.23}$$

for $n \in N_0$, then

$$u(n) \leq c \prod_{s=0}^{n-1} \left[1 + b(s) + \sum_{\tau=0}^{s-1} k(s,\tau) + \sum_{\tau=0}^{s-1} \left(\sum_{\sigma=0}^{\tau-1} h(s,\tau,\sigma) \right) \right], \tag{4.4.24}$$

for $n \in N_0$.

(c_2) Let $g(u)$ be as in Theorem 4.4.2, part (b_2). If

$$u(n) \leq c + \sum_{s=0}^{n-1} b(s) g(u(s)) + \sum_{s=0}^{n-1} \left(\sum_{\tau=0}^{s-1} k(s,\tau) g(u(\tau)) \right)$$

$$+ \sum_{s=0}^{n-1} \left(\sum_{\tau=0}^{s-1} \left(\sum_{\sigma=0}^{\tau-1} h(s,\tau,\sigma) g(u(\sigma)) \right) \right), \tag{4.4.25}$$

for $n \in N_0$, then for $0 \leq n \leq n_2; n, n_2 \in N_0$,

$$u(n) \leq G^{-1} [G(c)$$

$$+ \sum_{s=0}^{n-1} \left[b(s) + \sum_{\tau=0}^{s-1} k(s,\tau) + \sum_{\tau=0}^{s-1} \left(\sum_{\sigma=0}^{\tau-1} h(s,\tau,\sigma) \right) \right] \Bigg], \tag{4.4.26}$$

where G, G^{-1} are as in Theorem 4.4.2, part (b_2) and $n_2 \in N_0$ be chosen so that

$$G(c) + \sum_{s=0}^{n-1} \left[b(s) + \sum_{\tau=0}^{s-1} k(s,\tau) + \sum_{\tau=0}^{s-1} \left(\sum_{\sigma=0}^{\tau-1} h(s,\tau,\sigma) \right) \right] \in Dom\left(G^{-1}\right),$$

for all $n \in N_0$ lying in $0 \leq n \leq n_2$.

Proof. (c_1) Define a function $z(n)$ by the right hand side of (4.4.25). Then $z(0) = c$, $u(n) \leq z(n)$, $z(n)$ is nondecreasing for $n \in N_0$ and

$$\Delta z(n) = b(n) u(n) + \sum_{\tau=0}^{n-1} k(n,\tau) u(\tau) + \sum_{\tau=0}^{n-1} \left(\sum_{\sigma=0}^{\tau-1} h(n,\tau,\sigma) u(\sigma) \right)$$

$$\leq b(n) z(n) + \sum_{\tau=0}^{n-1} k(n,\tau) z(\tau) + \sum_{\tau=0}^{n-1} \left(\sum_{\sigma=0}^{\tau-1} h(n,\tau,\sigma) z(\sigma) \right)$$

$$\leq \left[b(n) + \sum_{\tau=0}^{n-1} k(n,\tau) + \sum_{\tau=0}^{n-1} \left(\sum_{\sigma=0}^{\tau-1} h(n,\tau,\sigma) \right) \right] z(n). \tag{4.4.27}$$

Now a suitable application of Corollary 1.2.2 given in [42, p. 12] to (4.4.27) yields

$$z(n) \leq c \prod_{s=0}^{n-1} \left[1 + b(s) + \sum_{\tau=0}^{s-1} k(s,\tau) + \sum_{\tau=0}^{s-1} \left(\sum_{\sigma=0}^{\tau-1} h(s,\tau,\sigma) \right) \right]. \tag{4.4.28}$$

Using (4.4.28) in $u(n) \leq z(n)$ we get the desired inequality in (4.4.24).

(c_2) The proof can be completed by following the proof of (c_1) and closely looking at the proof of Theorem 4.4.2, part (b_2). Here we omit the details.

Remark 4.4.2. We note that the inequalities in Theorems 4.4.2 and 4.4.3 parts (b_1) and (c_1) provides the growth estimates on the discrete versions of the integral inequalities due to Bykov and Salpagarov [9] given in Theorem 1.4.2, while the inequalities in (b_2) and (c_2) provides the growth estimates on the general versions of the inequalities given in [9], which can be used conveniently in certain applications.

The inequalities established in [73] are embodied in the following theorems.

In what follows, let $J_i = \{(n_1, ..., n_i) : (n_1, ..., n_i) \in N_0^i\}$ for $i = 1, ..., m$. For any functions $w(n) \in D(N_0, R_+)$, $k_i(n_1, ..., n_i) \in D(J_i, R_+)$ for $i = 1, ..., m$; first we give the following notations used to simplify the details of presentation:

$$B_i[w](n) = \sum_{n_1=0}^{n-1} \left(\sum_{n_2=0}^{n_1-1} \cdots \left(\sum_{n_i=0}^{n_{i-1}-1} k_i(n_1, ..., n_i) \right) \cdots \right),$$

$$G[w](n) = k_1(n) w(n) + \sum_{n_2=0}^{n-1} k_2(n,n_2) w(n_2) + \sum_{n_2=0}^{n-1} \left(\sum_{n_3=0}^{n_2-1} k_3(n,n_2,n_3) w(n_3) \right)$$

$$+ ... + \sum_{n_2=0}^{n-1} \left(\sum_{n_3=0}^{n_2-1} \cdots \left(\sum_{n_m=0}^{n_{m-1}-1} k_m(n,n_2,...,n_m) w(n_m) \right) \cdots \right).$$

Theorem 4.4.4. Let $u(n), a(n) \in D(N_0, R_+)$, $k_i(n_1, ..., n_i) \in D(J_i, R_+)$ for $i = 1, ..., m$ and $c \geq 0$ is a real constant.

(d_1) If

$$u(n) \leq c + \sum_{i=1}^{m} B_i[u](n), \tag{4.4.29}$$

for $n \in N_0$, then

$$u(n) \leq c \prod_{n_1=0}^{n-1} [1 + G[1](n_1)], \tag{4.4.30}$$

for $n \in N_0$.

(d_2) Let $a(n)$ be nondecreasing for $n \in N_0$. If

$$u(n) \leq a(n) + \sum_{i=1}^{m} B_i[u](n), \tag{4.4.31}$$

for $n \in N_0$, then

$$u(n) \leq a(n) \prod_{n_1=0}^{n-1} [1 + G[1](n_1)], \tag{4.4.32}$$

for $n \in N_0$.

Proof. (d_1) Define a function $z(n)$ by the right hand side of (4.4.29) i.e.,

$$z(n) = c + \sum_{n_1=0}^{n-1} k_1(n_1) u(n_1) + \sum_{n_1=0}^{n-1} \left(\sum_{n_2=0}^{n_1-1} k_2(n_1, n_2) u(n_2) \right)$$

$$+ \sum_{n_1=0}^{n-1} \left(\sum_{n_2=0}^{n_1-1} \left(\sum_{n_3=0}^{n_2-1} k_3(n_1, n_2, n_3) u(n_3) \right) \right) + \ldots$$

$$+ \sum_{n_1=0}^{n-1} \left(\sum_{n_2=0}^{n_1-1} \left(\sum_{n_3=0}^{n_2-1} \cdots \left(\sum_{n_m=0}^{n_{m-1}-1} k_m(n_1, n_2, n_3, \ldots, n_m) u(n_m) \right) \cdots \right) \right).$$

Then $z(0) = c$, $u(n) \leq z(n)$, $z(n)$ is nondecreasing for $n \in N_0$ and

$$\Delta z(n) = k_1(n) u(n) + \sum_{n_2=0}^{n-1} k_2(n, n_2) u(n_2) + \sum_{n_2=0}^{n-1} \left(\sum_{n_3=0}^{n_2-1} k_3(n, n_2, n_3) u(n_3) \right)$$

$$+ \ldots + \sum_{n_2=0}^{n-1} \left(\sum_{n_3=0}^{n_2-1} \cdots \left(\sum_{n_m=0}^{n_{m-1}-1} k_m(n, n_2, n_3, \ldots, n_m) u(n_m) \right) \cdots \right)$$

$$\leq \left[k_1(n) + \sum_{n_2=0}^{n-1} k_2(n, n_2) + \sum_{n_2=0}^{n-1} \left(\sum_{n_3=0}^{n_2-1} k_3(n, n_2, n_3) \right) \right.$$

$$\left. + \ldots + \sum_{n_2=0}^{n-1} \left(\sum_{n_3=0}^{n_2-1} \cdots \left(\sum_{n_m=0}^{n_{m-1}-1} k_m(n, n_2, n_3, \ldots, n_m) \right) \cdots \right) \right] z(n)$$

i.e.,

$$\Delta z(n) \leq G[1](n) z(n). \tag{4.4.33}$$

Now a suitable application of Theorem 1.2.1 given in [42] to (4.4.33) yields

$$z(n) \leq c \prod_{n_1=0}^{n-1} [1 + G[1](n_1)]. \tag{4.4.34}$$

Using (4.4.34) in $u(n) \leq z(n)$ we get the desired inequality in (4.4.30).

(d_2) The proof can be completed by closely looking at the proof of Theorem 1.2.4 given in [42] and by making use of the inequality established in (d_1) . We omit the details.

Theorem 4.4.5. Let $u(n)$, $k_i(n_1, ..., n_i)$ for $i = 1, ..., m$ be as in Theorem 4.4.4.

(e_1) Let $\phi(n) \in D(N_0, R_+)$ and $\Delta\phi(n) \geq 0$ for $n \in N_0$. If

$$u(n) \leq \phi(n) + \sum_{i=1}^{m} B_i[u](n), \tag{4.4.35}$$

for $n \in N_0$, then

$$u(n) \leq \phi(0) \prod_{n_1=0}^{n-1} [1 + G[1](n_1)] + \sum_{n_1=0}^{n-1} \Delta\phi(n_1) \prod_{\sigma=n_1+1}^{n-1} [1 + G[1](\sigma)], \tag{4.4.36}$$

for $n \in N_0$.

(e_2) Let $a(n), b(n) \in D(N_0, R_+)$. If

$$u(n) \leq a(n) + b(n) \sum_{n_1=0}^{m} B_i[u](n), \tag{4.4.37}$$

for $n \in N_0$, then

$$u(n) \leq a(n) + b(n) \sum_{n_1=0}^{n-1} G[a](n_1) \prod_{\sigma=n_1+1}^{n-1} [1 + G[b](\sigma)], \tag{4.4.38}$$

for $n \in N_0$.

Proof. (e_1) From the hypotheses on $\phi(n)$ we observe that $\phi(n)$ is nondecreasing for $n \in N_0$. Define a function $z(n)$ by the right hand side of (4.4.35). Then $z(0) = \phi(0)$, $u(n) \leq z(n)$, $z(n)$ is nondecreasing for $n \in N_0$ and as in the proof of Theorem 4.4.4, part (d_1) we have

$$\Delta z(n) \leq \Delta\phi(n) + G[1](n) z(n). \tag{4.4.39}$$

Now a suitable application of Theorem 1.2.1 given in [42] to (4.4.39) yields

$$z(n) \leq \phi(0) \prod_{n_1=0}^{n-1} [1 + G[1](n_1)] + \sum_{n_1=0}^{n-1} \Delta\phi(n_1) \prod_{\sigma=n_1+1}^{n-1} [1 + G[1](\sigma)]. \tag{4.4.40}$$

Using (4.4.40) in $u(n) \leq z(n)$ we get the required inequality in (4.4.36).

(e_2) Define a function $z(n)$ by

$$z(n) = \sum_{i=0}^{m} B_i[u](n). \tag{4.4.41}$$

Then as in the proof of Theorem 4.4.4, part (d_1), $z(0) = 0$, $z(n)$ is nondecreasing for $n \in N_0$; (4.4.37) can be restated as

$$u(n) \leq a(n) + b(n) z(n), \tag{4.4.42}$$

and

$$\Delta z(n) = k_1(n) u(n) + \sum_{n_2=0}^{n-1} k_2(n, n_2) u(n_2) + \sum_{n_2=0}^{n-1} \left(\sum_{n_3=0}^{n_2-1} k_3(n, n_2, n_3) u(n_3) \right)$$
$$+ \ldots + \sum_{n_2=0}^{n-1} \left(\sum_{n_3=0}^{n_2-1} \cdots \left(\sum_{n_m=0}^{n_{m-1}-1} k_m(n, n_2, n_3, \ldots, n_m) u(n_m) \right) \ldots \right)$$
$$\leq G[a](n) + G[b](n) z(n). \tag{4.4.43}$$

Now an application of Theorem 1.2.1 given in [42] to (4.4.43) yields

$$z(n) \leq \sum_{n_1=0}^{n-1} G[a](n_1) \prod_{\sigma=n_1+1}^{n-1} [1 + G[b](\sigma)]. \tag{4.4.44}$$

Using (4.4.44) in (4.4.42) we get the required inequality in (4.4.38).

Remark 4.4.3. The inequalities in Theorem 4.4.4 and 4.4.5 are motivated by the integral inequalities established by various investigators and given in [3, pp. 100-108]. For some useful singular finite difference inequalities, we refer the interested readers to the recent paper by Medveď [27] and some of the references cited therein.

4.5 Bounds on certain finite difference inequalities

The main goal of this section is to present some specific type of finite difference ineualities investigated by Pachpatte in [37,39,44,54,70,75]. The inequalities given here can be used in the analysis of certain finite difference and sum-difference equations.

Our first theorem deals with the finite difference inequalities proved in [70].

Theorem 4.5.1. Let $u(n), a(n), b(n), c(n), f(n), g(n) \in D(N_{\alpha,\beta}, R_+)$.

(a_1) Suppose that $\Delta a(n) \geq 0$ for $n \in N_{\alpha,\beta}$ and

$$u(n) \leq a(n) + \sum_{s=\alpha}^{n-1} b(s) u(s) + \sum_{s=\alpha}^{\beta} c(s) u(s), \tag{4.5.1}$$

for $n \in N_{\alpha,\beta}$. If

$$q_1 = \sum_{s=\alpha}^{\beta} c(s) \prod_{\tau=\alpha}^{s-1} [1 + b(\tau)] < 1, \tag{4.5.2}$$

then

$$u(n) \leq N_1 \prod_{s=\alpha}^{n-1} [1 + b(s)] + \sum_{s=\alpha}^{n-1} \Delta a(s) \prod_{\sigma=s+1}^{n-1} [1 + b(\sigma)], \tag{4.5.3}$$

for $n \in N_{\alpha,\beta}$, where

$$N_1 = \frac{1}{1 - q_1} \left[a(\alpha) + \sum_{s=\alpha}^{\beta} c(s) \sum_{\tau=\alpha}^{s-1} \Delta a(\tau) \prod_{\sigma=\tau+1}^{s-1} [1 + b(\sigma)] \right]. \tag{4.5.4}$$

(a_2) Suppose that

$$u(n) \leq a(n) + b(n) \sum_{s=\alpha}^{n-1} f(s) u(s) + \sum_{s=\alpha}^{\beta} g(s) u(s), \tag{4.5.5}$$

for $n \in N_{\alpha,\beta}$. If

$$q_2 = \sum_{s=\alpha}^{\beta} g(s) L_2(s) < 1, \tag{4.5.6}$$

then

$$u(n) \leq L_1(n) + N_2 L_2(n), \tag{4.5.7}$$

for $n \in N_{\alpha,\beta}$, where

$$L_1(n) = a(n) + b(n) \sum_{s=\alpha}^{n-1} f(s) a(s) \prod_{\sigma=s+1}^{n-1} [1 + f(\sigma) b(\sigma)], \tag{4.5.8}$$

$$L_2(n) = c(n) + b(n) \sum_{s=\alpha}^{n-1} f(s) c(s) \prod_{\sigma=s+1}^{n-1} [1 + f(\sigma) b(\sigma)], \tag{4.5.9}$$

and

$$N_2 = \frac{1}{1 - q_2} \sum_{s=\alpha}^{\beta} g(s) L_1(s). \tag{4.5.10}$$

(a_3) Let $r(n,s), \Delta r(n,s) \in D\left(N_{\alpha,\beta}^2, R_+\right)$ for $\alpha \leq s \leq n \leq \beta$ and

$$u(n) \leq a(n) + \sum_{s=\alpha}^{n-1} r(n,s) u(s) + \sum_{s=\alpha}^{\beta} c(s) u(s), \tag{4.5.11}$$

for $n \in N_{\alpha,\beta}$. If

$$q_3 = \sum_{s=\alpha}^{\beta} c(s) \prod_{\tau=\alpha}^{s-1} \left[1 + \bar{B}(\tau)\right] < 1, \tag{4.5.12}$$

then

$$u(n) \leq a(n) + N_3 \prod_{s=\alpha}^{n-1} \left[1 + \bar{B}(s)\right] + \sum_{s=\alpha}^{n-1} \bar{A}(s) \prod_{\sigma=s+1}^{n-1} \left[1 + \bar{B}(\sigma)\right], \tag{4.5.13}$$

for $n \in N_{\alpha,\beta}$, where

$$\bar{A}(n) = r(n+1,n) a(n) + \sum_{s=\alpha}^{n-1} \Delta_1 r(n,s) a(s), \tag{4.5.14}$$

$$\bar{B}(n) = r(n+1,n) + \sum_{s=\alpha}^{n-1} \Delta_1 r(n,s), \tag{4.5.15}$$

and

$$N_3 = \frac{1}{1 - q_3} \sum_{s=\alpha}^{\beta} c(s) \left[a(s) + \sum_{\tau=\alpha}^{s-1} \bar{A}(\tau) \prod_{\sigma=\tau+1}^{s-1} \left[1 + \bar{B}(\sigma)\right]\right]. \tag{4.5.16}$$

(a_1) Define a function $z(n)$ by the right hand side of (4.5.1). Then $u(n) \leq z(n)$,

$$z(\alpha) = a(\alpha) + \sum_{s=\alpha}^{\beta} c(s)\, u(s), \tag{4.5.17}$$

and

$$\begin{aligned} \Delta z(n) &= \Delta a(n) + b(n)\, u(n) \\ &\leq \Delta a(n) + b(n)\, z(n). \end{aligned} \tag{4.5.18}$$

Now a suitable applicarion of Theorem 1.2.1 given in [42] to (4.5.18) and using the fact that $u(n) \leq z(n)$ we have

$$u(n) \leq z(\alpha) \prod_{s=\alpha}^{n-1} [1 + b(s)] + \sum_{s=\alpha}^{n-1} \Delta a(s) \prod_{\sigma=s+1}^{n-1} [1 + b(\sigma)]. \tag{4.5.19}$$

From (4.5.17),(4.5.19) and in view of (4.5.2) we have

$$z(\alpha) \leq N_1. \tag{4.4.20}$$

Using (4.5.20) in (4.5.19) we get the required inequality in (4.5.3).

The proofs of (a_2) and (a_3) follows by closely looking at the proof of (a_1) and the proofs of Theorem 1.5.1, part (a_2) and Theorem 1.5.2, part (b_1). Here we omit the details.

Remark 4.5.1. By taking $c(n) = 0$ in (a_1) and $N_{\alpha,\beta}$ is replaced by N_0, we get the inequality given in Theorem 1.2.6 in [42]. The inequalities in (a_2) and (a_3) can be considered as the useful variants of the inequalities in Theorems 1.2.3 and 1.3.4 given in [42].

The next theorem contains the inequalities investigated in [54,75].

Theorem 4.5.2. Let $u(n) \in D(N_{\alpha,\beta}, R_+)$ and $k \geq 0$ be a real constant.

(b_1) Let $a(n,s), b(n,s), c(n,s) \in D(E, R_+)$; $a(n,s), b(n,s)$ be nondecreasing in n for each $s \in N_{\alpha,\beta}$ where $E = \left\{(n,s) \in N^2_{\alpha,\beta} : \alpha \leq s \leq n \leq \beta\right\}$ and

$$u(n) \leq k + \sum_{s=\alpha}^{n-1} a(n,s) \left[u(s) + \sum_{\sigma=\alpha}^{s-1} c(s,\sigma)\, u(\sigma)\right] + \sum_{s=\alpha}^{\beta} b(n,s)\, u(s), \tag{4.5.21}$$

for $n \in N_{\alpha,\beta}$. If

$$q(n) = \sum_{s=\alpha}^{\beta} b(n,s) \prod_{\xi=\alpha}^{n-1} [1 + B(n,\xi)] < 1, \tag{4.5.22}$$

for $n \in N_{\alpha,\beta}$, where

$$B(n,\xi) = a(n,\xi)\left[1 + \sum_{\sigma=\alpha}^{\xi-1} c(\xi,\sigma)\right], \tag{4.5.23}$$

for $(n,\xi) \in E$, then

$$u(n) \leq \frac{k}{1 - q(n)} \prod_{\xi=\alpha}^{n-1} [1 + B(n,\xi)], \tag{4.5.24}$$

for $n \in N_{\alpha,\beta}$.

(b_2) Let $f(n), g(n), h(n) \in D(N_{\alpha,\beta}, R_+)$ and

$$u(n) \leq k + \sum_{s=\alpha}^{n-1} f(s)\left[u(s) + \sum_{\sigma=\alpha}^{s-1} g(\sigma)u(\sigma) + \sum_{\sigma=\alpha}^{\beta} h(\sigma)u(\sigma)\right], \tag{4.5.25}$$

for $n \in N_{\alpha,\beta}$. If

$$r = \sum_{\sigma=\alpha}^{\beta} h(\sigma) \prod_{\tau=\alpha}^{\sigma-1} [1 + f(\tau) + g(\tau)] < 1, \tag{4.5.26}$$

then

$$u(n) \leq \frac{k}{1 - r} \prod_{s=\alpha}^{n-1} [1 + f(s) + g(s)], \tag{4.5.27}$$

for $n \in N_{\alpha,\beta}$.

Proof. (b_1) Fix any $m \in N_{\alpha,\beta}$, then for $\alpha \leq n \leq m$, from (4.5.21) we have

$$u(n) \leq k + \sum_{s=\alpha}^{n-1} a(m,s)\left[u(s) + \sum_{\sigma=\alpha}^{s-1} c(m,\sigma)u(\sigma)\right] + \sum_{s=\alpha}^{\beta} b(m,s)u(s). \tag{4.5.28}$$

Define a function $z(n,m)$, $\alpha \leq n \leq m$ by the right hand side of (4.5.28). Then for $\alpha \leq n \leq m$, $u(n) \leq z(n,m)$, $z(n,m)$ is nondecreasing in n,

$$z(\alpha,m) = k + \sum_{s=\alpha}^{\beta} b(m,s)u(s), \tag{4.5.29}$$

and

$$\Delta_1 z(n,m) = a(m,n)\left[u(n) + \sum_{\sigma=\alpha}^{n-1} c(n,\sigma)u(\sigma)\right]$$

$$\leq a(m,n)\left[1+\sum_{\sigma=\alpha}^{n-1} c(n,\sigma)\right] z(n,m),$$

i.e.,

$$z(n+1,m) \leq \left[1+a(m,n)\left[1+\sum_{\sigma=\alpha}^{n-1} c(n,\sigma)\right]\right] z(n,m), \tag{4.5.30}$$

for $\alpha \leq n \leq m$. By setting $n=\xi$ in (4.5.30) and subsituting $\xi = \alpha, \alpha+1, ..., m-1$ successively, we obtain

$$z(m,m) \leq z(\alpha,m)\prod_{\xi=\alpha}^{m-1}\left[1+a(m,\xi)\left[1+\sum_{\sigma=\alpha}^{\xi-1} c(\xi,\sigma)\right]\right]. \qquad (4.5.31])$$

Since m is arbitrary, from (4.5.31) and (4.5.29) with m replaced by n and using $u(n) \leq z(n,n)$ we have

$$u(n) \leq z(\alpha,n)\prod_{\xi=\alpha}^{n-1}\left[1+a(n,\xi)\left[1+\sum_{\sigma=\alpha}^{\xi-1} c(\xi,\sigma)\right]\right], \tag{4.5.32}$$

where

$$z(\alpha,n) = k+\sum_{s=\alpha}^{\beta} b(n,s)\,u(s), \tag{4.5.33}$$

Using (4.5.32) on the right hand side of (4.5.33) and in view of (4.5.22) it is easy to observe that

$$z(\alpha,n) \leq \frac{k}{1-q(n)}. \tag{4.5.34}$$

Using (4.5.34) in (4.5.32) and (4.5.23) we get (4.5.24).

(b_2) Define a function $z(n)$ by the right hand side of (4.5.25). Then $z(\alpha) = k$,, $u(n) \leq z(n)$ and

$$\Delta z(n) = f(n)\left[u(n)+\sum_{\sigma=\alpha}^{n-1} g(\sigma)\,u(\sigma)+\sum_{\sigma=\alpha}^{\beta} h(\sigma)\,u(\sigma)\right],$$

$$\leq f(n)\left[z(n)+\sum_{\sigma=\alpha}^{n-1} g(\sigma)\,z(\sigma)+\sum_{\sigma=\alpha}^{\beta} h(\sigma)\,z(\sigma)\right],$$

for $n \in N_{\alpha,\beta}$. Define a function $v(n)$ by

$$v(n) = z(n)+\sum_{\sigma=\alpha}^{n-1} g(\sigma)\,z(\sigma)+\sum_{\sigma=\alpha}^{\beta} h(\sigma)\,z(\sigma). \tag{4.5.35}$$

Then $z(n) \leq v(n)$, $\Delta z(n) \leq f(n)\, v(n)$,

$$v(\alpha) = k + \sum_{\sigma=\alpha}^{\beta} h(\sigma)\, z(\sigma), \tag{4.5.36}$$

and

$$\begin{aligned} \Delta v(n) &= \Delta z(n) + g(n)\, z(n) \\ &\leq [f(n) + g(n)]\, v(n). \end{aligned} \tag{4.5.37}$$

Now a suitable application of Theorem 1.2.1 given in [42] to (4.5.37) yields

$$v(n) \leq v(\alpha) \prod_{s=\alpha}^{n-1} [1 + f(s) + g(s)]. \tag{4.5.38}$$

Using (4.5.38) in $z(n) \leq v(n)$ we get

$$z(n) \leq v(\alpha) \prod_{s=\alpha}^{n-1} [1 + f(s) + g(s)], \tag{4.5.39}$$

for $n \in N_{\alpha,\beta}$. Using (4.5.39) on the right hand side of (4.5.36) and in view of (4.5.26) we observe that

$$v(\alpha) \leq \frac{k}{1-r}. \tag{4.5.40}$$

Using (4.5.40) in (4.5.39) and the fact that $u(n) \leq z(n)$ we get the required inequality in (4.5.27).

Remark 4.5.2. We note that, if we take in Theorem 4.5.2, part (b_1), $c(n,s) = 0$, then we get the inequality established in [52, Theorem 2]. Furthermore, in the various special cases of Theorem 4.5.2, we get new inequalities which can be used conveniently in certain situations.

In the following theorem, we present some of the inequalities established in [39,44].

Theorem 4.5.3. Let $u(n), a(n), b(n) \in D(N_0, R_+)$.

(c_1) Let $a(n)$ be nonincreasing for $n \in N_0$. If

$$u(n) \leq a(n) + \sum_{s=n+1}^{\infty} b(s)\, u(s), \tag{4.5.41}$$

for $n \in N_0$, then

$$u(n) \leq a(n) \prod_{s=n+1}^{\infty} [1 + b(s)], \tag{4.5.42}$$

for $n \in N_0$.

(c_2) Let $c(n) \in D(N_0, R_+)$. If

$$u(n) \leq a(n) + b(n) \sum_{s=n+1}^{\infty} c(s) u(s), \tag{4.5.43}$$

for $n \in N_0$, then

$$u(n) \leq a(n) + b(n) d(n) \prod_{s=n+1}^{\infty} [1 + c(s) b(s)], \tag{4.5.44}$$

for $n \in N_0$, where

$$d(n) = \sum_{s=n+1}^{\infty} c(s) a(s), \tag{4.5.45}$$

for $n \in N_0$.

(c_3) Let $L : N_0 \times R_+ \to R_+$ be a function which satisfies the condition

$$0 \leq L(n, u) - L(n, v) \leq M(n, v)(u - v), \tag{4.5.46}$$

for $n \in N_0$, $u \geq v \geq 0$, where $M : N_0 \times R_+ \to R_+$. If

$$u(n) \leq a(n) + b(n) \sum_{s=n+1}^{\infty} L(s, u(s)), \tag{4.5.47}$$

for $n \in N_0$, then

$$u(n) \leq a(n) + b(n) e(n) \prod_{s=n+1}^{\infty} [1 + M(s, a(s)) b(s)], \tag{4.5.48}$$

for $n \in N_0$, where

$$e(n) = \sum_{s=n+1}^{\infty} L(s, a(s)), \tag{4.5.49}$$

for $n \in N_0$

Proof. (c_1) Let $a(n) > 0$ for $n \in N_0$, then from (4.5.41) it is easy to observe that

$$\frac{u(n)}{a(n)} \leq 1 + \sum_{s=n+1}^{\infty} b(s) \frac{u(s)}{a(s)}. \tag{4.5.50}$$

Define a function $z(n)$ by the right hand side of (4.5.50), then $\frac{u(n)}{a(n)} \leq z(n)$ and

$$z(n) - z(n+1) = b(n+1) \frac{u(n+1)}{a(n+1)}$$

$$\leq b\,(n+1)\,z\,(n+1)\,. \tag{4.5.51}$$

From (4.5.51) we observe that

$$z\,(n) \leq [1 + b\,(n+1)]\,z\,(n+1)\,. \tag{4.5.52}$$

By setting $n = s$ in (4.5.52) and then substituting $s = n, n+1, ..., m-1$ ($m \geq n+1$ is arbitrary in N_0) successively, we obtain the estimate

$$z\,(n) \leq z\,(m) \prod_{s=n+1}^{m} [1 + b\,(s)]\,. \tag{4.5.53}$$

Noting that $\lim_{m \to \infty} z\,(m) = 1$ and by letting $m \to \infty$ in (4.5.53) we get

$$z\,(n) \leq \prod_{s=n+1}^{\infty} [1 + b\,(s)]\,. \tag{4.5.54}$$

Using (4.5.54) in $\frac{u(n)}{a(n)} \leq z\,(n)$ we get the desired inequality in (4.5.42). The proof of the case when $a\,(n) \geq 0$ can be completed as mentioned in the proof of Theorem 4.2.3, part (b_1).

(c_2) Define a function $z(n)$ by

$$z\,(n) = \sum_{s=n+1}^{\infty} c\,(s)\,u\,(s)\,, \tag{4.5.55}$$

for $n \in N_0$ Then (4.5.43) can be written as

$$u\,(n) \leq a\,(n) + b\,(n)\,z\,(n)\,. \tag{4.5.56}$$

From (4.5.55) and (4.5.56) we have

$$z\,(n) \leq d\,(n) + \sum_{s=n+1}^{\infty} c\,(s)\,b\,(s)\,z\,(s)\,, \tag{4.5.57}$$

where $d(n)$ is given by (4.5.45). Clearly $d(n)$ is real-valued, nonnegative and nonincreasing function for $n \in N_0$. Now a suitable application of the inequality in part (c_1) to (4.5.57) yields

$$z\,(n) \leq d\,(n) \prod_{s=n+1}^{\infty} [1 + c\,(s)\,b\,(s)]\,. \tag{4.5.58}$$

Using (4.5.58) in (4.5.56) we get the required inequality in (4.5.44).

(c_3) Define a function $z(n)$ by

$$z(n) = \sum_{s=n+1}^{\infty} L(s, u(s)), \tag{4.5.59}$$

then from (4.4.47) we have

$$u(n) \leq a(n) + b(n) z(n). \tag{4.5.60}$$

From (4.5.59), (4.5.60) and the hypotheses on L, we observe that

$$z(n) \leq \sum_{s=n+1}^{\infty} [L(s, a(s) + b(s) z(s)) - L(s, a(s)) + L(s, a(s))]$$

$$\leq e(n) + \sum_{s=n+1}^{\infty} M(s, a(s)) b(s) z(s), \tag{4.5.61}$$

where $e(n)$ is given by (4.5.49). Clearly $e(n)$ is real-valued, nonnegative and nonincreasing function for $n \in N_0$. Now an application of the inequality in part (c_1) to (4.5.61) yields

$$z(n) \leq e(n) \prod_{s=n+1}^{\infty} [1 + M(s, a(s)) b(s)]. \tag{4.5.62}$$

The desired inequality in (4.5.48) follows from (4.5.60) and (4.5.62).

Our last theorem in this section gives the inequalities proved in [37].

Theorem 4.5.4. Let $u(n), a(n), b(n) \in D(N_0, R_+)$ and $p > 1$ be a real constant.

(d_1) Let $f(n), g(n) \in D(N_0, R_+)$. If

$$u^p(n) \leq a(n) + b(n) \sum_{s=n+1}^{\infty} [f(s) u(s) + g(s)], \tag{4.5.63}$$

for $n \in N_0$, then

$$u(n) \leq \left[a(n) + b(n) A(n) \prod_{s=n+1}^{\infty} \left[1 + \frac{b(s)}{p} f(s)\right]\right]^{\frac{1}{p}}, \tag{4.5.64}$$

for $n \in N_0$, where

$$A(n) = \sum_{s=n+1}^{\infty} \left[f(s) \left(\frac{p-1}{p} + \frac{a(s)}{p}\right) + g(s)\right], \tag{4.5.65}$$

for $n \in N_0$.

(d_2) Let L, M be as in Theorem 4.5.3, part (c_3) and the condition (4.5.46) holds. If

$$u^p(n) \le a(n) + b(n) \sum_{s=n+1}^{\infty} L(s, u(s)), \tag{4.5.66}$$

for $n \in N_0$, then

$$u(n) \le \Big[a(n) + b(n) B(n) \times \prod_{s=n+1}^{\infty} \left[1 + M\left(s, \frac{p-1}{p} + \frac{a(s)}{p}\right) \frac{b(s)}{p}\right]\Big]^{\frac{1}{p}}, \tag{4.5.67}$$

for $n \in N_0$, where

$$B(n) = \sum_{s=n+1}^{\infty} L\left(s, \frac{p-1}{p} + \frac{a(s)}{p}\right), \tag{4.5.68}$$

for $n \in N_0$.

Proof. (d_1) Define a function $z(n)$ by

$$z(n) = \sum_{s=n+1}^{\infty} [f(s) u(s) + g(s)], \tag{4.5.69}$$

for $n \in N_0$. Then (4.5.63) can be written as

$$u^p(n) \le a(n) + b(n) z(n). \tag{4.5.70}$$

From (4.5.70) as in the proof of Theorem 1.3.1, part (a_1) we obtain

$$u(n) \le \frac{p-1}{p} + \frac{a(n)}{p} + \frac{b(n)}{p} z(n). \tag{4.5.71}$$

From (4.5.69) and (4.5.71) we have

$$z(n) \le \sum_{s=n+1}^{\infty} \left[f(s)\left(\frac{p-1}{p} + \frac{a(s)}{p} + \frac{b(s)}{p} z(s)\right) + g(s)\right]$$

$$= A(n) + \sum_{s=n+1}^{\infty} f(s) \frac{b(s)}{p} z(s), \tag{4.5.72}$$

where $A(n)$ is given by (4.5.65). Clearly $A(n)$ is real-valued, nonnegative and nonincreasing function for $n \in N_0$. Now an application of Theorem 4.5.3, part (c_3) to (4.5.72) yields

$$z(n) \le A(n) \prod_{s=n+1}^{\infty} \left[1 + f(s) \frac{b(s)}{p}\right]. \tag{4.5.73}$$

The desired inequality in (4.5.64) follows from (4.5.70) and (4.5.73).

(d_2) The proof can be completed by closely looking at the proof of (d_1) and the proof of Theorem 4.5.3, part (c_3). We omit the details.

4.6 Applications

The inequalities given in earlier sections are recently investigated and used in various contexts. In this section we present applications of some of the inequalities to study basic properties of solutions of certain finite difference and sum-difference equations, which we hope will be a source for future work.

4.6.1 Perturbed difference equations

Consider a system of finite difference equations

$$x(n+1) = A(n)x(n) + f(n, x(n)) + r(n), \quad x(0) = x_0, \tag{4.6.1}$$

as a perturbation of the linear system

$$y(n+1) = A(n)y(n), \quad y(0) = x_0 \tag{4.6.2}$$

where $n \in N_0$, x, y, f, r are the elements of R^m, the m dimensional Euclidean space, $A(n)$ is an $m \times m$ matrix with $\det A(n) \neq 0$, the functions r and f are defined on N_0 and $N_0 \times R^m$ respectively and x_0 is a given vector in R^m. The symbol $|.|$ will denote some convenient norm on R^m as well as a corresponding consistent matrix norm. We denote by $Y(n)$ the fundamental solution matrix of the system (4.6.2) such that $Y(0) = I$, the identity matrix. It is known that the solution $x(n)$ of (4.6.1) is equivalent to the sum-difference equation (see [42, p. 55])

$$x(n) = Y(n)Y^{-1}(0)x_0 + \sum_{s=0}^{n-1} Y(n)Y^{-1}(s+1)\{f(s, x(s)) + r(s)\}. \tag{4.6.3}$$

We assume that the fundamental solution matrix $Y(n)$ of (4.6.2) satisfies

$$\left|Y(n)Y^{-1}(s)\right| \leq M, \quad 0 \leq s \leq n; s, n \in N_0, \tag{4.6.4}$$

where M is a positive constant.

The following theorems illustrate the applications of Theorem 4.2.1 (see [57]).

Theorem 4.6.1. Suppose that the function f in (4.6.1) satisfies

$$|f(n,x)| \leq p(n)|x|, \tag{4.6.5}$$

for $n \in N_0$, $x \in R^m$, where $p(n) \in D(N_0, R_+)$.If $x(n)$ is any solution of equation (4.6.1) for $n \in N_0$, then

$$|x(n)| \leq M\left\{|x_0| + \sum_{s=0}^{n-1}(|r(s)| + |x_0| Mp(s)) \prod_{\sigma=s+1}^{n-1}[1 + Mp(\sigma)]\right\}, \tag{4.6.6}$$

for $n \in N_0$, where M is given as in (4.6.4).

Proof. By using the variation of constants formula any solution $x(n)$ of (4.6.1) is represented by (4.6.3). Using (4.6.4), (4.6.5) in (4.6.3) we obtain

$$|x(n)| \leq M|x_0| + M\left(\sum_{s=0}^{n-1}|r(s)| + \sum_{s=0}^{n-1} p(s)|x(s)|\right). \tag{4.6.7}$$

Now a suitable application of Theorem 4.2.1 to (4.6.7) yields the required estimation in (4.6.6).

Theorem 4.6.2. Suppose that the function f in (4.6.1) satisfies

$$|f(n,x) - f(n,y)| \leq p(n)|x - y|, \tag{4.6.8}$$

for $n \in N_0$, $x, y \in R^m$, where $p(n) \in D(N_0, R_+)$. Then the equation (4.6.1) has at most one solution on N_0.

Proof. Let $x_1(n)$ and $x_2(n)$ be two solutions of (4.6.1) on N_0, then we have

$$x_1(n) - x_2(n) = \sum_{s=0}^{n-1} Y(n) Y^{-1}(s+1)\{f(s, x_1(s)) - f(s, x_2(s))\}. \tag{4.6.9}$$

From (4.6.9), (4.6.4), (4.6.8) we obtain

$$|x_1(n) - x_2(n)| \leq \sum_{s=0}^{n-1} \left|Y(n) Y^{-1}(s+1)\right| |f(s, x_1(s)) - f(s, x_2(s))|$$

$$\leq M \sum_{s=0}^{n-1} p(s)|x_1(s) - x_2(s)|. \tag{4.6.10}$$

By a suitable application of Theorem 4.2.1 to (4.6.10) we have $|x_1(n) - x_2(n)| \leq 0$.Therefore $x_1(n) = x_2(n)$ i.e., there is at most one solution of the equation (4.6.1) on N_0.

4.6.2 Volterra type difference equations involving iterated sums

In this section we present applications of the inequality in Theorem 4.4.2, part (b_1) (see [67]) to study certain properties of solutions of nonlinear sum-difference equation of the form

$$y(n) = f(n) + \sum_{s=0}^{n-1} F(n, s, y(s)) + \sum_{s=0}^{n-1} \left(\sum_{\sigma=0}^{s-1} H(n, s, \sigma, y(\sigma)) \right), \tag{4.6.11}$$

for $n \in N_0$, where $y(n) \in D(N_0, R)$ is an unknown function, $f \in D(N_0, R)$; $F : E_1 \times R \to R$, $H : E_2 \times R \to R$ in which $E_1 = \{(n, s) \in N_0^2 : 0 \le s \le n < \infty\}$, $E_2 = \{(n, s, \sigma) \in N_0^3 : 0 \le \sigma \le s \le n < \infty\}$.

Theorem 4.6.3. Suppose that the functions f, F, H in equation (4.6.11) satisfy the conditions

$$|f(n)| \le c, \tag{4.6.12}$$

$$|F(n, s, y)| \le k(n, s) |y|, \tag{4.6.13}$$

$$|H(n, s, \sigma, y)| \le h(n, s, \sigma) |y|, \tag{4.6.14}$$

where $c \ge 0$ is a real constant and $k(n, s) \in D(E_1, R_+)$, $h(n, s, \sigma) \in D(E_2, R_+)$. If $y(n)$ is any solution of equation (4.6.11) on N_0, then

$$|y(n)| \le c \prod_{s=0}^{n-1} [1 + P(s) + Q(s)], \tag{4.6.15}$$

where $P(n), Q(n)$ are given by (4.4.15), (4.4.16) in which $\Delta_1 k(n, s) \in D(E_1, R_+)$, $\Delta_1 h(n, s, \sigma) \in D(E_2, R_+)$.

Proof. Let $y(n)$ be a solution of equation (4.6.11). Using (4.6.12)-(4.6.14) in (4.6.11) we have

$$|y(n)| \le c + \sum_{s=0}^{n-1} k(n, s) |y(s)| + \sum_{s=0}^{n-1} \left(\sum_{\sigma=0}^{s-1} h(n, s, \sigma) |y(\sigma)| \right). \tag{4.6.16}$$

Now an application of Theorem 4.4.2, part (b_1) to (4.6.16) yields the required estimate in (4.6.15).

Theorem 4.6.4. Suppose that the functions F, H in equation (4.6.11) satisfy the conditions

$$|F(n, s, y) - F(n, s, \bar{y})| \le k(n, s) |y - \bar{y}|, \tag{4.6.17}$$

$$|H(n, s, \sigma, y) - H(n, s, \sigma, \bar{y})| \le h(n, s, \sigma) |y - \bar{y}|, \tag{4.6.18}$$

where $k(n, s)$, $h(n, s, \sigma)$ are as in Theorem 4.6.3. Let $P(n), Q(n)$ be as in Theorem 4.6.3. Then the equation (4.6.11) has at most one solution on N_0.

Proof. Let $u(n)$ and $v(n)$ be two solutions of equation (4.6.11) on N_0. Using this fact and the conditions (4.6.17), (4.6.18) we have

$$|u(n) - v(n)| \leq \sum_{s=0}^{n-1} k(n,s)\,|u(s) - v(s)|$$

$$+\sum_{s=0}^{n-1}\left(\sum_{\sigma=0}^{s-1} h(n,s,\sigma)\,|u(\sigma) - v(\sigma)|\right). \qquad (4.6.19)$$

Now a suitable application of Theorem 4.4.2, part (b_1) (when $c = 0$) to (4.6.19) yields $u(n) = v(n)$ i.e., there is at most one solution of equation (4.6.11) on N_0

4.6.3 Volterra-Fredholm type sum-difference equations

In this section we present applications of the inequality in Theorem 4.5.1, part (a_2) to study certain properties of solutions of Volterra-Fredholm type sum-difference equation of the form

$$z(n) = e(n) + \sum_{s=\alpha}^{n-1} F(n,s,z(s)) + \sum_{s=\alpha}^{\beta} G(n,s,z(s)), \qquad (4.6.20)$$

for $n \in N_{\alpha,\beta}$, where $z(n) \in D(N_{\alpha,\beta}, R)$ is an unknown function, $e(n) \in D(N_{\alpha,\beta}, R)$; $F, G : E \times R \to R$ in which $E = \left\{(n,s) \in N^2_{\alpha,\beta} : \alpha \leq s \leq n \leq \beta\right\}$, see [70].

Theorem 4.6.5. Suppose that the functions e, F, G in equation (4.6.20) satisfy the conditions

$$|e(n)| \leq a(n), \qquad (4.6.21)$$

$$|F(n,s,z)| \leq b(n)\,f(s)\,|z|, \qquad (4.6.22)$$

$$|G(n,s,z)| \leq c(n)\,g(s)\,|z|, \qquad (4.6.23)$$

where $a(n), b(n), c(n), f(n), g(n) \in D(N_{\alpha,\beta}, R_+)$. Let q_2 be as in (4.5.6), Theorem 4.5.1, part (a_2).If $z(n)$ is a solution of equation (4.6.20) on $N_{\alpha,\beta}$, then

$$|z(n)| \leq L_1(n) + N_2 L_2(n), \qquad (4.6.24)$$

for $n \in N_{\alpha,\beta}$,where $L_1(n)$, $L_2(n)$, N_2 are as in Theorem 4.5.1, part (a_2)

Proof. Let $z(n)$ be a solution of equation (4.6.20) on $N_{\alpha,\beta}$. Using the fact that $z(n)$ is a solution of equation (4.6.20) and (4.6.21)-(4.6.23) we have

$$|z(n)| \leq a(n) + b(n) \sum_{s=\alpha}^{n-1} f(s) |z(s)| + c(n) \sum_{s=\alpha}^{\beta} g(s) |z(s)| . \tag{4.6.25}$$

Now an application of the inequality in Theorem 4.5.1, part (a_2) to (4.6.25) yields the required estimate in (4.6.24).

Theorem 4.6.6. Suppose that the funcrions F, G in equation (4.6.20) satisfy the conditions

$$|F(n, s, z) - F(n, s, \bar{z})| \leq b(n) f(s) |z - \bar{z}| , \tag{4.6.26}$$

$$|G(n, s, z) - G(n, s, \bar{z})| \leq c(n) g(s) |z - \bar{z}| , \tag{4.6.27}$$

where $b(n), c(n), f(n), g(n) \in D(N_{\alpha,\beta}, R_+)$. Let $q_2, L_1(n), L_2(n), N_2$ be as in Theorem 4.5.1, part (a_2). Then the equation (4.6.20) has at most one solution on $N_{\alpha,\beta}$.

Proof. Let $u(n)$ and $v(n)$ be two solutions of equation (4.6.20) on $N_{\alpha,\beta}$.Using the facts that $u(n)$ and $v(n)$ are the solutions of equation (4.6.20) and (4.6.26), (4.6.27) we have

$$|u(n) - v(n)| \leq b(n) \sum_{s=\alpha}^{n-1} f(s) |u(s) - v(s)| + c(n) \sum_{s=\alpha}^{\beta} g(s) |u(s) - v(s)|. \tag{4.6.28}$$

Now an application of the inequality given in Theorem 4.5.1, part (a_2) (with $a(n) = 0$ which in fact implies $L_1(n) = 0$, $N_2 = 0$) to (4.6.28) yields $u(n) = v(n)$ i.e., there is at most one solution of equation (4.6.20) on $N_{\alpha,\beta}$.

4.6.4 Fredholm type sum-difference equations

In this section we present applications given in [75] of the special version of the inequality in Theorem 4.5.2, part (b_2) to study the properties of solutions of the Fredholm type sum-difference equation

$$\Delta x(n) = F\left(n, x(n), \sum_{\sigma=\alpha}^{\beta} k(n, \sigma, x(\sigma))\right), \tag{4.6.29}$$

with the given initial condition

$$x(\alpha) = x_0 \tag{4.6.30}$$

where x, k, F are the elements of R^m an m-dimensional Euclidean space with norm $|.|$ and $k : E \times R^m \to R^m$, $F : N_{\alpha,\beta} \times R^m \times R^m \to R^m$, in which $E = \left\{ (n, s) \in N^2_{\alpha,\beta} : \alpha \leq s \leq n \leq \beta \right\}$.

Theorem 4.6.7. Assume that

$$|k(n, s, x)| \leq e(n) h(s) |x|, \tag{4.6.31}$$

$$|F(n, x, y)| \leq f(n) (|x| + |y|), \tag{4.6.32}$$

where $e(n), h(n), f(n) \in D(N_{\alpha,\beta}, R_+)$ and $e(n) \geq 1$. Let

$$r_0 = \sum_{\sigma=\alpha}^{\beta} h(\sigma) \prod_{\tau=\alpha}^{\sigma-1} [1 + e(\tau) f(\tau)] < 1. \tag{4.6.33}$$

If $x(n)$ is any solution of (4.6.29)-(4.6.30), then

$$|x(n)| \leq \frac{|x_0|}{1 - r_0} \prod_{s=\alpha}^{n-1} [1 + e(s) f(s)], \tag{4.6.34}$$

for $n \in N_{\alpha,\beta}$.

Proof. The solution $x(n)$ of (4.6.29)-(4.6.30) satisfies the following equivalent sum-difference equation

$$x(n) = x_0 + \sum_{s=\alpha}^{n-1} F\left(s, x(s), \sum_{\sigma=\alpha}^{\beta} k(s, \sigma, x(\sigma)) \right). \tag{3.6.35)}$$

Using (4.6.31), (4.6.32) in (4.6.35) we observe that

$$|x(n)| \leq |x_0| + \sum_{s=\alpha}^{n-1} f(s) \left(|x(s)| + \sum_{\sigma=\alpha}^{\beta} e(s) h(\sigma) |x(\sigma)| \right)$$

$$\leq |x_0| + \sum_{s=\alpha}^{n-1} f(s) e(s) \left(|x(s)| + \sum_{\sigma=\alpha}^{\beta} h(\sigma) |x(\sigma)| \right). \tag{4.6.36}$$

Now a suitable application of Theorem 4.5.2, part (b_2) (when $g(n) = 0$) to (4.6.36) yields (4.6.34).

Theorem 4.6.8. Let $x(n), y(n)$, $n \in N_{\alpha,\beta}$ be the solutions of (4.6.29) with initial conditions

$$x(\alpha) = x_0, \tag{4.6.37}$$

$$y(\alpha) = y_0, \tag{4.6.38}$$

respectively.Suppose that the functions k and F in equation (4.6.29) satisfy the conditions

$$|k(n,s,x) - k(n,s,y)| \leq e(n)\, h(s)\, |x-y|, \tag{4.6.39}$$

$$|F(n,x,y) - F(n,\bar{x},\bar{y})| \leq f(n)\,(|x-\bar{x}| + |y-\bar{y}|), \tag{4.6.40}$$

where $e(n), h(n), f(n)$ are given as in Theorem 4.6.7. Let r_0 be as given in (4.6.33). Then

$$|x(n) - y(n)| \leq \frac{|x_0 - y_0|}{1 - r_0} \prod_{s=\alpha}^{n-1} [1 + e(s)\, f(s)], \tag{4.6.41}$$

for $n \in N_{\alpha,\beta}$.

Proof. Using the facts that $x(n), y(n)$ are the solutions of (4.6.29)-(4.6.37), (4.6.29)-(4.6.38) respectively, we have

$$x(n) - y(n) = x_0 - y_0 + \sum_{s=\alpha}^{n-1} \left\{ F\left(s, x(s), \sum_{\sigma=\alpha}^{\beta} k(s,\sigma,x(\sigma))\right) \right.$$

$$\left. -F\left(s, y(s), \sum_{\sigma=\alpha}^{\beta} k(s,\sigma,y(\sigma))\right)\right\}. \tag{4.6.42}$$

Using (4.6.39), (4.6.40), (4.6.42) we observe that

$$|x(n) - y(n)| \leq |x_0 - y_0| + \sum_{s=\alpha}^{n-1} f(s)\, e(s) \left(|x(s) - y(s)| \right.$$

$$\left. + \sum_{\sigma=\alpha}^{\beta} h(\sigma)\, |x(\sigma) - y(\sigma)| \right). \tag{4.6.43}$$

Now a suitable application of Theorem 4.5.2, part (b_2) (when $g(n) = 0$) to (4.6.43) yields the desired estimate in (4.6.41), which shows the continuous dependence of solutions of equation (4.6.29) on given initial data.

Finally, we note that a variety of new methods and tools are developed by various investigators to study different types of finite difference equations. The inequalities and applications given above are recently investigated and further progress is expected.

4.7 Notes

Owing to the considerable applications,recently some new finite difference inequalities are developed to widen the scope of their applications. This chapter presents some basic finite difference inequalities recently developed in the literature. Sections 4.2-4.5 are devoted to the variety of new finite difference inequalities investigated by Pachpatte in [35,37,39,44,45,53,54,55,57,67,68,70,73 ,75]. I think that these inequalities places a new stepping stone to the vast literature on the subject and inspire further work in this area.In section 4.6,some applications are discussed to illustrate,how some of these inequalities can be used to study various types of finite and sum-difference equations.The number of applications of the inequalities given here is considerable and those presented in section 1.6 are taken from some of the above noted references.

Chapter 5

Finite difference inequalities in two variables

5.1 Introduction

The study of dynamics of physical systems governed by partial finite difference equations is equally important. Inspite of the great possibilities for applications, the theory of partial finite difference equations is developing rather slowely. Indeed, one need new theory and efficient techniques for its significant developments. Finite difference inequalities in two and more independent variables which provide explicit estimates on unknown functions have become very effective and powerful tools for studying qualitative behavior of solutions of partial finite difference equations. In the past few years a large number of new finite difference inequalities involving functions of two independent variables have been discovered and used in various applications, see [36,38,40,41,45,48,49,53,55,56,62,66,68,71,76]. In this chapter, our goal is to present some basic finite difference inequalities recently discovered and which can be used as handy tools in the study of different classes of partial finite and sum-difference equations. Applications of some of the inequalities are also presented.

5.2 Some basic finite difference inequalities

During the past few years some new finite difference inequalities have been developed in order to widen the scope of their applications. In this section we present some fundamental finite difference inequalities involving functions of two independent variables, recently investigated by Pachpatte in [56,68,55,40,45].

Our first theorem deals with the comparison inequalities related to certain partial finite difference equations proved in [56].

Theorem 5.2.1. Let $\bar{M}_0 = \{m_0, m_0+1, ...\}, \bar{N}_0 = \{n_0, n_0+1, ...\}$ where m_0, n_0 are integers and $\Delta_0 = \bar{M}_0 \times \bar{N}_0$.

(a_1) Let $f(m,n,r)$ be a function defined for $(m,n) \in \Delta_0$. $0 \le r < \infty$ and nondecreasing with respect to r for fixed $(m,n) \in \Delta_0$. Let $u(m,n)$ and $v(m,n)$ be two functions defined for $(m,n) \in \Delta_0$ and $u(m_0,n_0) \le v(m_0,n_0)$. Assume further that

$$u(m+1,n+1) \le f(m,n,u(m,n)), \tag{5.2.1}$$

$$v(m+1,n+1) \ge f(m,n,v(m,n)), \tag{5.2.2}$$

for $(m,n) \in \Delta_0$. Then

$$u(m,n) \le v(m,n), \tag{5.2.3}$$

for $(m,n) \in \Delta_0$.

(a_2) Suppose that the functions $W_1(m,n,r), W_2(m,n,r)$ be nonnegative and defined for $(m,n) \in \Delta_0$, $0 \le r < \infty$ and nondecreasing with respect to r for fixed $(m,n) \in \Delta_0$. Let $z(m,n)$ be a function defined for $(m,n) \in \Delta_0$ and

$$W_2(m,n,z(m,n)) \le z(m+1,n+1) \le W_1(m,n,z(m,n)), \tag{5.2.4}$$

for $(m,n) \in \Delta_0$. Let $u(m,n)$ and $v(m,n)$ be solutions of the difference equations

$$u(m+1,n+1) = W_1(m,n,u(m,n)), u(m_0,n_0) = u_0, \tag{5.2.5}$$

$$v(m+1,n+1) = W_2(m,n,v(m,n)), v(m_0,n_0) = v_0, \tag{5.2.6}$$

and suppose that $v_0 \le z(m_0,n_0) \le u_0$. Then

$$v(m,n) \le z(m,n) \le u(m,n), \tag{5.2.7}$$

for $(m,n) \in \Delta_0$.

(a_3) Let $x(m,n)$ and $y(m,n)$ be solutions of the difference equations

$$x(m+1,n+1) = g(m,n,x(m,n)), x(m_0,n_0) = x_0, \tag{5.2.8}$$

and

$$y(m+1,n+1) = h(m,n,y(m,n)), y(m_0,n_0) = y_0, \tag{5.2.9}$$

where $x(m,n), y(m,n), g(m,n,r), h(m,n,r)$ are defined for $(m,n) \in \Delta_0$, $0 \le r < \infty$. Let the functions $W_1(m,n,r)$ and $W_2(m,n,r)$ be as in (a_2). Suppose that the functions g and h in (5.2.8) and (5.2.9) satisfy the condition

$$W_2(m,n,|x-y|) \le |g(m,n,x) - h(m,n,y)| \le W_1(m,n,|x-y|), \tag{5.2.10}$$

for $(m,n) \in \Delta_0$. Let $u(m,n)$ and $v(m,n)$ be solutions of the equations (5.2.5) and (5.2.6) for $(m,n) \in \Delta_0$ and assume that $v_0 \le |x_0 - y_0| \le u_0$. Then

$$v(m,n) \le |x(m,n) - y(m,n)| \le u(m,n), \tag{5.2.11}$$

for $(m,n) \in \Delta_0$.

Proof. (a_1) Since $u(m_0,n_0) \le v(m_0,n_0)$, from the nondecreasing character of f we obtain

$$\begin{aligned} u(m_0+1,n_0+1) &\le f(m_0,n_0,u(m_0,n_0)) \\ &\le f(m_0,n_0,v(m_0,n_0)) \\ &\le v(m_0+1,n_0+1). \end{aligned}$$

If the inequality (5.2.3) is fulfilled for $m = m_0+i$, $n = n_0+i$ $(i = 2,3,...,k)$, it follows by the nondecreasing character of f that

$$\begin{aligned} u(m_0+k+1,n_0+k+1) &\le f(m_0+k,n_0+k,u(m_0+k,n_0+k)) \\ &\le f(m_0+k,n_0+k,v(m_0+k,n_0+k)) \\ &\le v(m_0+k+1,n_0+k+1). \end{aligned}$$

Hence by mathematical induction we obtain (5.2.3).

(a_2) Applying the inequality in part (a_1) to the second part of (5.2.4) and (5.2.5) we obtain the right half of the inequality (5.2.7). A similar argument yields the left half of the inequality (5.2.7).

(a_3) Let $z(m,n) = |x(m,n) - y(m,n)|$. Then $z(m_0,n_0) = |x(m_0,n_0) - y(m_0,n_0)| \le u(m_0,n_0)$. On account of the nondecreasing nature of $W_1(m,n,r)$ we obtain

$$\begin{aligned} z(m_0+1,n_0+1) &= |x(m_0+1,n_0+1) - y(m_0+1,n_0+1)| \\ &= |g(m_0,n_0,x(m_0,n_0)) - h(m_0,n_0,y(m_0,n_0))| \\ &\le W_1(m_0,n_0,|x(m_0,n_0) - y(m_0,n_0)|) \\ &\le W_1(m_0,n_0,u(m_0,n_0)) \\ &= u(m_0+1,n_0+1). \end{aligned}$$

If the inequality $z(m,n) \le u(m,n)$ is fulfilled for $m = m_0+i, n = n_0+i(i = 2,3,...k)$, then it follows by the nondecreasing nature of $W_1(m,n,r)$ that

$$\begin{aligned} &z(m_0+k+1,n_0+k+1) = |x(m_0+k+1,n_0+k+1) - y(m_0+k+1, \\ &n_0+k+1)| \\ &= |g(m_0+k,n_0+k,x(m_0+k,n_0+k)) - h(m_0+k,n_0+k,y(m_0+k,n_0+k))| \\ &\quad \le W_1(m_0+k,n_0+k,|x(m_0+k,n_0+k) - y(m_0+k,n_0+k)|) \\ &\quad \le W_1(m_0+k,n_0+k,u(m_0+k,n_0+k)) \\ &\quad = u(m_0+k+1,n_0+k+1). \end{aligned}$$

Hence by mathematical induction we obtain $|x(m,n) - y(m,n)| \le u(m,n)$ for $(m,n) \in \Delta_0$. The proof of the left half of the inequality (5.2.11) is similar.

Explicit representation of the solution $u(m,n)$ (or $v(m,n)$) of a comparision equation of the form (5.2.5) (or (5.2.6)) is not always possible. Therefore in applications this solution is often replaced by an upper (or lower) bound for it. The following theorems deals with some such inequalities proved in [40,45,55,68].

Theorem 5.2.2. Let $u(m,n), a(m,n) \in D(N_0^2, R_+), k(m,n,\sigma,\tau), \Delta_1 k(m, n,\sigma,\tau), \Delta_2 k(m,n,\sigma,\tau), \Delta_2\Delta_1 k(m,n,\sigma,\tau) \in D(E, R_+)$ where
$E = \{(m,n,\sigma,\tau) \in N_0^4 : 0 \le \sigma \le m < \infty, 0 \le \tau \le n < \infty\}$.

(b_1) Let $g \in C(R_+, R_+)$ be a nondecreasing function with $g(u) > 0$ for $u > 0$. If

$$u(m,n) \le c + \sum_{\sigma=0}^{m-1}\sum_{\tau=0}^{n-1} k(m,n,\sigma,\tau) g(u(\sigma,\tau)), \tag{5.2.12}$$

for $m,n \in N_0$, where $c \ge 0$ is a real constant, then for $0 \le m \le m_1, 0 \le n \le n_1$; $m, m_1, n, n_1 \in N_0$,

$$u(m,n) \le G^{-1}\left[G(c) + \sum_{s=0}^{m-1}\sum_{t=0}^{n-1} Q(s,t)\right], \tag{5.2.13}$$

where

$$Q(m,n) = k(m+1,n+1,m,n) + \sum_{\sigma=0}^{m-1} \Delta_1 k(m,n+1,\sigma,n)$$

$$+\sum_{\tau=0}^{n-1} \Delta_2 k(m+1,n,m,\tau) + \sum_{\sigma=0}^{m-1}\sum_{\tau=0}^{n-1} \Delta_2\Delta_1 k(m,n,\sigma,\tau), \tag{5.2.14}$$

$$G(r) = \int_{r_0}^{r} \frac{ds}{g(s)}, r > 0, \tag{5.2.15}$$

$r_0 > 0$ is arbitrary, G^{-1} is the inverse of G and $m_1, n_1 \in N_0$ are chosen so that

$$G(c) + \sum_{s=0}^{m-1}\sum_{t=0}^{n-1} Q(s,t) \in Dom\left(G^{-1}\right),$$

for all m and n lying in $0 \le m \le m_1$ and $0 \le n \le n_1$.

(b_2) Let g, G, G^{-1} be as in (b_1) and suppose in addition $g(u)$ is subadditive. If

$$u(m,n) \le a(m,n) + \sum_{\sigma=0}^{m-1}\sum_{\tau=0}^{n-1} k(m,n,\sigma,\tau) g(u(\sigma,\tau)), \tag{5.2.16}$$

for $m, n \in N_0$, then for $0 \leq m \leq m_2, 0 \leq n \leq n_2$; $m, m_2, n, n_2 \in N_0$,

$$u(m,n) \leq a(m,n) + G^{-1}\left[G(A(m,n)) + \sum_{s=0}^{m-1}\sum_{t=0}^{n-1} Q(s,t)\right], \tag{5.2.17}$$

where $Q(m, n)$ is given by (5.2.14),

$$A(m,n) = \sum_{\sigma=0}^{m-1}\sum_{\tau=0}^{n-1} k(m,n,\sigma,\tau) g(a(\sigma,\tau)), \tag{5.2.18}$$

$m_2, n_2 \in N_0$ are chosen so that

$$G(A(m,n)) + \sum_{s=0}^{m-1}\sum_{t=0}^{n-1} Q(s,t) \in Dom\left(G^{-1}\right),$$

for all m and n lying in $0 \leq m \leq m_2$, $0 \leq n \leq n_2$.

Proof. (b_1) Let $c > 0$ and define a function $z(m,n)$ by the right hand side of (5.2.12). Then $z(0,n) = z(m,0) = c$, $u(m,n) \leq z(m,n)$, $z(m,n)$ is positive and nondecreasing for $m, n \in N_0$ and

$$\begin{aligned}
\Delta_1 z(m,n) &= z(m+1,n) - z(m,n) \\
&= \sum_{\sigma=0}^{m}\sum_{\tau=0}^{n-1} k(m+1,n,\sigma,\tau) g(u(\sigma,\tau)) \\
&- \sum_{\sigma=0}^{m-1}\sum_{\tau=0}^{n-1} k(m+1,n,\sigma,\tau) g(u(\sigma,\tau)) \\
&+ \sum_{\sigma=0}^{m-1}\sum_{\tau=0}^{n-1} k(m+1,n,\sigma,\tau) g(u(\sigma,\tau)) \\
&- \sum_{\sigma=0}^{m-1}\sum_{\tau=0}^{n-1} k(m,n,\sigma,\tau) g(u(\sigma,\tau)) \\
&= \sum_{\tau=0}^{n-1} k(m+1,n,m,\tau) g(u(m,\tau)) \\
&+ \sum_{\sigma=0}^{m-1}\sum_{\tau=0}^{n-1} \Delta_1 k(m,n,\sigma,\tau) g(u(\sigma,\tau)).
\end{aligned} \tag{5.2.19}$$

From (5.2.19) we have

$$\Delta_2\Delta_1 z(m,n) = \Delta_1 z(m,n+1) - \Delta_1 z(m,n)$$

$$= \sum_{\tau=0}^{n} k(m+1,n+1,m,\tau) g(u(m,\tau))$$
$$+ \sum_{\sigma=0}^{m-1} \sum_{\tau=0}^{n} \Delta_1 k(m,n+1,\sigma,\tau)\, g(u(\sigma,\tau))$$
$$- \sum_{\tau=0}^{n-1} k(m+1,n,m,\tau)\, g(u(m,\tau))$$
$$- \sum_{\sigma=0}^{m-1} \sum_{\tau=0}^{n-1} \Delta_1 k(m,n,\sigma,\tau)\, g(u(\sigma,\tau))$$
$$= \sum_{\tau=0}^{n} k(m+1,n+1,m,\tau) g(u(m,\tau))$$
$$- \sum_{\tau=0}^{n-1} k(m+1,n+1,m,\tau)\, g(u(m,\tau))$$
$$+ \sum_{\tau=0}^{n-1} k(m+1,n+1,m,\tau)\, g(u(m,\tau))$$
$$- \sum_{\tau=0}^{n-1} k(m+1,n,m,\tau)\, g(u(m,\tau))$$
$$+ \sum_{\sigma=0}^{m-1} \sum_{\tau=0}^{n} \Delta_1 k(m,n+1,\sigma,\tau)\, g(u(\sigma,\tau))$$
$$- \sum_{\sigma=0}^{m-1} \sum_{\tau=0}^{n-1} \Delta_1 k(m,n,\sigma,\tau)\, g(u(\sigma,\tau))$$
$$= k(m+1,n+1,m,n)\, g(u(m,n)) + \sum_{\tau=0}^{n-1} \Delta_2 k(m+1,n,m,\tau) g(u(m,\tau))$$
$$+ \sum_{\sigma=0}^{m-1} \left\{ \sum_{\tau=0}^{n} \Delta_1 k(m,n+1,\sigma,\tau)\, g(u(\sigma,\tau)) - \sum_{\tau=0}^{n-1} \Delta_1 k(m,n,\sigma,\tau)\, g(u(\sigma,\tau)) \right\}$$
$$= k(m+1,n+1,m,n)\, g(u(m,n)) + \sum_{\tau=0}^{n-1} \Delta_2 k(m+1,n,m,\tau) g(u(m,\tau))$$
$$+ \sum_{\sigma=0}^{m-1} \left\{ \sum_{\tau=0}^{n} \Delta_1 k(m,n+1,\sigma,\tau)\, g(u(\sigma,\tau)) - \sum_{\tau=0}^{n-1} \Delta_1 k(m,n+1,\sigma,\tau)\, g(u(\sigma,\tau)) \right.$$
$$\left. + \sum_{\tau=0}^{n-1} \Delta_1 k(m,n+1,\sigma,\tau)\, g(u(\sigma,\tau)) - \sum_{\tau=0}^{n-1} \Delta_1 k(m,n,\sigma,\tau)\, g(u(\sigma,\tau)) \right\}$$

$$= k(m+1,n+1,m,n)\,g(u(m,n)) + \sum_{\tau=0}^{n-1} \Delta_2 k(m+1,n,m,\tau) g(u(m,\tau))$$

$$+ \sum_{\sigma=0}^{n-1} \Delta_1 k(m,n+1,\sigma,n)\,g(u(\sigma,n))$$

$$+ \sum_{\sigma=0}^{m-1} \sum_{\tau=0}^{n-1} \Delta_2\Delta_1 k(m,n,\sigma,\tau) g(u(\sigma,\tau))$$

$$\leq k(m+1,n+1,m,n)\,g(z(m,n)) + \sum_{\tau=0}^{n-1} \Delta_2 k(m+1,n,m,\tau) g(z(m,\tau))$$

$$+ \sum_{\sigma=0}^{n-1} \Delta_1 k(m,n+1,\sigma,n)\,g(z(\sigma,n))$$

$$+ \sum_{\sigma=0}^{m-1} \sum_{\tau=0}^{n-1} \Delta_2\Delta_1 k(m,n,\sigma,\tau) g(z(\sigma,\tau))$$

$$\leq Q(m,n)\,g(z(m,n)). \tag{5.2.20}$$

The rest of the proof can be completed by following the proof of Theorem 5.2.1 given in [42, p. 388].

(b_2) Define a function $z(m,n)$ by

$$z(m,n) = \sum_{\sigma=0}^{m-1} \sum_{\tau=0}^{n-1} k(m,n,\sigma,\tau)\,g(u(\sigma,\tau)). \tag{5.2.21}$$

From (5.2.21) and using the fact that $u(m,n) \leq a(m,n) + z(m,n)$ and hypotheses on g we have

$$z(m,n) \leq \sum_{\sigma=0}^{m-1} \sum_{\tau=0}^{n-1} k(m,n,\sigma,\tau)\,g(a(\sigma,\tau) + z(\sigma,\tau))$$

$$\leq A(m,n) + \sum_{\sigma=0}^{m-1} \sum_{\tau=0}^{n-1} k(m,n,\sigma,\tau)\,g(z(\sigma,\tau)), \tag{5.2.22}$$

where $A(m,n)$ is given by (5.2.18). The remaining proof can be completed by closely looking at the proof of Theorem 5.2.2 given in [42].

Remark 5.2.1. We note that the inequalities in (b_1) and (b_2) are the further generalizations of the inequality given in Theorem 5.2.1 in [42], which can be used in more general situations.

Theorem 5.2.3. Let $u(m,n)$, $k(m,n,\sigma,\tau), \Delta_1 k(m,n,\sigma,\tau)$, $\Delta_2 k(m,n,\sigma,\tau)$, $\Delta_2\Delta_1 k(m,n,\sigma,\tau)$ and c be as in Theorem 5.2.2.

(c_1) If

$$u^2(m,n) \le c + \sum_{\sigma=0}^{m-1}\sum_{\tau=0}^{n-1} k(m,n,\sigma,\tau)u(\sigma,\tau), \tag{5.2.23}$$

for $m,n \in N_0$, then

$$u(m,n) \le \sqrt{c} + \frac{1}{2}\sum_{s=0}^{m-1}\sum_{t=0}^{n-1} Q(s,t), \tag{5.2.24}$$

for $m,n \in N_0$, where $Q(m,n)$ is given by (5.2.14)

(c_2) Let $g(u)$ be as in Theorem 5.2.2, part (b_1). If

$$u^2(m,n) \le c + \sum_{\sigma=0}^{m-1}\sum_{\tau=0}^{n-1} k(m,n,\sigma,\tau)u(\sigma,\tau)\,g(u(\sigma,\tau)), \tag{5.2.25}$$

for $m,n \in N_0$, then for $0 \le m \le m_3, 0 \le n \le n_3; m, m_3, n, n_3 \in N_0$,

$$u(m,n) \le G^{-1}\left[G\left(\sqrt{c}\right) + \frac{1}{2}\sum_{s=0}^{m-1}\sum_{t=0}^{n-1} Q(s,t)\right], \tag{5.2.26}$$

where $Q(m,n)$ is given by (5.2.14), G, G^{-1} are as defined in Theorem 5.2.2, part (b_1) and m_3, n_3 are chosen so that

$$G\left(\sqrt{c}\right) + \frac{1}{2}\sum_{s=0}^{m-1}\sum_{t=0}^{n-1} Q(s,t) \in Dom\left(G^{-1}\right),$$

for all $m,n \in N_0$ lying in $0 \le m \le m_3, 0 \le n \le n_3$.

Proof. Let $c > 0$ and define a function $z(m,n)$ by the right hand side of (5.2.23). Then $z(0,n) = z(m,0) = c$, $u(m,n) \le \sqrt{z(m,n)}$, $z(m,n)$ is positive and nondecreasing for $m,n \in N_0$ and following the proof of Theorem 5.2.2, part (b_1) we get

$$\Delta_2\Delta_1 z(m,n) \le Q(m,n)\sqrt{z(m,n)}. \tag{5.2.27}$$

The rest of the proof follows by using the arguments as in the proof of Theorem 5.4.1 given in [42].

(c_2) The proof can be completed by following the proof of (c_1) given above and the proof of Theorem 5.4.3 in [42]. We omit the details.

Remark 5.2.2. In the special case when $k(m,n,\sigma,\tau) = a(\sigma,\tau)$, the inequalities in (c_1) and (c_2) reduces to the corresponding inequalities in Theorem 5.4.1 and Theorem 5.4.3 given in [42].

Theorem 5.2.4. Let $u(m,n), a(m,n), b(m,n), g(m,n), h(m,n) \in D\left(N_0^2, R_+\right)$ and $p > 1$ is a real constant.

(d_1) If

$$u^p(m,n) \le a(m,n) + b(m,n) \sum_{\sigma=0}^{m-1} \sum_{\tau=0}^{n-1} \left[g(s,t)\, u^p(s,t) + h(s,t)\, u(s,t)\right], \quad (5.2.28)$$

for $m, n \in N_0$, then

$$u(m,n) \le \{a(m,n) + b(m,n)\, e(m,n)$$

$$\times \prod_{s=0}^{m-1} \left[1 + \sum_{t=0}^{n-1} \left(g(s,t) + \frac{h(s,t)}{p}\right) b(s,t)\right]\Bigg\}^{\frac{1}{p}}, \quad (5.2.29)$$

for $m, n \in N_0$, where

$$e(m,n) = \sum_{s=0}^{m-1} \sum_{t=0}^{n-1} \left[g(s,t)\, a(s,t) + h(s,t)\left(\frac{p-1}{p} + \frac{a(s,t)}{p}\right)\right], \quad (5.2.30)$$

for $m, n \in N_0$.

(d_2) Let $L : N_0^2 \times R_+ \to R_+$ be a function which satisfies the condition

$$0 \le L(m,n,u) - L(m,n,v) \le M(m,n,v)(u-v),$$

for $u \ge v \ge 0$, where $M : N_0^2 \times R_+ \to R_+$. If

$$u^p(m,n) \le a(m,n) + b(m,n) \sum_{\sigma=0}^{m-1} \sum_{\tau=0}^{n-1} L(s,t,u(s,t)), \quad (5.2.31)$$

for $m, n \in N_0$, then

$$u(m,n) \le \{a(m,n) + b(m,n)\, \bar{e}(m,n)$$

$$\times \prod_{s=0}^{m-1} \left[1 + \sum_{t=0}^{n-1} M\left(s,t,\frac{p-1}{p} + \frac{a(s,t)}{p}\right) \frac{b(s,t)}{p}\right]\Bigg\}^{\frac{1}{p}}, \quad (5.2.32)$$

for $m, n \in N_0$, where

$$\bar{e}(m,n) = \sum_{s=0}^{m-1} \sum_{t=0}^{n-1} L\left(s,t,\frac{p-1}{p} + \frac{a(s,t)}{p}\right), \quad (5.2.33)$$

for $m, n \in N_0$.

Proof. (d_1) Define a function $z(m,n)$ by

$$z(m,n) = \sum_{s=0}^{m-1} \sum_{t=0}^{n-1} \left[g(s,t)\, u^p(s,t) + h(s,t)\, u(s,t) \right], \tag{5.2.34}$$

then $z(m,0) = z(0,n) = 0$ and (5.2.28) can be written as

$$u^p(m,n) \leq a(m,n) + b(m,n)\, z(m,n). \tag{5.2.35}$$

From (5.2.35) as in the proof of Theorem 2.3.3, part (c_1) we obtain

$$u(m,n) \leq \frac{p-1}{p} + \frac{a(m,n)}{p} + \frac{b(m,n)}{p} z(m,n). \tag{5.2.36}$$

From (5.2.34)-(5.2.36) we observe that

$$z(m,n) \leq \sum_{s=0}^{m-1} \sum_{t=0}^{n-1} \Big[g(s,t) \left(a(s,t) + b(s,t)\, z(s,t) \right)$$

$$+ h(s,t) \left(\frac{p-1}{p} + \frac{a(s,t)}{p} + \frac{b(s,t)}{p} z(s,t) \right) \Big]$$

$$= e(m,n) + \sum_{s=0}^{m-1} \sum_{t=0}^{n-1} b(s,t) \left(g(s,t) + \frac{h(s,t)}{p} \right) z(s,t), \tag{5.2.37}$$

where $e(m,n)$ is given by (5.2.30). Clearly $e(m,n)$ is nonnegative and nondecreasing function for $m,n \in N_0$. Now an application of Theorem 4.2.2 given in [42] to (5.2.37) yields

$$z(m,n) \leq e(m,n) \prod_{s=0}^{m-1} \left[1 + \sum_{t=0}^{n-1} b(s,t) \left(g(s,t) + \frac{h(s,t)}{p} \right) \right]. \tag{5.2.38}$$

The required inequality in (5.2.29) follows from (5.2.35) and (5.2.38).

(d_2) Define a function $z(m,n)$ by

$$z(m,n) = \sum_{s=0}^{m-1} \sum_{t=0}^{n-1} L(s,t,u(s,t)), \tag{5.2.39}$$

then as in the proof of Part (d_1) above, from (5.2.31) we see that the inequalities (5.2.35), (5.2.36) hold.From (5.2.39),(5.2.36) and the assumptions on L it follows that

$$z(m,n) \leq \sum_{s=0}^{m-1} \sum_{t=0}^{n-1} \left\{ L\left(s,t,\frac{p-1}{p} + \frac{a(s,t)}{p} + \frac{b(s,t)}{p} z(s,t) \right) \right.$$

$$-L\left(s,t,\frac{p-1}{p}+\frac{a(s,t)}{p}\right)+L\left(s,t,\frac{p-1}{p}+\frac{a(s,t)}{p}\right)\Bigg\}$$

$$\leq \bar{e}(m,n)+\sum_{s=0}^{m-1}\sum_{t=0}^{n-1}M\left(s,t,\frac{p-1}{p}+\frac{a(s,t)}{p}\right)\frac{b(s,t)}{p}z(s,t), \tag{5.2.40}$$

where $\bar{e}(m,n)$ is given by (5.2.33).Clearly $\bar{e}(m,n)$ is nonnegative and nondecreasing function for $m,n\in N_0$. An application of Theorem 4.2.2 given in [42] to (5.2.40) yields

$$z(m,n)\leq \bar{e}(m,n)\prod_{s=0}^{m-1}\left[1+\sum_{t=0}^{n-1}M\left(s,t,\frac{p-1}{p}+\frac{a(s,t)}{p}\right)\frac{b(s,t)}{p}\right]. \tag{5.2.41}$$

From (5.2.35) and (5.2.41) the desired inequality in (5.2.32) follows.

Remark 5.2.3. We note that the inequalities given in (d_1) and (d_2) are of more general type and in the various special cases, one can obtain new inequalities which can also be used as tools in certain applications.

Theorem 5.2.5. Let $u(m,n), f(m,n)\in D\left(N_0^2,R_+\right)$, $h(m,n,\sigma,\tau)\in D(E,R_+)$ and $c\geq 0$, $p>1$ be real constants, where $E=\left\{(m,n,\sigma,\tau)\in N_0^4 : 0\leq\sigma\leq m<\infty,\ 0\leq\tau\leq n<\infty\right\}$

(e_1) If

$$u^p(m,n)\leq c+\sum_{s=0}^{m-1}\sum_{t=0}^{n-1}\Bigg[f(s,t)\,g(u(s,t))$$

$$+\sum_{\sigma=0}^{s-1}\sum_{\tau=0}^{t-1}h(s,t,\sigma,\tau)\,g(u(\sigma,\tau))\Bigg], \tag{5.2.42}$$

for $m,n\in N_0$, then for $0\leq m\leq m_4, 0\leq n\leq n_4; m,m_4,n,n_4\in N_0$,

$$u(m,n)\leq\left\{H^{-1}\left[H(c)+B(m,n)\right]\right\}^{\frac{1}{p}}, \tag{5.2.43}$$

where

$$B(m,n)=\sum_{s=0}^{m-1}\sum_{t=0}^{n-1}\left[f(s,t)+\sum_{\sigma=0}^{s-1}\sum_{\tau=0}^{t-1}h(s,t,\sigma,\tau)\right], \tag{5.2.44}$$

$$H(r)=\int_{r_0}^{r}\frac{ds}{g\left(s^{\frac{1}{p}}\right)}, r>0, \tag{5.2.45}$$

$r_0>0$ is arbitrary, H^{-1} is the inverse of H and $m_4,n_4\in N_0$ are chosen so that

$$H(c)+B(m,n)\in Dom\left(H^{-1}\right),$$

for all m,n lying in $0\leq m\leq m_4, 0\leq n\leq n_4$.

(e_2) If

$$u^p(m,n) \le c + \sum_{s=0}^{m-1}\sum_{t=0}^{n-1}\left[f(s,t)\,u(s,t) + \sum_{\sigma=0}^{s-1}\sum_{\tau=0}^{t-1} h(s,t,\sigma,\tau)u(\sigma,\tau)\right], \quad (5.2.46)$$

for $m, n \in N_0$, then

$$u(m,n) \le \left\{c^{\frac{p-1}{p}} + \left(\frac{p-1}{p}\right) B(m,n)\right\}^{\frac{1}{p-1}}, \quad (5.2.47)$$

for $m, n \in N_0$, where $B(m,n)$ is given by (5.2.44).

Proof. (e_1) Let $c > 0$ and define a function $z(m,n)$ by the right hand side of (5.2.42).Then $z(0,n) = z(m,0) = c$, $u(m,n) \le \{z(m,n)\}^{\frac{1}{p}}$, $z(m,n)$ is positive and nondecreasing for $m, n \in N_0$ and

$$\begin{aligned} &z(m+1,n) - z(m,n) \\ &= \sum_{t=0}^{n-1}\left[f(m,t)\,g(u(m,t)) + \sum_{\sigma=0}^{m-1}\sum_{\tau=0}^{t-1} h(m,t,\sigma,\tau)\,g(u(\sigma,\tau))\right] \\ &\le \sum_{t=0}^{n-1}\left[f(m,t)\,g\left(\{z(m,t)\}^{\frac{1}{p}}\right) + \sum_{\sigma=0}^{m-1}\sum_{\tau=0}^{t-1} h(m,t,\sigma,\tau)\,g\left(\{z(\sigma,\tau)\}^{\frac{1}{p}}\right)\right] \\ &\le g\left(\{z(m,n)\}^{\frac{1}{p}}\right)\sum_{t=0}^{n-1}\left[f(m,t) + \sum_{\sigma=0}^{m-1}\sum_{\tau=0}^{t-1} h(m,t,\sigma,\tau)\right]. \end{aligned} \quad (5.2.48)$$

From (5.2.45), (5.2.48) we observe that

$$\begin{aligned} H(z(m+1,n)) - H(z(m,n)) &= \int_{z(m,n)}^{z(m+1,n)} \frac{ds}{g\left(s^{\frac{1}{p}}\right)} \\ &\le \frac{z(m+1,n) - z(m,n)}{g\left(\{z(m,n)\}^{\frac{1}{p}}\right)} \\ &\le \sum_{t=0}^{n-1}\left[f(m,t) + \sum_{\sigma=0}^{m-1}\sum_{\tau=0}^{t-1} h(m,t,\sigma,\tau)\right]. \end{aligned} \quad (5.2.49)$$

Keeping n fixed in (5.2.49), setting $m = s$ and summing over s from 0 to $m-1$ we obtain

$$H(z(m,n)) \le H(c) + B(m,n). \quad (5.2.50)$$

Now substituting the bound on $z(m,n)$ from (5.2.50) in $u(m,n) \le \{z(m,n)\}^{\frac{1}{p}}$, we obtain the required inequality in (5.2.43). The proof of the case when $c \ge 0$ can be completed as mentioned in the proof of Theorem 4.2.3, part (b_1). The domain $0 \le m \le m_4, 0 \le n \le n_4$ is obvious.

(e_2) The proof is similar to that of Theorem 1.3.4. We omit the details.

Remark 5.2.4. We note that the inequality in (e_1) is a Bihari type discrete inequality in two independent variables and if we take $p = 2, h = 0$ in (e_2), then we get a slight variant of the inequality in Theorem 5.4.1 given in [42].

5.3 Further finite difference inequalities

In view of the important applications,a great deal of attention has been given to establish finite difference inequalities which provide explicit bounds on unknown functions. In this section, we offer some more finite difference inequalities investigated by Pachpatte in [38,53,66] which provide a natural and effective means in certain applications.

We begin with the following theorem which contains the inequalities proved in [38].

Theorem5.3.1. Let $u(m,n), a(m,n), b(m,n), p(m,n), g(m,n), h(m,n) \in D(N_0^2, R_+)$. Let $L : N_0^2 \times R_+ \to R_+$ be a function which satisfies the condition

$$0 \le L(m,n,u) - L(m,n,v) \le M(m,n,v)(u-v), \tag{5.3.1}$$

for $u \ge v \ge 0$, where $M : N_0^2 \times R_+ \to R_+$.

(a_1) Let $a(m,n)$ be nondecreasing in m. If

$$u(m,n) \le a(m,n) + p(m,n) \sum_{s=0}^{m-1} b(s,n) u(s,n)$$

$$+ \sum_{s=0}^{m-1} \sum_{t=0}^{n-1} L(s,t,u(s,t)), \tag{5.3.2}$$

for $m, n \in N_0$, then

$$u(m,n) \le f(m,n) [a(m,n) + e(m,n)$$

$$\times \prod_{s=0}^{m-1} \left[1 + \sum_{t=0}^{n-1} M(s,t,f(s,t)a(s,t)) f(s,t)\right]\Bigg], \tag{5.3.3}$$

for $m, n \in N_0$, where

$$f(m,n) = 1 + p(m,n) \sum_{s=0}^{m-1} b(s,n) \prod_{\sigma=s+1}^{m-1} [1 + b(\sigma,n) p(\sigma,n)], \tag{5.3.4}$$

$$e(m,n) = \sum_{s=0}^{m-1}\sum_{t=0}^{n-1} L(s,t,f(s,t)\,a(s,t)), \tag{5.3.5}$$

for $m, n \in N_0$.

(a_2) Let $a(m,n)$ be as in (a_1). If

$$u(m,n) \le a(m,n) + \sum_{s=0}^{m-1} g(s,n)\left(u(s,n) + \sum_{\sigma=0}^{s-1} h(\sigma,n)\,u(\sigma,n)\right)$$

$$+ \sum_{s=0}^{m-1}\sum_{t=0}^{n-1} L(s,t,u(s,t)), \tag{5.3.6}$$

for $m, n \in N_0$, then

$$u(m,n) \le k(m,n)\,[a(m,n) + \bar{e}(m,n)$$

$$\times \prod_{s=0}^{m-1}\left[1 + \sum_{t=0}^{n-1} M(s,t,k(s,t)\,a(s,t))\,k(s,t)\right]\Bigg], \tag{5.3.7}$$

for $m, n \in N_0$, where

$$k(m,n) = 1 + \sum_{s=0}^{m-1} g(s,n) \prod_{\sigma=0}^{s-1} [1 + g(\sigma,n) + h(\sigma,n)], \tag{5.3.8}$$

$$\bar{e}(m,n) = \sum_{s=0}^{m-1}\sum_{t=0}^{n-1} L(s,t,k(s,t)\,a(s,t)), \tag{5.3.9}$$

for $m, n \in N_0$.

Proof. (a_1) Define a function $z(m,n)$ by

$$z(m,n) = a(m,n) + \sum_{s=0}^{m-1}\sum_{t=0}^{n-1} L(s,t,u(s,t)). \tag{5.3.10}$$

Then (5.3.2) can be restated as

$$u(m,n) \le z(m,n) + p(m,n) \sum_{s=0}^{m-1} b(s,n)\,u(s,n). \tag{5.3.11}$$

Clearly $z(m,n)$ is nonnegative and nondecreasing function for $m \in N_0$. Treating (5.3.11) as an one dimensional inequality for any fixed $n \in N_0$ and a suitable application of Theorem 1.2.4 given in [42] to (5.3.11) yields

$$u(m,n) \le z(m,n)\,f(m,n), \tag{5.3.12}$$

where $f(m,n)$ is defined by (5.3.4). From (5.3.10) and (5.3.12) we have

$$u(m,n) \le f(m,n)\left[a(m,n) + r(m,n)\right], \tag{5.3.13}$$

where

$$r(m,n) = \sum_{s=0}^{m-1}\sum_{t=0}^{n-1} L(s,t,u(s,t)). \tag{5.3.14}$$

Using (5.3.13),(5.3.1) in (5.3.14) we observe that

$$r(m,n) \le \sum_{s=0}^{m-1}\sum_{t=0}^{n-1} \{L(s,t,f(s,t)[a(s,t)+r(s,t)])$$

$$-L(s,t,f(s,t)a(s,t)) + L(s,t,f(s,t)a(s,t))\}$$

$$\le e(m,n) + \sum_{s=0}^{m-1}\sum_{t=0}^{n-1} M(s,t,f(s,t)a(s,t))f(s,t)r(s,t), \tag{5.3.15}$$

where $e(m,n)$ is defined by (5.3.5). It is easy to observe that $e(m,n)$ is nonnegative and nondecreasing for $m,n \in N_0$. Now a suitable application of Theorem 4.2.2 given in [42] to (5.3.15) yields

$$r(m,n) \le e(m,n) \prod_{s=0}^{m-1}\left[1 + \sum_{t=0}^{n-1} M(s,t,f(s,t)a(s,t))f(s,t)\right]. \tag{5.3.16}$$

Now using (5.3.16) in (5.3.13) we get the desired inequality in (5.3.3).

(a_2) Define a function $z(m,n)$ by (5.3.10), then (5.3.6) can be written as

$$u(m,n) \le z(m,n) + \sum_{s=0}^{m-1} g(s,n)\left(u(s,n) + \sum_{\sigma=0}^{s-1} h(\sigma,n)u(\sigma,n)\right). \tag{5.3.17}$$

Clearly $z(m,n)$ is nonnegative and nondecreasing function for $m \in N_0$. Treating (5.3.17) as one-dimensional inequality for any fixed $n \in N_0$ and a suitable application of Theorem 1.4.2 given in [42] to (5.3.17) yields

$$u(m,n) \le z(m,n)k(m,n), \tag{5.3.18}$$

where $k(m,n)$ is defined by (5.3.8). Now by following the proof of (a_1) we obtain the desired inequality in (5.3.7).

Remark 5.3.1. If we take $p(m.n) = 0$ in (a_1), $g(m,n) = 0$ in (a_2), then $f(m,n) = 1 = k(m,n)$, $e(m,n) = \bar{e}(m,n) = \sum_{s=0}^{m-1}\sum_{t=0}^{n-1} L(s,t,a(s,t)) = e_0(m,n)$ (say) and the bounds obtained in (5.3.3) and (5.3.7) reduces to

$$u(m,n) \le a(m,n) + e_0(m,n) \prod_{s=0}^{m-1}\left[1 + \sum_{t=0}^{n-1} M(s,t,a(s,t))\right]. \tag{5.3.19}$$

For some such inequaliies and their applications, see [42].

The inequalities embodied in the following theorem are established in [53].

Theorem 5.3.2. Let $u(m,n), f(m,n), a(m,n) \in D(N_0^2, R_+)$, $k(m,n,\sigma,\tau)$, $\Delta_1 k(m,n,\sigma,\tau)$, $\Delta_2 k(m,n,\sigma,\tau)$, $\Delta_2\Delta_1 k(m,n,\sigma,\tau) \in D(E, R_+)$ and $c \geq 0$ be a real constant, where $E = \{(m,n,\sigma,\tau) \in N_0^4 : 0 \leq \sigma \leq m < \infty, 0 \leq \tau \leq n < \infty\}$.

(b_1) If

$$u(m,n) \leq c + \sum_{s=0}^{m-1}\sum_{t=0}^{n-1} f(s,t)\left[u(s,t) + \sum_{\sigma=0}^{s-1}\sum_{\tau=0}^{t-1} k(s,t,\sigma,\tau)u(\sigma,\tau)\right], \quad (5.3.20)$$

for $m, n \in N_0$, then

$$u(m,n) \leq c\left[1 + \sum_{s=0}^{m-1}\sum_{t=0}^{n-1} f(s,t)\prod_{\xi=0}^{s-1}\left[1 + \sum_{\eta=0}^{t-1}[f(\xi,\eta) + Q(\xi,\eta)]\right]\right], \quad (5.3.21)$$

for $m, n \in N_0$, where $Q(m,n)$ is defined by (5.2.14).

(b_2) If

$$u(m,n) \leq a(m,n) + \sum_{s=0}^{m-1}\sum_{t=0}^{n-1}\left[u(s,t)\right.$$

$$\left. + \sum_{\sigma=0}^{s-1}\sum_{\tau=0}^{t-1} k(s,t,\sigma,\tau)\,u(\sigma,\tau)\right], \quad (5.3.22)$$

for $m, n \in N_0$, then

$$u(m,n) \leq a(m,n) + H(m,n)\left[1 + \sum_{s=0}^{m-1}\sum_{t=0}^{n-1} f(s,t)\right.$$

$$\left. \times \prod_{\xi=0}^{s-1}\left[1 + \sum_{\eta=0}^{t-1}[f(\xi,\eta) + Q(\xi,\eta)]\right]\right], \quad (5.3.23)$$

for $m, n \in N_0$, where

$$H(m,n) = \sum_{s=0}^{m-1}\sum_{t=0}^{n-1} f(s,t)\left[a(s,t) + \sum_{\sigma=0}^{s-1}\sum_{\tau=0}^{t-1} k(s,t,\sigma,\tau)a(\sigma,\tau)\right], \quad (5.3.24)$$

for $m, n \in N_0$ and $Q(m,n)$ is defined by (5.2.14).

Proof. (b_1) Let $c > 0$ and define a function $z(m,n)$ by the right hand side of (5.3.20). Then $z(0,n) = z(m,0) = c$, $u(m,n) \le z(m,n)$ and

$$\Delta_2\Delta_1 z(m,n) = f(m,n)\left[u(m,n) + \sum_{\sigma=0}^{m-1}\sum_{\tau=0}^{n-1} k(m,n,\sigma,\tau)u(\sigma,\tau)\right]$$

$$\le f(m,n)\left[z(m,n) + \sum_{\sigma=0}^{m-1}\sum_{\tau=0}^{n-1} k(m,n,\sigma,\tau)z(\sigma,\tau)\right]. \tag{5.3.25}$$

Define a function $v(m,n)$ by

$$v(m,n) = z(m,n) + \sum_{\sigma=0}^{m-1}\sum_{\tau=0}^{n-1} k(m,n,\sigma,\tau)z(\sigma,\tau). \tag{5.3.26}$$

Then $v(m,0) = z(m,0) = c$, $v(0,n) = z(0,n) = c$, $\Delta_2\Delta_1 z(m,n) \le f(m,n)$ $v(m,n)$, $z(m,n) \le v(m,n)$ and following the proof of Theorem 5.2.2, part (b_1) and using the fact that $z(m,n)$ is nondecreasing for $m,n \in N_0$ we observe that

$$\Delta_2\Delta_1 v(m,n) \le \Delta_2\Delta_1 z(m,n) + Q(m,n)\, z(m,n)$$

$$\le [f(m,n) + Q(m,n)]\, v(m,n), \tag{5.3.27}$$

where $Q(m,n)$ is defined by (5.2.14). The rest of the proof can be completed as in the proof of Theorem 4.3.1 given in [42].

(b_2) The proof can be completed by following the proof of Theorem 4.3.3, part (a_4) given in [42]. We omit the details.

Remark 5.3.2. By taking $k(m,n,\sigma,\tau) = k(\sigma,\tau)$, the inequality in (b_1) reduces to the inequality in Theorem 4.3.1 given in [42]. The inequality in (b_2) is of more general type and can be used conveniently in certain situations.

In the following theorems we present the inequalities investigated in [66] which can be used in some applications.

Theorem 5.3.3. Let $E_1 = \{(m,n,s,t) \in N_0^4 : 0 \le s \le m < \infty, 0 \le t \le n < \infty\}$ and $E_2 = \{(m,n,s,t,\sigma,\tau) \in N_0^6 : 0 \le \sigma \le s \le m < \infty, 0 \le \tau \le t \le n < \infty\}$. Let $u(m,n) \in D(N_0^2, R_+)$; $k(m,n,s,t), \Delta_1 k(m,n,s,t), \Delta_2 k(m,n,s,t), \Delta_2\Delta_1 k$ $(m,n,s,t) \in D(E_1, R_+)$; $h(m,n,s,t,\sigma,\tau), \Delta_1 h(m,n,s,t,\sigma,\tau), \Delta_2 h(m,n,s,t,$ $\sigma,\tau), \Delta_2\Delta_1 h(m,n,s,t,\sigma,\tau) \in D(E_2, R_+)$ and $c \ge 0$ be a real constant.

(c_1) If

$$u(m,n) \leq c + \sum_{s=0}^{m-1} \sum_{t=0}^{n-1} k(m,n,s,t)\, u(s,t)$$

$$+ \sum_{s=0}^{m-1} \sum_{t=0}^{n-1} \left(\sum_{\sigma=0}^{s-1} \sum_{\tau=0}^{t-1} h(m,n,s,t,\sigma,\tau)\, u(\sigma,\tau) \right), \tag{5.3.28}$$

for $m, n \in N_0$, then

$$u(m,n) \leq c \prod_{x=0}^{m-1} \left[1 + \sum_{y=0}^{n-1} \left[A(x,y) + B(x,y) \right] \right], \tag{5.3.29}$$

for $m, n \in N_0$, where

$$A(x,y) = k(x+1, y+1, x, y) + \sum_{s=0}^{x-1} \Delta_1 k(x, y+1, s, y)$$

$$+ \sum_{t=0}^{y-1} \Delta_2 k(x+1, y, x, t) + \sum_{s=0}^{x-1} \sum_{t=0}^{y-1} \Delta_2 \Delta_1 k(x, y, s, t), \tag{5.3.30}$$

$$B(x,y) = \sum_{\sigma=0}^{x-1} \sum_{\tau=0}^{y-1} h(x+1, y+1, x, y, \sigma, \tau)$$

$$+ \sum_{s=0}^{x-1} \left(\sum_{\sigma=0}^{s-1} \sum_{\tau=0}^{y-1} \Delta_1 h(x, y+1, s, y, \sigma, \tau) \right)$$

$$+ \sum_{t=0}^{y-1} \left(\sum_{\sigma=0}^{x-1} \sum_{\tau=0}^{t-1} \Delta_2 h(x+1, y, x, t, \sigma, \tau) \right)$$

$$+ \sum_{s=0}^{x-1} \sum_{t=0}^{y-1} \left(\sum_{\sigma=0}^{s-1} \sum_{\tau=0}^{t-1} \Delta_2 \Delta_1 h(x, y, s, t, \sigma, \tau) \right). \tag{5.3.31}$$

(c_2) Ler $g \in C(R_+, R_+)$ be a nondecreasing function with $g(u) > 0$ for $u > 0$. If

$$u(m,n) \leq c + \sum_{s=0}^{m-1} \sum_{t=0}^{n-1} k(m,n,s,t)\, g(u(s,t))$$

$$+ \sum_{s=0}^{m-1} \sum_{t=0}^{n-1} \left(\sum_{\sigma=0}^{s-1} \sum_{\tau=0}^{t-1} h(m,n,s,t,\sigma,\tau)\, g(u(\sigma,\tau)) \right), \tag{5.3.32}$$

for $m,n \in N_0$, then for $0 \le m \le m_1, 0 \le n \le n_1; m, m_1, n, n_1 \in N_0$,

$$u(m,n) \le G^{-1}\left[G(c) + \sum_{x=0}^{m-1}\sum_{y=0}^{n-1}[A(x,y) + B(x,y)]\right], \tag{5.3.33}$$

where $A(x,y), B(x,y)$ are given by (5.3.30), (5.3.31),

$$G(r) = \int_{r_0}^{r} \frac{dw}{g(w)}, r > o, \tag{5.3.34}$$

$r_0 > o$ is arbitrary, G^{-1} is the inverse of G and $m_1, n_1 \in N_0$ be chosen so that

$$G(c) + \sum_{x=0}^{m-1}\sum_{y=0}^{n-1}[A(x,y) + B(x,y)] \in Dom\left(G^{-1}\right),$$

for all $m,n \in N_0$ such that $0 \le m \le m_1, 0 \le n \le n_1$.

Theorem 5.3.4. Let $u(m,n), k(m,n,s,t)$, $h(m,n,s,t,\sigma,\tau)$, c be as in Theorem 5.3.3 and $b(m,n) \in D\left(N_0^2, R_+\right)$.

(d_1) If

$$u(m,n) \le c + \sum_{s=0}^{m-1}\sum_{t=0}^{n-1} b(s,t)\,u(s,t) + \sum_{s=0}^{m-1}\sum_{t=0}^{n-1}\left(\sum_{\sigma=0}^{s-1}\sum_{\tau=0}^{t-1} k(s,t,\sigma,\tau)\,u(\sigma,\tau)\right)$$

$$+\sum_{s=0}^{m-1}\sum_{t=0}^{n-1}\left(\sum_{\sigma=0}^{s-1}\sum_{\tau=0}^{t-1}\left(\sum_{\xi=0}^{\sigma-1}\sum_{\eta=0}^{\tau-1} h(s,t,\sigma,\tau,\xi,\eta)\,u(\xi,\eta)\right)\right), \tag{5.3.35}$$

for $m,n \in N_0$, then

$$u(m,n) \le c\prod_{s=0}^{m-1}\left[1 + \sum_{t=0}^{n-1} Q(s,t)\right], \tag{5.3.36}$$

for $m,n \in N_0$, where

$$Q(m,n) = b(m,n) + \sum_{\sigma=0}^{m-1}\sum_{\tau=0}^{n-1} k(m,n,\sigma,\tau)$$

$$+\sum_{\sigma=0}^{m-1}\sum_{\tau=0}^{n-1}\left(\sum_{\xi=0}^{\sigma-1}\sum_{\eta=0}^{\tau-1} h(m,n,\sigma,\tau,\xi,\eta)\right). \tag{5.3.37}$$

(d_2) Let $g(u)$ be as in Theorem 5.3.3, part (c_2). If

$$u(m,n) \le c + \sum_{s=0}^{m-1}\sum_{t=0}^{n-1} b(s,t)\, g(u(s,t)) + \sum_{s=0}^{m-1}\sum_{t=0}^{n-1}\left(\sum_{\sigma=0}^{s-1}\sum_{\tau=0}^{t-1} k(s,t,\sigma,\tau)\, g(u(\sigma,\tau))\right)$$

$$+ \sum_{s=0}^{m-1}\sum_{t=0}^{n-1}\left(\sum_{\sigma=0}^{s-1}\sum_{\tau=0}^{t-1}\left(\sum_{\xi=0}^{\sigma-1}\sum_{\eta=0}^{\tau-1} h(s,t,\sigma,\tau,\xi,\eta)\, g(u(\xi,\eta))\right)\right), \qquad (5.3.38)$$

for $m, n \in N_0$, then for $0 \le m \le m_2, 0 \le n \le n_2$; $m, m_2, n, n_2 \in N_0$,

$$u(m,n) \le G^{-1}\left[G(c) + \sum_{s=0}^{m-1}\sum_{t=0}^{n-1} Q(s,t)\right], \qquad (5.3.39)$$

where $Q(x,y)$ is given by (5.3.37), G, G^{-1} are as in Theorem 5.3.3, part (c_2) and $m_2, n_2 \in N_0$ be chosen so that

$$G(c) + \sum_{s=0}^{m-1}\sum_{t=0}^{n-1} Q(s,t) \in Dom\left(G^{-1}\right),$$

for all $m, n \in N_0$ such that $0 \le m \le m_2, 0 \le n \le n_2$.

Proofs of Theorems 5.3.3 and 5.3.4. (c_1) Let $c > 0$ and define a function $z(m,n)$ by the right hand side of (5.3.28). Then $z(m,n) > 0$, $z(0,n) = z(m,0) = c$ and

$$\Delta_1 z(m,n) = z(m+1,n) - z(m,n)$$

$$= \sum_{s=0}^{m}\sum_{t=0}^{n-1} k(m+1,n,s,t)\, u(s,t) + \sum_{s=0}^{m}\sum_{t=0}^{n-1}\left(\sum_{\sigma=0}^{s-1}\sum_{\tau=0}^{t-1} h(m+1,n,s,t,\sigma,\tau)\, u(\sigma,\tau)\right)$$

$$- \sum_{s=0}^{m-1}\sum_{t=0}^{n-1} k(m,n,s,t)\, u(s,t) - \sum_{s=0}^{m-1}\sum_{t=0}^{n-1}\left(\sum_{\sigma=0}^{s-1}\sum_{\tau=0}^{t-1} h(m,n,s,t,\sigma,\tau)\, u(\sigma,\tau)\right)$$

$$= \sum_{t=0}^{n-1} k(m+1,n,m,t)\, u(m,t) + \sum_{s=0}^{m-1}\sum_{t=0}^{n-1} k(m+1,n,s,t)\, u(s,t)$$

$$- \sum_{s=0}^{m-1}\sum_{t=0}^{n-1} k(m,n,s,t)\, u(s,t)$$

$$+ \sum_{t=0}^{n-1}\left(\sum_{\sigma=0}^{m-1}\sum_{\tau=0}^{t-1} h(m+1,n,s,t,\sigma,\tau)\, u(\sigma,\tau)\right)$$

$$+ \sum_{s=0}^{m-1}\sum_{t=0}^{n-1}\left(\sum_{\sigma=0}^{s-1}\sum_{\tau=0}^{t-1} h(m+1,n,s,t,\sigma,\tau)\, u(\sigma,\tau)\right)$$

$$-\sum_{s=0}^{m-1}\sum_{t=0}^{n-1}\left(\sum_{\sigma=0}^{s-1}\sum_{\tau=0}^{t-1} h(m,n,s,t,\sigma,\tau)\,u(\sigma,\tau)\right)$$

$$=\sum_{t=0}^{n-1} k(m+1,n,m,t)\,u(m,t)+\sum_{s=0}^{m-1}\sum_{t=0}^{n-1}\Delta_1 k(m,n,s,t)\,u(s,t)$$

$$+\sum_{t=0}^{n-1}\left(\sum_{\sigma=0}^{m-1}\sum_{\tau=0}^{t-1} h(m+1,n,m,t,\sigma,\tau)\,u(\sigma,\tau)\right)$$

$$+\sum_{s=0}^{m-1}\sum_{t=0}^{n-1}\left(\sum_{\sigma=0}^{s-1}\sum_{\tau=0}^{t-1}\Delta_1 h(m,n,s,t,\sigma,\tau)\,u(\sigma,\tau)\right). \tag{5.3.40}$$

From (5.3.40) and using the facts that $u(m,n)\le z(m,n)$, $z(m,n)$ is nondecreasing for $m,n\in N_0$, we have

$$\Delta_2\Delta_1 z(m,n)=\Delta_1 z(m,n+1)-\Delta_1 z(m,n)$$

$$=\sum_{t=0}^{n} k(m+1,n+1,m,t)\,u(m,t)$$

$$+\sum_{s=0}^{m-1}\sum_{t=0}^{n}\Delta_1 k(m,n+1,s,t)\,u(s,t)$$

$$+\sum_{t=0}^{n}\left(\sum_{\sigma=0}^{m-1}\sum_{\tau=0}^{t-1} h(m+1,n+1,m,t,\sigma,\tau)\,u(\sigma,\tau)\right)$$

$$+\sum_{s=0}^{m-1}\sum_{t=0}^{n}\left(\sum_{\sigma=0}^{s-1}\sum_{\tau=0}^{t-1}\Delta_1 h(m,n+1,s,t,\sigma,\tau)u(\sigma,\tau)\right)$$

$$-\sum_{t=0}^{n-1} k(m+1,n,m,t)\,u(m,t)$$

$$-\sum_{s=0}^{m-1}\sum_{t=0}^{n-1}\Delta_1 k(m,n,s,t)\,u(s,t)$$

$$-\sum_{t=0}^{n-1}\left(\sum_{\sigma=0}^{m-1}\sum_{\tau=0}^{t-1} h(m+1,n,m,t,\sigma,\tau)\,u(\sigma,\tau)\right)$$

$$-\sum_{s=0}^{m-1}\sum_{t=0}^{n-1}\left(\sum_{\sigma=0}^{s-1}\sum_{\tau=0}^{t-1}\Delta_1 h(m,n,s,t,\sigma,\tau)u(\sigma,\tau)\right)$$

$$=k(m+1,n+1,m,n)\,u(m,n)$$

$$+\sum_{t=0}^{n-1} k(m+1,n+1,m,t)\,u(m,t)$$

$$+\sum_{s=0}^{m-1} \Delta_1 k(m, n+1, s, n)\, u(s, n)$$

$$+\sum_{s=0}^{m-1} \sum_{t=0}^{n-1} \Delta_1 k(m, n+1, s, t)\, u(s, t)$$

$$-\sum_{t=0}^{n-1} k(m+1, n, m, t)\, u(m, t)$$

$$-\sum_{s=0}^{m-1} \sum_{t=0}^{n-1} \Delta_1 k(m, n, s, t)\, u(s, t)$$

$$+\sum_{\sigma=0}^{m-1} \sum_{\tau=0}^{n-1} h(m+1, n+1, m, n, \sigma, \tau) u(\sigma, \tau)$$

$$+\sum_{t=0}^{n-1} \left(\sum_{\sigma=0}^{m-1} \sum_{\tau=0}^{t-1} h(m+1, n+1, m, t, \sigma, \tau)\, u(\sigma, \tau) \right)$$

$$+\sum_{s=0}^{m-1} \left(\sum_{\sigma=0}^{s-1} \sum_{\tau=0}^{n-1} \Delta_1 h(m, n+1, s, n, \sigma, \tau)\, u(\sigma, \tau) \right)$$

$$+\sum_{s=0}^{m-1} \sum_{t=0}^{n-1} \left(\sum_{\sigma=0}^{s-1} \sum_{\tau=0}^{t-1} \Delta_1 h(m, n+1, s, t, \sigma, \tau)\, u(\sigma, \tau) \right)$$

$$-\sum_{t=0}^{n-1} \left(\sum_{\sigma=0}^{m-1} \sum_{\tau=0}^{t-1} h(m+1, n, m, t, \sigma, \tau)\, u(\sigma, \tau) \right)$$

$$-\sum_{s=0}^{m-1} \sum_{t=0}^{n-1} \left(\sum_{\sigma=0}^{s-1} \sum_{\tau=0}^{t-1} \Delta_1 h(m, n, s, t, \sigma, \tau)\, u(\sigma, \tau) \right)$$

$$= k(m+1, n+1, m, n)\, u(m, n)$$

$$+\sum_{s=0}^{m-1} \Delta_1 k(m, n+1, s, n)\, u(s, n)$$

$$+\sum_{t=0}^{n-1} \Delta_2 k(m+1, n, m, t)\, u(m, t)$$

$$+\sum_{s=0}^{m-1} \sum_{t=0}^{n-1} \Delta_2 \Delta_1 k(m, n, s, t)\, u(s, t)$$

$$+\sum_{\sigma=0}^{m-1} \sum_{\tau=0}^{n-1} h(m+1, n+1, m, n, \sigma, \tau) u(\sigma, \tau)$$

$$+\sum_{s=0}^{m-1}\left(\sum_{\sigma=0}^{s-1}\sum_{\tau=0}^{n-1}\Delta_1 h\,(m,n+1,s,n,\sigma,\tau)\,u\,(\sigma,\tau)\right)$$

$$+\sum_{t=0}^{n-1}\left(\sum_{\sigma=0}^{m-1}\sum_{\tau=0}^{t-1}\Delta_2 h\,(m+1,n,m,t,\sigma,\tau)\,u\,(\sigma,\tau)\right)$$

$$+\sum_{s=0}^{m-1}\sum_{t=0}^{n-1}\left(\sum_{\sigma=0}^{s-1}\sum_{\tau=0}^{t-1}\Delta_2\Delta_1 h\,(m,n,s,t,\sigma,\tau)\,u\,(\sigma,\tau)\right)$$

$$\leq [A\,(m,n)+B\,(m,n)]\,z\,(m,n)\,. \tag{5.3.41}$$

Now by following the proof of Theorem 4.2.1 given in [42] we get

$$z\,(m,n)\leq c\prod_{x=0}^{m-1}\left[1+\sum_{y=0}^{n-1}[A\,(x,y)+B\,(x,y)]\right], \tag{5.3.42}$$

for $m,n\in N_0$. Using (5.3.42) in $u\,(m,n)\leq z\,(m,n)$ we get the required inequality in (5.3.29). If $c\geq 0$, we carry out the above procedure with $c+\varepsilon$ instead of c, where $\varepsilon>0$ is an arbitrary small constant, and subsequently pass to the limit as $\varepsilon\to 0$ to obtain (5.3.29).

(c_2) Let $c>0$ and define a function $z(m,n)$ by the right hand side of (5.3.32). Then $z(m,n)>0$, $z(m,0)=z(0,n)=c$, $u\,(m,n)\leq z\,(m,n)$ and $z(m,n)$ is nondecreasing for $m,n\in N_0$. By following the arguments as in the proof of (c_1) upto (5.3.41) with suitable modifications we get

$$\Delta_2\Delta_1 z\,(m,n)\leq [A\,(m,n)+B\,(m,n)]\,g\,(z\,(m,n))\,. \tag{5.3.43}$$

The remaining proof can be completed as in the proof of Theorem 5.2.1 given in [42].

(d_1) Let $c>0$ and define a function $z(m,n)$ by the right hand side of (5.3.35). Then $z(m,n)>0$, $z(m,0)=z(0,n)=c$, $u\,(m,n)\leq z\,(m,n)$ and $z(m,n)$ is nondecreasing for $m,n\in N_0$ and

$$\Delta_2\Delta_1 z\,(m,n)=b\,(m,n)\,u\,(m,n)+\sum_{\sigma=0}^{m-1}\sum_{\tau=0}^{n-1}k\,(m,n,\sigma,\tau)\,u\,(\sigma,\tau)$$

$$+\sum_{\sigma=0}^{m-1}\sum_{\tau=0}^{n-1}\left(\sum_{\xi=0}^{\sigma-1}\sum_{\eta=0}^{\tau-1}h\,(m,n,\sigma,\tau,\xi,\eta)\,u\,(\xi,\eta)\right)$$

$$\leq Q\,(m,n)\,z\,(m,n)\,. \tag{5.3.44}$$

The rest of the proof can be completed by following the proof of Theorem 4.2.1 given in [42].

(d_2) The proof can be completed by following the proof of (d_1) and closely looking at the proof of Theorem 5.2.1 given in [42]. Here we leave the details to the reader.

Remark 5.3.3. We note that the inequalities in Theorems 5.3.3 and 5.3.4 can be considered as two independent variable discrete generalizations of the integral inequalities established by Bykov and Salpagarov in [9] (see also [12]). The important feature of these inequalities lies in their successful utilizations to the situations for which the other available inequalities do not apply directly.

5.4 Estimates on certain finite difference inequalities I

In [36,41,48,49] Pachpatte has investigated a number of new finite difference inequalities involving functions of two independent variables. In this section we shall give some of the inequalities established in the above papers which find applications in the study of some specific types of finite difference equations.

We start with the following theorem which deals with the inequalities proved in [36].

Theorem 5.4.1. Let $u(m,n), a(m,n), b(m,n) \in D\left(N_0^2, R_+\right)$.

(a_1) Let $a(m,n)$ be nondecreasing in m and nonincreasing in n. If

$$u(m,n) \leq a(m,n) + \sum_{s=0}^{m-1} \sum_{t=n+1}^{\infty} b(s,t)\, u(s,t), \tag{5.4.1}$$

for $m,n \in N_0$, then

$$u(m,n) \leq a(m,n) \prod_{s=0}^{m-1} \left[1 + \sum_{t=n+1}^{\infty} b(s,t)\right], \tag{5.4.2}$$

for $m,n \in N_0$.

(a_2) Let $a(m,n)$ be nonincreasing in each variable m and n. If

$$u(m,n) \leq a(m,n) + \sum_{s=m+1}^{\infty} \sum_{t=n+1}^{\infty} b(s,t)\, u(s,t), \tag{5.4.3}$$

for $m, n \in N_0$, then

$$u(m,n) \le a(m,n) \prod_{s=m+1}^{\infty} \left[1 + \sum_{t=n+1}^{\infty} b(s,t)\right], \tag{5.4.4}$$

for $m, n \in N_0$.

Proof. (a_1) First we assume that $a(m,n) > 0$ for $m, n \in N_0$. From (5.4.1) it is easy to observe that

$$\frac{u(m,n)}{a(m,n)} \le 1 + \sum_{s=0}^{m-1} \sum_{t=n+1}^{\infty} b(s,t) \frac{u(s,t)}{a(s,t)}. \tag{5.4.5}$$

Define a function $z(m,n)$ by

$$z(m,n) = 1 + \sum_{s=0}^{m-1} \sum_{t=n+1}^{\infty} b(s,t) \frac{u(s,t)}{a(s,t)}, \tag{5.4.6}$$

then $\frac{u(m,n)}{a(m,n)} \le z(m,n)$ and

$$\begin{aligned} &[z(m+1,n) - z(m,n)] - [z(m+1,n+1) - z(m,n+1]\\ &= b(m,n+1) \frac{u(m,n+1)}{a(m,n+1)}\\ &\le b(m,n+1) z(m,n+1). \end{aligned} \tag{5.4.7}$$

From (5.4.7) and using the facts that $z(m,n) > 0$, $z(m,n+1) \le z(m,n)$ for $m, n \in N_0$, we observe that

$$\begin{aligned} &\frac{[z(m+1,n) - z(m,n)]}{z(m,n)} - \frac{[z(m+1,n+1) - z(m,n+1)]}{z(m,n+1)}\\ &\le b(m,n+1). \end{aligned} \tag{5.4.8}$$

Keeping m fixed in (5.4.8), set $n = t$ and sum over $t = n, n+1, ..., r-1$ $(r \ge n+1$ is arbitrary in N_0) to obtain

$$\begin{aligned} &\frac{[z(m+1,n) - z(m,n)]}{z(m,n)} - \frac{[z(m+1,r) - z(m,r)]}{z(m,r)}\\ &\le \sum_{t=n+1}^{r} b(m,t). \end{aligned} \tag{5.4.9}$$

Noting that $\lim_{r \to \infty} z(m,r) = \lim_{r \to \infty} z(m+1,r) = 1$ and by letting $r \to \infty$ in (5.4.9) we get

$$\frac{[z(m+1,n) - z(m,n)]}{z(m,n)} \le \sum_{t=n+1}^{\infty} b(m,t),$$

i.e.,

$$z(m+1,n) \leq \left[1 + \sum_{t=n+1}^{\infty} b(m,t)\right] z(m,n). \tag{5.4.10}$$

Now keeping n fixed in (5.4.10) and setting $m = s$ and substituting $s = 0, 1, 2, ..., m-1$ successively and using the fact that $z(0,n) = 1$ we get

$$z(m,n) \leq \prod_{s=0}^{m-1} \left[1 + \sum_{t=n+1}^{\infty} b(s,t)\right]. \tag{5.4.11}$$

Using (5.4.11) in (5.4.5) we get the required inequality in (5.4.2). If $a(m,n)$ is nonnegative, we carry out the above procedure with $a(m,n) + \varepsilon$ instead of $a(m,n)$, where $\varepsilon > 0$ is an arbitrary small constant , and subsequently pass to the limit as $\varepsilon \to 0$ to obtain (5.4.2).

(a_2) We first assume that $a(m,n) > 0$ for $m, n \in N_0$. From (5.4.3) it is easy to observe that

$$\frac{u(m,n)}{a(m,n)} \leq 1 + \sum_{s=m+1}^{\infty} \sum_{t=n+1}^{\infty} b(s,t) \frac{u(s,t)}{a(s,t)}. \tag{5.4.12}$$

Define a function $z(m,n)$ by the right hand side of (5.4.12), then $\frac{u(m,n)}{a(m,n)} \leq z(m,n)$ and

$$\begin{aligned}
&[z(m,n) - z(m+1,n)] - [z(m,n+1) - z(m+1,n+1]\\
&= b(m+1,n+1) \frac{u(m+1,n+1)}{a(m+1,n+1)}\\
&\leq b(m+1,n+1) z(m+1,n+1).
\end{aligned} \tag{5.4.13}$$

From (5.4.13) and using the facts that $z(m,n) > 0$, $z(m+1,n+1) \leq z(m+1,n)$ for $m, n \in N_0$, we observe that

$$\begin{aligned}
&\frac{[z(m,n) - z(m+1,n)]}{z(m+1,n)} - \frac{[z(m,n+1) - z(m+1,n+1)]}{z(m+1,n+1)}\\
&\leq b(m+1,n+1).
\end{aligned} \tag{5.4.14}$$

Keeping m fixed in (5.4.14), set $n = t$ and sum over $t = n, n+1, ..., q-1$ ($q \geq n+1$ is arbitrary in N_0) to obtain

$$\begin{aligned}
&\frac{[z(m,n) - z(m+1,n)]}{z(m+1,n)} - \frac{[z(m,q) - z(m+1,q)]}{z(m+1,q)}\\
&\leq \sum_{t=n+1}^{q} b(m+1,t).
\end{aligned} \tag{5.4.15}$$

Noting that $\lim_{q \to \infty} z(m,q) = \lim_{q \to \infty} z(m+1,q) = 1$ and by letting $q \to \infty$ in (5.4.15) we get

$$\frac{[z(m,n) - z(m+1,n)]}{z(m+1,n)} \leq \sum_{t=n+1}^{\infty} b(m+1,t),$$

i.e.

$$z(m,n) \leq \left[1 + \sum_{t=n+1}^{\infty} b(m+1,t)\right] z(m+1,n). \tag{5.4.16}$$

Now keeping n fixed in (5.4.16) and by setting $m = s$ and by substituting $s = m, m+1, ..., p-1$ ($p \geq m+1$ is arbitrary in N_0) successively, we obtain

$$z(m,n) \leq z(p,n) \prod_{s=m+1}^{p} \left[1 + \sum_{t=n+1}^{\infty} b(s,t)\right]. \tag{5.4.17}$$

Noting that $\lim_{p \to \infty} z(p,n) = 1$, and letting $p \to \infty$ in (5.4.17) we get

$$z(m,n) \leq \prod_{s=m+1}^{\infty} \left[1 + \sum_{t=n+1}^{\infty} b(s,t)\right]. \tag{5.4.18}$$

Using (5.4.18) in (5.4.12) we get the required inequality in (5.4.4). The case, when $a(m,n)$ is nonnegative can be completed as mentioned in the proof of part (a_1).

In the following theorems we present the inequalities investigated in [36].

Theorem 5.4.2. Let $u(m,n), b(m,n) \in D(N_0^2, R_+)$ and $c \geq 0$ be a real constant. Let $g \in C(R_+, R_+)$ be a nondecreasing function with $g(u) > 0$ for $u > 0$.

(b_1) If

$$u(m,n) \leq c + \sum_{s=0}^{m-1} \sum_{t=n+1}^{\infty} b(s,t) g(u(s,t)), \tag{5.4.19}$$

for $m, n \in N_0$, then for $0 \leq m \leq m_1, 0 \leq n \leq n_1; m, m_1, n, n_1 \in N_0$,

$$u(m,n) \leq G^{-1}\left[G(c) + \sum_{s=0}^{m-1} \sum_{t=n+1}^{\infty} b(s,t)\right], \tag{5.4.20}$$

where

$$G(r)=\int_{r_0}^{r}\frac{ds}{g(s)},r>0, \tag{5.4.21}$$

$r_0>0$ is arbitrary, G^{-1} is the inverse of G and $m_1,n_1\in N_0$ are chosen so that

$$G(c)+\sum_{s=0}^{m-1}\sum_{t=n+1}^{\infty}b(s,t)\in Dom\left(G^{-1}\right),$$

for all $m,n\in N_0$ such that $0\le m\le m_1, 0\le n\le n_1$.

(b_2) If

$$u(m,n)\le c+\sum_{s=m+1}^{\infty}\sum_{t=n+1}^{\infty}b(s,t)g(u(s,t)), \tag{5.4.22}$$

for $m,n\in N_0$, then for $0\le m\le m_2, 0\le n\le n_2; m,m_2,n,n_2\in N_0$,

$$u(m,n)\le G^{-1}\left[G(c)+\sum_{s=m+1}^{\infty}\sum_{t=n+1}^{\infty}b(s,t)\right], \tag{5.4.23}$$

where G, G^{-1} are as in (b_1) and m_2,n_2 are chosen so that

$$G(c)+\sum_{s=m+1}^{\infty}\sum_{t=n+1}^{\infty}b(s,t)\in Dom\left(G^{-1}\right),$$

for all $m,n\in N_0$ such that $0\le m\le m_2, 0\le n\le n_2$.

Theorem 5.4.3. Let $u(m,n),a(m,n),b(m,n)\in D\left(N_0^2,R_+\right)$ and $L:N_0^2\times R_+\to R_+$ be a function which satisfies the condition

$$0\le L(m,n,u)-L(m,n,v)\le M(m,n,v)(u-v), \tag{5.4.24}$$

for $u\ge v\ge 0$,, where $M:N_0^2\times R_+\to R_+$.

(c_1) If

$$u(m,n)\le a(m,n)+b(m,n)\sum_{s=0}^{m-1}\sum_{t=n+1}^{\infty}L(s,t,u(s,t)), \tag{5.4.25}$$

for $m,n\in N_0$, then

$$u(m,n)\le a(m,n)+b(m,n)e(m,n)$$

$$\times \prod_{s=0}^{m-1} \left[1 + \sum_{t=n+1}^{\infty} M(s,t,a(s,t))\, b(s,t)\right], \tag{5.4.26}$$

for $m, n \in N_0$, where

$$e(m,n) = \sum_{s=0}^{m-1} \sum_{t=n+1}^{\infty} L(s,t,a(s,t)), \tag{5.4.27}$$

for $m, n \in N_0$.

(c_2) If

$$u(m,n) \le a(m,n) + b(m,n) \sum_{s=m+1}^{\infty} \sum_{t=n+1}^{\infty} L(s,t,u(s,t)), \tag{5.4.28}$$

for $m, n \in N_0$, then

$$u(m,n) \le a(m,n) + b(m,n)\, \bar{e}(m,n)$$

$$\times \prod_{s=m+1}^{\infty} \left[1 + \sum_{t=n+1}^{\infty} M(s,t,a(s,t))\, b(s,t)\right], \tag{5.4.29}$$

for $m, n \in N_0$, where

$$\bar{e}(m,n) = \sum_{s=m+1}^{\infty} \sum_{t=n+1}^{\infty} L(s,t,a(s,t)), \tag{5.4.30}$$

for $m, n \in N_0$.

Theorem 5.4.4. Let $u(m,n), a(m,n), b(m,n) \in D(N_0^2, R_+)$ and $L : N_0^2 \times R_+ \to R_+$ be a function which satisfies the condition

$$0 \le L(m,n,u) - L(m,n,v) \le M(m,n,v)\, \psi^{-1}(u-v), \tag{5.4.31}$$

for $u \ge v \ge 0$, where $M : N_0^2 \times R_+ \to R_+$ and $\psi : R_+ \to R_+$ be a continuous and strictly increasing function with $\psi(0) = 0$, ψ^{-1} is the inverse function of ψ and $\psi^{-1}(xy) \le \psi^{-1}(x)\, \psi^{-1}(y)$ for $x, y \in R_+$.

(d_1) If

$$u(m,n) \le a(m,n) + b(m,n)\, \psi\left(\sum_{s=0}^{m-1} \sum_{t=n+1}^{\infty} L(s,t,u(s,t))\right), \tag{5.4.32}$$

for $m, n \in N_0$, then

$$u(m,n) \le a(m,n) + b(m,n)$$

$$\times \psi\left(e(m,n) \prod_{s=0}^{m-1} \left[1 + \sum_{t=n+1}^{\infty} M(s,t,a(s,t))\, \psi^{-1}(b(s,t))\right]\right), \tag{5.4.33}$$

for $m, n \in N_0$ where $e(m,n)$ is defined by (5.4.27).

(d_2) If

$$u(m,n) \le a(m,n) + b(m,n)\,\psi\left(\sum_{s=m+1}^{\infty}\sum_{t=n+1}^{\infty} L(s,t,u(s,t))\right), \quad (5.4.34)$$

for $m, n \in N_0$, then

$$u(m,n) \le a(m,n) + b(m,n)$$

$$\times\psi\left(\bar{e}(m,n)\prod_{s=m+1}^{\infty}\left[1 + \sum_{t=n+1}^{\infty} M(s,t,a(s,t))\,\psi^{-1}(b(s,t))\right]\right), \quad (5.4.35)$$

for $m, n \in N_0$, where $\bar{e}(m,n)$ is defined by (5.4.30).

Proofs of Theorems 5.4.2-5.4.4. We give the details of the proofs of (b_1), (c_1), (d_1) only. The proofs of (b_2), (c_2), (d_2) can be completed similarly with suitable modifications.

(b_1) Let $c > 0$ and define a function $z(m,n)$ by the right hand side of (5.4.19). Then $u(m,n) \le z(m,n)$ and

$$[z(m+1,n) - z(m,n)] - [z(m+1,n+1) - z(m,n+1]$$

$$= b(m,n+1)\,g(u(m,n+1))$$

$$\le b(m,n+1)\,g(z(m,n+1)). \quad (5.4.36)$$

From (5.4.36) and using the facts that $z(m,n) > 0$, $z(m,n+1) \le z(m,n)$ for $m, n \in N_0$, we observe that

$$\frac{[z(m+1,n) - z(m,n)]}{g(z(m,n))} - \frac{[z(m+1,n+1) - z(m,n+1]}{g(z(m,n+1))}$$

$$\le b(m,n+1). \quad (5.4.37)$$

Now by following the similar arguments as in the proof of Theorem 5.4.1, part (a_1) below (5.4.8) upto (5.4.10) we have

$$\frac{[z(m+1,n) - z(m,n)]}{g(z(m,n))} \le \sum_{t=n+1}^{\infty} b(m,t). \quad (5.4.38)$$

From (5.4.21) and (5.4.38) we have

$$G(z(m+1,n)) - G(z(m,n)) = \int_{z(m,n)}^{z(m+1,n)} \frac{ds}{g(s)}$$

$$\le \frac{1}{g(z(m,n))}[z(m+1,n) - z(m,n)]$$

$$\leq \sum_{t=n+1}^{\infty} b(m,t). \tag{5.4.39}$$

Keeping n fixed in (5.4.39), setting $m = s$ and taking the sum over $s = 0, 1, 2, ..., m-1$ and using the fact that $z(0,n) = c$ we obtain

$$G(z(m,n)) - G(c) \leq \sum_{s=0}^{m-1} \sum_{t=n+1}^{\infty} b(s,t). \tag{5.4.40}$$

The required inequality in (5.4.20) follows from (5.4.40) and the fact that $u(m,n) \leq z(m,n)$. The proof of the case when $c \geq 0$ can be completed as mentioned in the proof of Theorem 5.4.1, part (a_1). The subdomain $0 \leq m \leq m_1, 0 \leq n \leq n_1$ is obvious.

(c_1) Define a function $z(m,n)$ by

$$z(m,n) = \sum_{s=0}^{m-1} \sum_{t=n+1}^{\infty} L(s,t,u(s,t)). \tag{5.4.41}$$

Then (5.4.25) can be restated as

$$u(m,n) \leq a(m,n) + b(m,n)\, z(m,n). \tag{5.4.42}$$

From (5.4.41), (5.4.42) and (5.4.24) we observe that

$$u(m,n) \leq \sum_{s=0}^{m-1} \sum_{t=n+1}^{\infty} \{L(s,t,a(s,t) + b(s,t)\, z(s,t))$$

$$-L(s,t,a(s,t)) + L(s,t,a(s,t))\}$$

$$\leq e(m,n) + \sum_{s=0}^{m-1} \sum_{t=n+1}^{\infty} M(s,t,a(s,t))\, b(s,t)\, z(s,t), \tag{5.4.43}$$

where $e(m,n)$ is defined by (5.4.27). Clearly $e(m,n)$ is real-valued, nonnegative,nondecreasing in m and nonincreasing in n for $m,n \in N_0$. Now an application of Theorem 5.4.1, part (a_1) to (5.4.43) yields

$$z(m,n) \leq e(m,n) \prod_{s=0}^{m-1} \left[1 + \sum_{t=n+1}^{\infty} M(s,t,a(s,t))\, b(s,t)\right]. \tag{5.4.44}$$

The desired inequality in (5.4.26) follows from (5.4.42) and (5.4.44).

(d_1) Define a function $z(m,n)$ by (5.4.41), then from (5.4.32) we have

$$u(m,n) \leq a(m,n) + b(m,n)\,\psi(z(m,n)). \tag{5.4.45}$$

From (5.4.41), (5.4.45), (5.4.31) and the hypotheses on ψ we observe that

$$z(m,n) \leq \sum_{s=0}^{m-1} \sum_{t=n+1}^{\infty} \{L(s,t,a(s,t) + b(s,t)\,\psi(z(s,t)))$$

$$-L(s,t,a(s,t)) + L(s,t,a(s,t))\}$$

$$\leq e(m,n) + \sum_{s=0}^{m-1} \sum_{t=n+1}^{\infty} M(s,t,a(s,t))\,\psi^{-1}(b(s,t))z(s,t),$$

where $e(m,n)$ is defined by (5.4.27). Now, by following the last arguments as in the proof of (c_1) given above we get the desired inequality in (5.4.33).

Remark 5.4.1. We note that the inequalities in Theorem5.4.2 are the useful versions of the more general inequalities given in [36, Theorem 2] and in the various special cases the inequalities in Theorems 5.4.3 and 5.4.4 can also be useful in certain applications.

The discrete analogues of Theorems 2.5.3-2.5.5 established in [41] are embodied in the following theorems.

Theorem 5.4.5. Let $u(m,n), a(m,n), b(m,n), c(m,n) \in D(N_0^2, R_+)$.

(e_1) If

$$u(m,n) \leq a(m,n) + b(m,n) \sum_{s=0}^{m-1} \sum_{t=n+1}^{\infty} c(s,t)\,u(s,t), \tag{5.4.46}$$

for $m, n \in N_0$, then

$$u(m,n) \leq a(m,n) + b(m,n)\,f(m,n) \prod_{s=0}^{m-1} \left[1 + \sum_{t=n+1}^{\infty} c(s,t)\,b(s,t)\right], \tag{5.4.47}$$

for $m, n \in N_0$, where

$$f(m,n) = \sum_{s=0}^{m-1} \sum_{t=n+1}^{\infty} c(s,t)\,a(s,t), \tag{5.4.48}$$

for $m, n \in N_0$.

(e_2) If

$$u(m,n) \le a(m,n) + b(m,n) \sum_{s=m+1}^{\infty} \sum_{t=n+1}^{\infty} c(s,t)\, u(s,t), \tag{5.4.49}$$

for $m, n \in N_0$, then

$$u(m,n) \le a(m,n) + b(m,n)\, \bar{f}(m,n)$$

$$\times \prod_{s=m+1}^{\infty} \left[1 + \sum_{t=n+1}^{\infty} c(s,t)\, b(s,t) \right], \tag{5.4.50}$$

for $m, n \in N_0$, where

$$\bar{f}(m,n) = \sum_{s=m+1}^{\infty} \sum_{t=n+1}^{\infty} c(s,t)\, a(s,t), \tag{5.4.51}$$

for $m, n \in N_0$.

Theorem 5.4.6. Let $u(m,n), a(m,n), b(m,n), c(m,n) \in D\left(N_0^2, R_+\right)$.

(p_1) Assume that $a(m,n)$ is nondecreasing in m for $m \in N_0$. If

$$u(m,n) \le a(m,n) + \sum_{s=0}^{m-1} b(s,n)\, u(s,n) + \sum_{s=0}^{m-1} \sum_{t=n+1}^{\infty} c(s,t)\, u(s,t), \tag{5.4.52}$$

for $m, n \in N_0$, then

$$z(m,n) \le q(m,n) \left[a(m,n) + F(m,n) \right.$$

$$\left. \times \prod_{s=0}^{m-1} \left[1 + \sum_{t=n+1}^{\infty} c(s,t)\, q(s,t) \right] \right], \tag{5.4.53}$$

for $m, n \in N_0$, where

$$q(m,n) = \prod_{s=0}^{m-1} \left[1 + b(s,n) \right], \tag{5.4.54}$$

$$F(m,n) = \sum_{s=0}^{m-1} \sum_{t=n+1}^{\infty} c(s,t)\, q(s,t)\, a(s,t), \tag{5.4.55}$$

for $m, n \in N_0$.

(p_2) Assume that $a(m,n)$ is nonincreasing in m for $m \in N_0$. If

$$u(m,n) \leq a(m,n) + \sum_{s=m+1}^{\infty} b(s,n)\, u(s,n)$$

$$+ \sum_{s=m+1}^{\infty} \sum_{t=n+1}^{\infty} c(s,t)\, u(s,t), \tag{5.4.56}$$

for $m, n \in N_0$, then

$$u(m,n) \leq \bar{q}(m,n) \left[a(m,n) + \bar{F}(m,n)\right.$$

$$\left.\times \prod_{s=0}^{m-1} \left[1 + \sum_{t=n+1}^{\infty} c(s,t)\, \bar{q}(s,t)\right]\right], \tag{5.4.57}$$

for $m, n \in N_0$, where

$$\bar{q}(m,n) = \prod_{s=m+1}^{\infty} \left[1 + b(s,n)\right], \tag{5.4.58}$$

$$\bar{F}(m,n) = \sum_{s=m+1}^{\infty} \sum_{t=n+1}^{\infty} c(s,t)\, \bar{q}(s,t)\, a(s,t), \tag{5.4.59}$$

for $m, n \in N_0$.

Theorem 5.4.7. Let $u(m,n), a(m,n), b(m,n) \in D\left(N_0^2, R_+\right)$. Let L, M be as in Theorem 5.4.3 and the condition (5.4.24) holds.

(q_1) Assume that $a(m,n)$ is nondecreasing in m for $m \in N_0$. If

$$u(m,n) \leq a(m,n) + \sum_{s=0}^{m-1} b(s,n)\, u(s,n) + \sum_{s=0}^{m-1} \sum_{t=n+1}^{\infty} L(s,t,u(s,t)), \tag{5.4.60}$$

for $m, n \in N_0$, then

$$u(m,n) \leq q(m,n) \left[a(m,n) + H(m,n)\right.$$

$$\left.\times \prod_{s=0}^{m-1} \left[1 + \sum_{t=n+1}^{\infty} M(s,t,q(s,t)\, a(s,t))\, q(s,t)\right]\right], \tag{5.4.61}$$

for $m, n \in N_0$, where $q(m,n)$ is defined by (5.4.54) and

$$H(m,n) = \sum_{s=0}^{m-1} \sum_{t=n+1}^{\infty} L(s,t,q(s,t)\, a(s,t)), \tag{5.4.62}$$

for $m, n \in N_0$.

(q_2) Assume that $a(m,n)$ is nonincreasing in m for $m \in N_0$. If

$$u(m,n) \le a(m,n) + \sum_{s=m+1}^{\infty} b(s,n)\, u(s,n)$$

$$+ \sum_{s=m+1}^{\infty} \sum_{t=n+1}^{\infty} L(s,t,u(s,t)), \tag{5.4.63}$$

for $m, n \in N_0$, then

$$u(m,n) \le \bar{q}(m,n) \Big[a(m,n) + \bar{H}(m,n)$$

$$\times \prod_{s=m+1}^{\infty} \left[1 + \sum_{t=n+1}^{\infty} M(s,t,\bar{q}(s,t)\, a(s,t))\, \bar{q}(s,t)\right]\Big], \tag{5.4.64}$$

for $m, n \in N_0$, where $\bar{q}(m,n)$ is defined by (5.4.58) and

$$\bar{H}(m,n) = \sum_{s=m+1}^{\infty} \sum_{t=n+1}^{\infty} L(s,t,\bar{q}(s,t)\, a(s,t)), \tag{5.4.65}$$

for $m, n \in N_0$.

Proofs of Theorems 5.4.5-5.4.7. We give the proofs of $(e_1), (p_2), (q_1)$; the proofs of $(e_2), (p_1), (q_2)$ can be completed similarly.

(e_1) Define a function $z(m,n)$ by

$$z(m,n) = \sum_{s=0}^{m-1} \sum_{t=n+1}^{\infty} c(s,t)\, u(s,t), \tag{5.4.66}$$

then (5.4.46) can be restated as

$$u(m,n) \le a(m,n) + b(m,n)\, z(m,n). \tag{5.4.67}$$

From (5.4.66) and (5.4.67) we have

$$z(m,n) \le \sum_{s=0}^{m-1} \sum_{t=n+1}^{\infty} c(s,t) \left[a(s,t) + b(s,t)\, z(s,t)\right]$$

$$\le f(m,n) + \sum_{s=0}^{m-1} \sum_{t=n+1}^{\infty} c(s,t)\, b(s,t)\, z(s,t), \tag{5.4.68}$$

where $f(m,n)$ is defined by (5.4.48). Clearly, $f(m,n)$ is real-valued, nonnegative function, nondecreasing in m and nonincreasing in n for $m, n \in N_0$. Now, an application of Theorem 5.4.1, part (a_1) to (5.4.68) yields

$$z(m,n) \le f(m,n) \prod_{s=0}^{m-1} \left[1 + \sum_{t=n+1}^{\infty} c(s,t)\, b(s,t)\right]. \tag{5.4.69}$$

The required inequality in (5.4.47) follows from (5.4.67) and (5.4.69).

(p_2) Define a function $w(m,n)$ by

$$w(m,n) = a(m,n) + \sum_{s=m+1}^{\infty} \sum_{t=n+1}^{\infty} c(s,t)\, u(s,t), \tag{5.4.70}$$

then (5.4.56) can be restated as

$$u(m,n) \leq w(m,n) + \sum_{s=m+1}^{\infty} b(s,n)\, u(s,n). \tag{5.4.71}$$

Clearly, $w(m,n)$ is real-valued, nonnegative and nonincreasing in m for $m \in N_0$. Keeping n fixed in (5.4.71) and applying Theorem 4.5.3, part (c_1) to (5.4.71), we obtain

$$u(m,n) \leq w(m,n)\, \bar{q}(m,n), \tag{5.4.72}$$

where $\bar{q}(m,n)$ is defined by (5.4.58). From (5.4.72) and (5.4.70) we have

$$u(m,n) \leq \bar{q}(m,n)\, [a(m,n) + v(m,n)], \tag{5.4.73}$$

where

$$v(m,n) = \sum_{s=m+1}^{\infty} \sum_{t=n+1}^{\infty} c(s,t)\, u(s,t). \tag{5.4.74}$$

From (5.4.74) and (5.4.73), it is easy to see that

$$v(m,n) \leq \bar{F}(m,n) + \sum_{s=m+1}^{\infty} \sum_{t=n+1}^{\infty} c(s,t)\, \bar{q}(s,t)\, v(s,t), \tag{5.4.75}$$

where $\bar{F}(m,n)$ is defined by (5.4.59). Clearly, $\bar{F}(m,n)$ is real-valued nonnegative function, nonincreasing in each variable m and n for $m,n \in N_0$. An application of Theorem 5.4.1, part (a_2) to (5.4.75) yields

$$v(m,n) \leq \bar{F}(m,n) \prod_{s=m+1}^{\infty} \left[1 + \sum_{t=n+1}^{\infty} c(s,t)\, \bar{q}(s,t)\right]. \tag{5.4.76}$$

Using (5.4.76) in (5.4.73) we get the required inequality in (5.4.57).

(q_1) Define a function $z(m,n)$ by

$$z(m,n) = a(m,n) + \sum_{s=0}^{m-1} \sum_{t=n+1}^{\infty} L(s,t,u(s,t)), \tag{5.4.77}$$

then (5.4.60) can be restated as

$$u(m,n) \leq z(m,n) + \sum_{s=0}^{m-1} b(s,n)\, u(s,n). \tag{5.4.78}$$

Clearly, $z(m,n)$ is real-valued, nonnegative and nondecreasing function in m for $m \in N_0$. Keeping n fixed in (5.4.78) and applying Corollary 1.2.5 given in [42, p. 15] to (5.4.78) we get

$$u(m,n) \le z(m,n)\, q(m,n), \tag{5.4.79}$$

where $q(m,n)$ is defined by (5.4.54). From (5.4.79) and (5.4.77) we have

$$u(m,n) \le q(m,n)\,[a(m,n) + v(m,n)], \tag{5.4.80}$$

where

$$v(m,n) = \sum_{s=0}^{m-1} \sum_{t=n+1}^{\infty} L(s,t,u(s,t)). \tag{5.4.81}$$

From (5.4.81), (5.4.80) and the hypotheses on L we observe that

$$\begin{aligned} v(m,n) &\le \sum_{s=0}^{m-1} \sum_{t=n+1}^{\infty} \{L(s,t,q(s,t)\,[a(s,t) + v(s,t)]) - L(s,t,q(s,t)\,a(s,t)) \\ &\quad + L(s,t,q(s,t)\,a(s,t))\} \\ &\le H(m,n) + \sum_{s=0}^{m-1} \sum_{t=n+1}^{\infty} M(s,t,q(s,t)\,a(s,t))\,q(s,t)\,v(s,t), \end{aligned} \tag{5.4.82}$$

where $H(m,n)$ is defined by (5.4.62). Clearly, $H(m,n)$ is real-valued, nonnegative, nondecreasing function in m and nonincreasing in n for $m,n \in N_0$. Now, applying Theorem 5.4.1, part (a_1) to (5.4.82) and substituting the bound on $v(m,n)$ in (5.4.80), we get the required inequality in (5.4.61).

The next theorem contains the inequalities obtained in [49].

Theorem 5.4.8. Let $u(m,n), a(m,n) \in D(N_0^2, R_+)$ and $c \ge 0$ be a real constant.

(r_1) If

$$u^2(m,n) \le c + \sum_{s=0}^{m-1} \sum_{t=n+1}^{\infty} a(s,t)\,u(s,t), \tag{5.4.83}$$

for $m,n \in N_0$, then

$$u(m,n) \le \sqrt{c} + \frac{1}{2} \sum_{s=0}^{m-1} \sum_{t=n+1}^{\infty} a(s,t), \tag{5.4.84}$$

for $m,n \in N_0$.

(r_2) Let g, G, G^{-1} be as in Theorem 5.4.2, part (b_1).If

$$u^2(m,n) \leq c + \sum_{s=0}^{m-1} \sum_{t=n+1}^{\infty} a(s,t) u(s,t) g(u(s,t)), \tag{5.4.85}$$

for $m, n \in N_0$, then for $0 \leq m \leq m_3, 0 \leq n \leq n_3; m, m_3, n, n_3 \in N_0$,

$$u(m,n) \leq G^{-1}\left[G(\sqrt{c}) + \frac{1}{2}\sum_{s=0}^{m-1} \sum_{t=n+1}^{\infty} a(s,t)\right], \tag{5.4.86}$$

and $m_3, n_3 \in N_0$ are chosen so that

$$G(\sqrt{c}) + \frac{1}{2}\sum_{s=0}^{m-1} \sum_{t=n+1}^{\infty} a(s,t) \in Dom\left(G^{-1}\right),$$

for all $m, n \in N_0$ such that $0 \leq m \leq m_3, 0 \leq n \leq n_3$.

(r_3) Let L, M be as in Theorem 5.4.3 and the condition (5.4.24) holds. If

$$u^2(m,n) \leq c + \sum_{s=0}^{m-1} \sum_{t=n+1}^{\infty} a(s,t) u(s,t) L(s,t,u(s,t)), \tag{5.4.87}$$

for $m, n \in N_0$, then

$$u(m,n) \leq \sqrt{c} + h(m,n) \prod_{s=0}^{m-1}\left[1 + \frac{1}{2}\sum_{t=n+1}^{\infty} a(s,t) M\left(s,t,\sqrt{c}\right)\right], \tag{5.4.88}$$

for $m, n \in N_0$, where

$$h(m,n) = \frac{1}{2}\sum_{s=0}^{m-1} \sum_{t=n+1}^{\infty} a(s,t) L\left(s,t,\sqrt{c}\right), \tag{5.4.89}$$

for $m, n \in N_0$.

Proof. (r_1) Let $c > 0$ and define a function $z(m,n)$ by the right hand side of (5.4.83), then $u(m,n) \leq \sqrt{z(m,n)}$ and

$$\begin{aligned} &[z(m+1,n) - z(m,n)] - [z(m+1,n+1) - z(m,n+1)] \\ &= a(m,n+1) u(m,n+1) \\ &\leq a(m,n+1)\sqrt{z(m,n+1)}. \end{aligned} \tag{5.4.90}$$

By using the facts that $\sqrt{z(m,n)} > 0$, $\sqrt{z(m,n+1)} \leq \sqrt{z(m,n)}$, $\sqrt{z(m,n+1)} \leq \sqrt{z(m+1,n+1)}$, $\sqrt{z(m+1,n+1)} \leq \sqrt{z(m+1,n)}$, we observe that

$$\left[\sqrt{z(m+1,n)} - \sqrt{z(m,n)}\right] - \left[\sqrt{z(m+1,n+1)} - \sqrt{z(m,n+1)}\right]$$

$$= \frac{[z(m+1,n) - z(m,n)]}{\sqrt{z(m+1,n)} + \sqrt{z(m,n)}} - \frac{[z(m+1,n+1) - z(m,n+1)]}{\sqrt{z(m+1,n+1)} + \sqrt{z(m,n+1)}}$$

$$\leq \frac{[z(m+1,n) - z(m,n)]}{\sqrt{z(m+1,n+1)} + \sqrt{z(m,n+1)}} - \frac{[z(m+1,n+1) - z(m,n+1)]}{\sqrt{z(m+1,n+1)} + \sqrt{z(m,n+1)}}$$

$$= \frac{[z(m+1,n) - z(m,n)] - [z(m+1,n+1) - z(m,n+1)]}{\sqrt{z(m+1,n+1)} + \sqrt{z(m,n+1)}}$$

$$\leq \frac{[z(m+1,n) - z(m,n)] - [z(m+1,n+1) - z(m,n+1)]}{\sqrt{z(m,n+1)} + \sqrt{z(m,n+1)}}$$

$$\leq \frac{1}{2} a(m,n+1). \tag{5.4.91}$$

Here, we have used (5.4.90) to get (5.4.91). Now, keeping m fixed in (5.4.91), set $n = t$ and sum over $t = n, n+1, ..., q-1$ ($q \geq n+1$ is arbitrary in N_0) to obtain

$$\left[\sqrt{z(m+1,n)} - \sqrt{z(m,n)}\right] - \left[\sqrt{z(m+1,q)} - \sqrt{z(m,q)}\right]$$

$$\leq \frac{1}{2} \sum_{t=n+1}^{q} a(m,t). \tag{5.4.92}$$

Noting that $\lim_{q \to \infty} \sqrt{z(m+1,q)} = \sqrt{z(m,q)} = \sqrt{c}$, and by letting $q \to \infty$ in (5.4.92) we get

$$\sqrt{z(m+1,n)} - \sqrt{z(m,n)} \leq \frac{1}{2} \sum_{t=n+1}^{\infty} a(m,t). \tag{5.4.93}$$

Keeping n fixed in (5.4.93), set $m = s$ and sum over $s = 0, 1, 2, ..., m-1$ and use the fact that $z(0,n) = c$, to obtain

$$\sqrt{z(m,n)} \leq \sqrt{c} + \frac{1}{2} \sum_{s=0}^{m-1} \sum_{t=n+1}^{\infty} a(s,t). \tag{5.4.94}$$

The desired inequality in (5.4.84) follows by using the fact that $u(m,n) \leq \sqrt{z(m,n)}$. If $c \geq 0$, we carry out the above procedure with $c + \varepsilon$ instead of c, where $\varepsilon > 0$ is an arbitrary small constant, and subsequently pass to the limit as $\varepsilon \to 0$ to obtain (5.4.84).

(r_2) Let $c > 0$ and define a function $z(m,n)$ by the right hand side of (5.4.85). Then by following the same arguments as in the proof of (r_1) upto (5.4.91) with suitable changes we get

$$\left[\sqrt{z(m+1,n)} - \sqrt{z(m,n)}\right] - \left[\sqrt{z(m+1,n+1)} - \sqrt{z(m,n+1)}\right]$$

$$\leq \frac{1}{2} a(m, n+1) g\left(\sqrt{z(m, n+1)}\right). \tag{5.4.95}$$

From (5.4.95) and using the fact that $g\left(\sqrt{z(m,n+1)}\right) \leq g\left(\sqrt{z(m,n)}\right)$ we observe that

$$\frac{\left[\sqrt{z(m+1,n)} - \sqrt{z(m,n)}\right]}{g\left(\sqrt{z(m,n)}\right)} - \frac{\left[\sqrt{z(m+1,n+1)} - \sqrt{z(m,n+1)}\right]}{g\left(\sqrt{z(m,n+1)}\right)}$$

$$\leq \frac{1}{2} a(m, n+1). \tag{5.4.96}$$

Keeping m fixed in (5.4.96), set $n = t$ and sum over $t = n, n+1, \ldots, q-1$ ($q \geq n+1$ is arbitrary in N_0) to obtain the estimate

$$\frac{\left[\sqrt{z(m+1,n)} - \sqrt{z(m,n)}\right]}{g\left(\sqrt{z(m,n)}\right)} - \frac{\left[\sqrt{z(m+1,q)} - \sqrt{z(m,q)}\right]}{g\left(\sqrt{z(m,q)}\right)}$$

$$\leq \frac{1}{2} \sum_{t=n+1}^{q} a(m,t) \tag{5.4.97}$$

Noting that $\lim_{q \to \infty} \sqrt{z(m+1,q)} = \sqrt{z(m,q)} = \sqrt{c}$ and by letting $q \to \infty$ in (5.4.97) we get

$$\frac{\left[\sqrt{z(m+1,n)} - \sqrt{z(m,n)}\right]}{g\left(\sqrt{z(m,n)}\right)} \leq \frac{1}{2} \sum_{t=n+1}^{\infty} a(m,t). \tag{5.4.98}$$

From (5.4.21) and (5.4.98) we have

$$G\left(\sqrt{z(m+1,n)}\right) - G\left(\sqrt{z(m,n)}\right) = \int_{\sqrt{z(m,n)}}^{\sqrt{z(m+1,n)}} \frac{ds}{g(s)}$$

$$\leq \frac{\left[\sqrt{z(m+1,n)} - \sqrt{z(m,n)}\right]}{g\left(\sqrt{z(m,n)}\right)}$$

$$\leq \frac{1}{2} \sum_{t=n+1}^{\infty} a(m,t). \tag{5.4.99}$$

Now keeping n fixed in (5.4.99), set $m = s$ and sum both sides over $s = 0, 1, 2, \ldots, m-1$ and use the fact that $z(0,n) = c$ to obtain

$$G\left(\sqrt{z(m,n)}\right) \leq G\left(\sqrt{c}\right) + \frac{1}{2} \sum_{s=0}^{m-1} \sum_{t=n+1}^{\infty} a(s,t). \qquad (5.4.100))$$

The required inequality in (5.4.86) follows from (5.4.100) and $u(m,n) \leq \sqrt{z(m,n)}$. The case $c \geq 0$ can be completed as mentioned in the proof of part (r_1). The subdomain $0 \leq m \leq m_3, 0 \leq n \leq n_3$ is obvious.

(r_3) Let $c > 0$ and define a function $z(m,n)$ by the right hand side of (5.4.87). Then $u(m,n) \leq \sqrt{z(m,n)}$ and by following the proof of (r_1) upto (5.4.91) with suitable changes we get

$$\left[\sqrt{z(m+1,n)} - \sqrt{z(m,n)}\right] - \left[\sqrt{z(m+1,n+1)} - \sqrt{z(m,n+1)}\right]$$

$$\leq \frac{1}{2} a(m,n+1) L\left(m, n+1, \sqrt{z(m,n+1)}\right). \tag{5.4.101}$$

Further by following the arguments as in the proof of (r_1) below (5.4.91) upto (5.4.94) with suitable changes we get

$$\sqrt{z(m,n)} \leq \sqrt{c} + \frac{1}{2} \sum_{s=0}^{m-1} \sum_{t=n+1}^{\infty} a(s,t) L\left(s, t, \sqrt{z(s,t)}\right). \tag{5.4.102}$$

Define a function $v(m,n)$ by

$$v(m,n) = \frac{1}{2} \sum_{s=0}^{m-1} \sum_{t=n+1}^{\infty} a(s,t) L\left(s, t, \sqrt{z(s,t)}\right). \tag{5.4.103}$$

From (5.4.103), (5.4.102) and the hypotheses on L we observe that

$$v(m,n) \leq \frac{1}{2} \sum_{s=0}^{m-1} \sum_{t=n+1}^{\infty} a(s,t) \left\{L\left(s,t,\sqrt{c} + v(s,t)\right) - L\left(s,t,\sqrt{c}\right) + L\left(s,t,\sqrt{c}\right)\right\}$$

$$\leq h(m,n) + \frac{1}{2} \sum_{s=0}^{m-1} \sum_{t=n+1}^{\infty} a(s,t) M\left(s,t,\sqrt{c}\right) v(s,t), \tag{5.4.104}$$

where $h(m,n)$ is defined by (5.4.89). Clearly, $h(m,n)$ is a real-valued nonnegative function, nondecreasing in m and nonincreasing in n for $m, n \in N_0$. An application of Theorem 5.4.1, part (a_1) to (5.4.104) yields

$$v(m,n) \leq h(m,n) \prod_{s=0}^{m-1} \left[1 + \frac{1}{2} \sum_{t=n+1}^{\infty} a(s,t) M\left(s,t,\sqrt{c}\right)\right]. \tag{5.4.105}$$

The required inequality in (5.4.88) follows by using the fact that $u(m,n) \leq \sqrt{z(m,n)}$ and (5.4.102). The proof of the case when $c \geq 0$ can be completed as mentioned in the proof of (r_1).

Our final theorem in this section deals with the inequalities proved in [48].

Theorem 5.4.9. Let $u(m,n), a(m,n), b(m,n), c(m,n) \in D(N_0^2, R_+)$ and $p > 1$ be a real constant.

(s_1) If

$$u^p(m,n) \leq a(m,n) + b(m,n) \sum_{s=0}^{m-1} \sum_{t=n+1}^{\infty} c(s,t)\, u(s,t), \tag{5.4.106}$$

for $m, n \in N_0$, then

$$u(m,n) \leq \{a(m,n) + b(m,n)\, E(m,n)$$

$$\times \prod_{s=0}^{m-1} \left[1 + \sum_{t=n+1}^{\infty} c(s,t) \frac{b(s,t)}{p}\right]\Bigg\}^{\frac{1}{p}}, \tag{5.4.107}$$

for $m, n \in N_0$, where

$$E(m,n) = \sum_{s=0}^{m-1} \sum_{t=n+1}^{\infty} c(s,t) \left(\frac{p-1}{p} + \frac{a(s,t)}{p}\right), \tag{5.4.108}$$

for $m, n \in N_0$.

(s_2) Let L, M be as in Theorem 5.4.3 and the condition (5.4.24) holds. If

$$u^p(m,n) \leq a(m,n) + b(m,n) \sum_{s=0}^{m-1} \sum_{t=n+1}^{\infty} L(s,t,u(s,t)), \tag{5.4.109}$$

for $m, n \in N_0$, then

$$u(m,n) \leq \{a(m,n) + b(m,n)\, \bar{E}(m,n)$$

$$\times \prod_{s=0}^{m-1} \left[1 + \sum_{t=n+1}^{\infty} M\left(s,t,\frac{p-1}{p} + \frac{a(s,t)}{p}\right) \frac{b(s,t)}{p}\right]\Bigg\}^{\frac{1}{p}}, \tag{5.4.110}$$

for $m, n \in N_0$, where

$$\bar{E}(m,n) = \sum_{s=0}^{m-1} \sum_{t=n+1}^{\infty} L\left(s,t,\frac{p-1}{p} + \frac{a(s,t)}{p}\right), \tag{5.4.111}$$

for $m, n \in N_0$.

Proof. (s_1) Define a function $z(m,n)$ by

$$z(m,n) = \sum_{s=0}^{m-1} \sum_{t=n+1}^{\infty} c(s,t)\, u(s,t). \tag{5.4.112}$$

Then (5.4.106) can be written as

$$u^p(m,n) \leq a(m,n) + b(m,n)\, z(m,n). \tag{5.4.113}$$

From (5.4.113) as in the proof of Theorem 2.3.3, part (c_1) we get

$$u(m,n) \leq \frac{p-1}{p} + \frac{a(m,n)}{p} + \frac{b(m,n)}{p} z(m,n). \tag{5.4.114}$$

From (5.4.112) and (5.4.114) we have

$$z(m,n) \leq \sum_{s=0}^{m-1} \sum_{t=n+1}^{\infty} c(s,t) \left(\frac{p-1}{p} + \frac{a(s,t)}{p} + \frac{b(s,t)}{p} z(s,t) \right)$$

$$\leq E(m,n) + \sum_{s=0}^{m-1} \sum_{t=n+1}^{\infty} c(s,t)\, \frac{b(s,t)}{p} z(s,t), \tag{5.4.115}$$

where $E(m,n)$is defined by (5.4.108). Clearly, $E(m,n)$ is real-valued, nonnegative function, nondecreasing in m and nonincreasing in n for $m,n \in N_0$. An application of Theorem 5.4.1, part (a_1) to (5.4.115) yields

$$z(m,n) \leq E(m,n) \prod_{s=0}^{m-1} \left[1 + \sum_{t=n+1}^{\infty} c(s,t)\, \frac{b(s,t)}{p} \right]. \tag{5.4.116}$$

The required inequality in (5.4.107) follows from (5.4.113) and (5.4.116).

(s_2) The proof follows by closely looking at the proof of (s_1) given above and the proof ofTheorem 5.4.3, part (c_1). Here we omit the details.

Remark 5.4.2. We note that one can very easily obtain explicit bounds on the inequalities given in (5.4.83), (5.4.85), (5.4.87), (5.4.106), (5.4.109) by replacing the double sum $\sum_{s=0}^{m-1} \sum_{t=n+1}^{\infty}$ by $\sum_{s=m+1}^{\infty} \sum_{t=n+1}^{\infty}$. Here we leave the details of such results to the readers to fill in where neede

5.5 Estimates on certain finite difference inequalities II

In this section we shall give some more finite difference inequalities, recently established by Pachpatte in [62,71,76], which can be used conveniently in certain new applications for which the inequalities given earlier do not apply directly.

Our first theorem deals with the inequalities investigated in [76].

Theorem 5.5.1. Let $u(m,n), a(m,n), b(m,n), c(m,n), f(m,n), g(m,n) \in D(N_0^2, R_+)$.

(a_1) Suppose that

$$u(m,n) \leq a(m,n) + b(m,n) \sum_{s=0}^{m-1} \sum_{t=0}^{n-1} f(s,t) u(s,t)$$

$$+c(m,n) \sum_{s=0}^{\infty} \sum_{t=0}^{\infty} g(s,t) u(s,t), \tag{5.5.1}$$

for $m, n \in N_0$. If

$$q_1 = \sum_{s=0}^{\infty} \sum_{t=0}^{\infty} g(s,t) Q_1(s,t) < 1, \tag{5.5.2}$$

then

$$u(m,n) \leq P_1(m,n) + N_1 Q_1(m,n), \tag{5.5.3}$$

for $m, n \in N_0$, where

$$P_1(m,n) = a(m,n) + b(m,n) L_1(m,n) \sum_{s=0}^{m-1} \sum_{t=0}^{n-1} f(s,t) a(s,t), \tag{5.5.4}$$

$$Q_1(m,n) = c(m,n) + b(m,n) L_1(m,n) \sum_{s=0}^{m-1} \sum_{t=0}^{n-1} f(s,t) c(s,t), \tag{5.5.5}$$

$$L_1(m,n) = \prod_{s=0}^{m-1} \left[1 + \sum_{t=0}^{n-1} f(s,t) b(s,t)\right], \tag{5.5.6}$$

and

$$N_1 = \frac{1}{1-q_1} \sum_{s=0}^{\infty} \sum_{t=0}^{\infty} g(s,t) P_1(s,t). \tag{5.5.7}$$

(a_2) Suppose that

$$u(m,n) \le a(m,n) + b(m,n) \sum_{s=m+1}^{\infty} \sum_{t=n+1}^{\infty} f(s,t)\, u(s,t)$$

$$+c(m,n) \sum_{s=0}^{\infty} \sum_{t=0}^{\infty} g(s,t)\, u(s,t), \tag{5.5.8}$$

for $m, n \in N_0$. If

$$q_2 = \sum_{s=0}^{\infty} \sum_{t=0}^{\infty} g(s,t)\, Q_2(s,t) < 1, \tag{5.5.9}$$

then

$$u(m,n) \le P_2(m,n) + N_2 Q_2(m,n), \tag{5.5.10}$$

for $m, n \in N_0$, where

$$P_2(m,n) = a(m,n) + b(m,n)\, L_2(m,n) \sum_{s=m+1}^{\infty} \sum_{t=n+1}^{\infty} f(s,t)\, a(s,t), \tag{5.5.11}$$

$$Q_2(m,n) = c(m,n) + b(m,n)\, L_2(m,n) \sum_{s=m+1}^{\infty} \sum_{t=n+1}^{\infty} f(s,t)\, c(s,t), \tag{5.5.12}$$

$$L_2(m,n) = \prod_{s=m+1}^{\infty} \left[1 + \sum_{t=n+1}^{\infty} f(s,t)\, b(s,t)\right], \tag{5.5.13}$$

and

$$N_2 = \frac{1}{1-q_2} \sum_{s=0}^{\infty} \sum_{t=0}^{\infty} g(s,t)\, P_2(s,t). \tag{5.5.14}$$

(a_3) Suppose that

$$u(m,n) \le a(m,n) + b(m,n) \sum_{s=0}^{m-1} \sum_{t=n+1}^{\infty} f(s,t)\, u(s,t)$$

$$+c(m,n) \sum_{s=0}^{\infty} \sum_{t=0}^{\infty} g(s,t)\, u(s,t), \tag{5.5.15}$$

for $m, n \in N_0$. If

$$q_3 = \sum_{s=0}^{\infty} \sum_{t=0}^{\infty} g(s,t)\, Q_3(s,t) < 1, \tag{5.5.16}$$

then

$$u(m,n) \leq P_3(m,n) + N_3 Q_3(m,n), \tag{5.5.17}$$

for $m, n \in N_0$, where

$$P_3(m,n) = a(m,n) + b(m,n) L_3(m,n) \sum_{s=0}^{m-1} \sum_{t=n+1}^{\infty} f(s,t) a(s,t), \tag{5.5.18}$$

$$Q_3(m,n) = c(m,n) + b(m,n) L_3(m,n) \sum_{s=0}^{m-1} \sum_{t=n+1}^{\infty} f(s,t) c(s,t), \tag{5.5.19}$$

$$L_3(m,n) = \prod_{s=0}^{m-1} \left[1 + \sum_{t=n+1}^{\infty} f(s,t) b(s,t)\right], \tag{5.5.20}$$

and

$$N_3 = \frac{1}{1-q_3} \sum_{s=0}^{\infty} \sum_{t=0}^{\infty} g(s,t) P_3(s,t). \tag{5.5.21}$$

Proof. (a_1) Let

$$v(m,n) = \sum_{s=0}^{m-1} \sum_{t=0}^{n-1} f(s,t) u(s,t), \tag{5.5.22}$$

$$r = \sum_{s=0}^{\infty} \sum_{t=0}^{\infty} g(s,t) u(s,t). \tag{5.5.23}$$

Then (5.5.1) can be restated as

$$u(m,n) \leq a(m,n) + b(m,n) v(m,n) + c(m,n) r. \tag{5.5.24}$$

From (5.5.22) and (5.5.24) we have

$$v(m,n) \leq d(m,n) + \sum_{s=0}^{m-1} \sum_{t=0}^{n-1} f(s,t) b(s,t) v(s,t), \tag{5.5.25}$$

where

$$d(m,n) = \sum_{s=0}^{m-1} \sum_{t=0}^{n-1} [f(s,t) a(s,t) + r f(s,t) c(s,t)]. \tag{5.5.26}$$

Clearly, $d(m,n)$ is real-valued,nonnegative function and nondecreasing in both the variables m and n for $m, n \in N_0$. Now an application of Theorem 4.2.2 given in [42] to (5.5.25) yields

$$v(m,n) \leq d(m,n) L_1(m,n). \tag{5.5.27}$$

Using (5.5.27) in (5.5.24) we have

$$u(m,n) \le a(m,n) + r\,c(m,n) + b(m,n)\,d(m,n)\,L_1(m,n)$$
$$= P_1(m,n) + r\,Q_1(m,n)\,. \tag{5.5.28}$$

Now from (5.5.28), (5.5.23) and (5.5.2) we have

$$r \le \sum_{s=0}^{\infty}\sum_{t=0}^{\infty} g(s,t)\left\{P_1(s,t) + r\,Q_1(s,t)\right\}$$

i.e.,

$$r\left\{1 - \sum_{s=0}^{\infty}\sum_{t=0}^{\infty} g(s,t)\,Q_1(s,t)\right\} \le \sum_{s=0}^{\infty}\sum_{t=0}^{\infty} g(s,t)\,P_1(s,t),$$

which implies

$$r \le N_1. \tag{5.5.29}$$

Using (5.5.29) in (5.5.28) we get (5.5.3).

(a_2) Let

$$v(m,n) = \sum_{s=m+1}^{\infty}\sum_{t=n+1}^{\infty} f(s,t)\,u(s,t)\,, \tag{5.5.30}$$

and r be as in (5.5.23). The proof can be completed by following the proof of (a_1) and using the inequality in Theorem 5.4.1, part (a_2).

(a_3) Let

$$v(m,n) = \sum_{s=0}^{m-1}\sum_{t=n+1}^{\infty} f(s,t)\,u(s,t)\,, \tag{5.5.31}$$

and r be as in (5.5.23). The proof follows by the similar arguments as in (a_1) and using the inequality in Theorem 5.4.1, part (a_1).

The next theorem contains the inequality established in [71].

Theorem 5.5.2. Let $u(m,n) \in D\left(N_0^2, R_+\right)$ and $a(m,n,s,t)\,, b(m,n,s,t) \in D(E, R_+)$ be nondecreasing in m, n for each $s, t \in N_0$, where $E = \{(m,n,s,t) \in N_0^4 : 0 \le s \le m < \infty, 0 \le t \le n < \infty\}$. $0 \le s \le m < \infty, 0 \le t \le n < \infty\}$. Suppose that

$$u(m,n) \le c + \sum_{s=0}^{m-1}\sum_{t=0}^{n-1} a(m,n,s,t)\,u(s,t) + \sum_{s=0}^{\infty}\sum_{t=0}^{\infty} b(m,n,s,t)\,u(s,t), \tag{5.5.32}$$

for $m, n \in N_0$, where $c \geq 0$ is a real constant. If

$$r(m,n) = \sum_{s=0}^{\infty}\sum_{t=0}^{\infty} b(m,n,s,t) \prod_{\sigma=0}^{s-1}\left[1+\sum_{\tau=0}^{t-1} a(s,t,\sigma,\tau)\right] < 1, \tag{5.5.33}$$

for $m, n \in N_0$, then

$$u(m,n) \leq \frac{c}{1-r(m,n)} \prod_{s=0}^{m-1}\left[1+\sum_{t=0}^{n-1} a(m,n,s,t)\right], \tag{5.5.34}$$

for $m, n \in N_0$.

Proof. Fix $(x,y) \in N_0^2$. Then for $0 \leq m \leq x, 0 \leq n \leq y; (m,n) \in N_0^2$ we have

$$u(m,n) \leq c + \sum_{s=0}^{m-1}\sum_{t=0}^{n-1} a(x,y,s,t)\, u(s,t) + \sum_{s=0}^{\infty}\sum_{t=0}^{\infty} b(x,y,s,t)\, u(s,t). \tag{5.5.35}$$

Let

$$k(x,y) = c + \sum_{s=0}^{\infty}\sum_{t=0}^{\infty} b(x,y,s,t)\, u(s,t), \tag{5.5.36}$$

then (5.5.35) can be restated as

$$u(m,n) \leq k(x,y) + \sum_{s=0}^{m-1}\sum_{t=0}^{n-1} a(x,y,s,t)\, u(s,t), \tag{5.5.37}$$

for $0 \leq m \leq x, 0 \leq n \leq y$. Now an application of Theorem 4.2.1 given in [42] to (5.5.37) yields

$$u(m,n) \leq k(x,y) \prod_{\sigma=0}^{m-1}\left[1+\sum_{\tau=0}^{n-1} a(x,y,\sigma,\tau)\right], \tag{5.5.38}$$

for $0 \leq m \leq x, 0 \leq n \leq y$.Since $(x,y) \in N_0^2$ is arbitrary, from (5.5.38) and (5.5.36) with (x,y) replaced by (m,n) we have

$$u(m,n) \leq k(m,n) \prod_{\sigma=0}^{m-1}\left[1+\sum_{\tau=0}^{n-1} a(m,n,\sigma,\tau)\right], \tag{5.5.39}$$

where

$$k(m,n) = c + \sum_{s=0}^{\infty}\sum_{t=0}^{\infty} b(m,n,s,t)\, u(s,t), \tag{5.5.40}$$

for all $(m,n) \in N_0^2$. Using (5.5.39) on the right hand side of (5.5.40) we have

$$k(m,n) \leq c + \sum_{s=0}^{\infty}\sum_{t=0}^{\infty} b(m,n,s,t) \left\{ k(m,n) \prod_{\sigma=0}^{m-1}\left[1+\sum_{\tau=0}^{n-1} a(s,t,\sigma,\tau)\right]\right\},$$

which in view of (5.5.33) implies

$$k(m,n) \le \frac{c}{1-r(m,n)}. \tag{5.5.41}$$

Using (5.5.41) in (5.5.39) we get the required inequality in (5.5.34).

Remark 5.5.1. We note that the inequality given in Theorem 5.5.2 is of more general type and in the special cases when $(i) b(m,n,s,t) = 0$, $(ii) a(m,n,s,t) = 0$, it can also be used more effectively in the situations for which the other available inequalities do not apply directly.

Tn the following theorem we present the inequality proved in [62].

Theorem 5.2.3. Let $u(m,n), p(m,n), f(m,n), g(m,n), h(m,n) \in D(N_0^2, R_+)$, and $c \ge 0$ be a real constant and suppose that

$$u(m,n) \le c + \sum_{s=0}^{m-1} p(s,n)\, u(s,n)$$

$$+ \sum_{s=0}^{m-1}\sum_{t=0}^{n-1} f(s,t) \Big[u(s,t)$$

$$+ \sum_{\sigma=0}^{s-1}\sum_{\tau=0}^{t-1} g(\sigma,\tau) u(\sigma,\tau) + \sum_{\sigma=0}^{\infty}\sum_{\tau=0}^{\infty} h(\sigma,\tau) u(\sigma,\tau) \Big], \tag{5.5.42}$$

for $(m,n) \in N_0^2$. If

$$r = \sum_{\sigma=0}^{\infty}\sum_{\tau=0}^{\infty} h(\sigma,\tau) B(\sigma,\tau)$$

$$\times \prod_{\xi=0}^{\sigma-1} \left[1 + \sum_{\eta=0}^{\tau-1} B(\xi,\eta) \left[f(\xi,\eta) + g(\xi,\eta) \right] \right] < 1, \tag{5.5.43}$$

where

$$B(\sigma,\tau) = \prod_{s=0}^{\sigma-1} [1 + p(s,\tau)], \tag{5.5.44}$$

for $(\sigma,\tau) \in N_0^2$, then

$$u(m,n) \le \frac{c}{1-r} B(m,n) \prod_{s=0}^{m-1} \left[1 + \sum_{t=0}^{n-1} B(s,t) \left[f(s,t) + g(s,t) \right] \right], \tag{5.5.45}$$

for $(m,n) \in N_0^2$.

Proof. Let $c > 0$ and define a function $z(m,n)$ by

$$z(m,n) = c + \sum_{s=0}^{m-1}\sum_{t=0}^{n-1} f(s,t)\left[u(s,t) + \sum_{\sigma=0}^{s-1}\sum_{\tau=0}^{t-1} g(\sigma,\tau)u(\sigma,\tau) + \sum_{\sigma=0}^{\infty}\sum_{\tau=0}^{\infty} h(\sigma,\tau)u(\sigma,\tau)\right]. \tag{5.5.46}$$

Then (5.5.42) can be restated as

$$u(m,n) \le z(m,n) + \sum_{s=0}^{m-1} p(s,n)\,u(s,n). \tag{5.5.47}$$

It is easy to observe that the function $z(m,n)$ is real-valued, positive and nondecreasing for $(m,n) \in N_0^2$. Now treating n fixed in (5.5.47) and applying the inequality given in Corollary 1.2.5 in [42] to (5.5.47) we get

$$u(m,n) \le B(m,n)\,z(m,n), \tag{5.5.48}$$

for $(m,n) \in N_0^2$, where $B(m,n)$ is defined by (5.5.44). From (5.5.46), (5.5.48) and the fact that $B(m,n) \ge 1$, we observe that

$$\begin{aligned} z(m,n) \le{}& c + \sum_{s=0}^{m-1}\sum_{t=0}^{n-1} f(s,t)\left[B(s,t)\,z(s,t) + \sum_{\sigma=0}^{s-1}\sum_{\tau=0}^{t-1} g(\sigma,\tau)B(\sigma,\tau)\,z(\sigma,\tau)\right.\\ &\left.+ \sum_{\sigma=0}^{\infty}\sum_{\tau=0}^{\infty} h(\sigma,\tau)B(\sigma,\tau)\,z(\sigma,\tau)\right]\\ \le{}& c + \sum_{s=0}^{m-1}\sum_{t=0}^{n-1} f(s,t)\,B(s,t)\left[z(s,t) + \sum_{\sigma=0}^{s-1}\sum_{\tau=0}^{t-1} g(\sigma,\tau)B(\sigma,\tau)\,z(\sigma,\tau)\right.\\ &\left.+ \sum_{\sigma=0}^{\infty}\sum_{\tau=0}^{\infty} h(\sigma,\tau)B(\sigma,\tau)\,z(\sigma,\tau)\right]. \end{aligned} \tag{5.5.49}$$

Define a function $v(m,n)$ by the right hand side of (5.5.49). Then $v(m,n) > 0$, $v(0,n) = v(m,0) = c$, $z(m,n) \le v(m,n)$ and

$$\begin{aligned} \Delta_2\Delta_1 v(m,n) ={}& f(m,n)\,B(m,n)\left[z(m,n) + \sum_{\sigma=0}^{m-1}\sum_{\tau=0}^{n-1} g(\sigma,\tau)\,B(\sigma,\tau)\,z(\sigma,\tau)\right.\\ &\left.+ \sum_{\sigma=0}^{\infty}\sum_{\tau=0}^{\infty} h(\sigma,\tau)\,B(\sigma,\tau)\,z(\sigma,\tau)\right]\\ \le{}& f(m,n)\,B(m,n)\left[v(m,n) + \sum_{\sigma=0}^{m-1}\sum_{\tau=0}^{n-1} g(\sigma,\tau)\,B(\sigma,\tau)\,v(\sigma,\tau)\right. \end{aligned}$$

$$+\sum_{\sigma=0}^{\infty}\sum_{\tau=0}^{\infty} h(\sigma,\tau) B(\sigma,\tau) v(\sigma,\tau)\Bigg]. \tag{5.5.50}$$

Define a function $w(m,n)$ by

$$w(m,n) = v(m,n) + \sum_{\sigma=0}^{m-1}\sum_{\tau=0}^{n-1} g(\sigma,\tau) B(\sigma,\tau) v(\sigma,\tau)$$

$$+\sum_{\sigma=0}^{\infty}\sum_{\tau=0}^{\infty} h(\sigma,\tau) B(\sigma,\tau) v(\sigma,\tau),$$

then $w(m,n) > 0$, $v(m,n) \le w(m,n)$, $\Delta_2\Delta_1 v(m,n) \le f(m,n) B(m,n) w(m,n)$,

$$w(0,n) = w(m,0) = c + \sum_{\sigma=0}^{\infty}\sum_{\tau=0}^{\infty} h(\sigma,\tau) B(\sigma,\tau) v(\sigma,\tau) = \bar{L}\ (say), \tag{5.5.51}$$

and

$$\Delta_2\Delta_1 w(m,n) = \Delta_2\Delta_1 v(m,n) + g(m,n) B(m,n) v(m,n)$$

$$\le f(m,n) B(m,n) w(m,n) + g(m,n) B(m,n) v(m,n)$$

$$\le B(m,n)[f(m,n) + g(m,n)] w(m,n). \tag{5.5.52}$$

Now, by following the proof of Theorem 4.2.1 given in [42], the inequality (5.5.52) implies the estimate

$$w(m,n) \le \bar{L} \prod_{s=0}^{m-1}\left[1 + \sum_{t=0}^{n-1} B(s,t)[f(s,t) + g(s,t)]\right]. \tag{5.5.53}$$

From (5.5.51), (5.5.53) and (5.5.43) we observe that

$$\bar{L} \le \frac{c}{1-r}. \tag{5.5.54}$$

Using (5.5.54) in (5.5.53) and the facts that $z(m,n) \le v(m,n)$, $u(m,n) \le B(m,n) z(m,n)$ we get the required inequality in (5.5.45). The proof of the case when $c \ge 0$ can be completed as mentioned in the proof of Theorem 5.4.8, part (r_1).

Remark 5.5.2. We note that,in the special cases when $(i) p(m,n) = 0$, $(ii) g(m,n) = 0$, $(iii) h(m,n) = 0$, the inequality in Theorem 5.5.3 reduces to the new inequalities which can be used as tools in different applications.

5.6 Applications

In this section we present applications of some of the inequalities given in earlier sections which deals with some fundamental properties of solutions of various types of finite difference equations in two independent variables. The inequalities given above are recently developed and hope will provide a fruitful source for future research.

5.6.1 Partial finite difference equations

First, consider the following partial finite difference equation

$$z(m+1, n+1) = F(m, n, z(m, n)), \tag{5.6.1}$$

with the given initial condition

$$z(m_0, n_0) = z_0, \tag{5.6.2}$$

for $(m, n) \in \Delta_0 = \bar{M}_0 \times \bar{N}_0$, $F : \Delta_0 \times R \to R$, where $\bar{M}_0$, $\bar{N}_0$ are as defined in Theorem 5.2.1.

As an application of the inequality given in Theorem 5.2.1, part (a_2) we present the following theorem which deals with the dependency of solutions of equation (5.6.1) on given initial values (see [56]).

Theorem 5.6.1. Suppose that the finction F in (5.6.1) satisfies

$$|F(m, n, x) - F(m, n, y)| \leq w(m, n, |x - y|), \tag{5.6.3}$$

for $(m, n) \in \Delta_0$, $x, y \in R$, where $w(m, n, r) : \Delta_0 \times R_+ \to R_+$ is a nondecreasing function with respect to r for fixed $(m, n) \in \Delta_0$. Let $z(m, n, m_0, n_0, z_i)$ $(i = 1, 2)$ be solutions of (5.6.1) with the given initial conditions

$$z(m_0, n_0, m_0, n_0, z_i) = z_i, \tag{5.6.4}$$

for $i = 1, 2$. Let $r(m, n)$ be a solution of the equation

$$r(m+1, n+1) = w(m, n, r(m, n)), \ r(m_0, n_0) = r_0, \tag{5.6.5}$$

for $(m, n) \in \Delta_0$ and $|z_1 - z_2| \leq r_0$. Then

$$|z(m, n, m_0, n_0, z_1) - z(m, n, m_0, n_0, z_2)| \leq r(m, n), \tag{5.6.6}$$

for $(m, n) \in \Delta_0$.

Proof. Let $p(m,n) = |z(m,n,m_0,n_0,z_1) - z(m,n,m_0,n_0,z_2)|$. Then

$$\begin{aligned}p(m+1,n+1) &= |z(m+1,n+1,m_0,n_0,z_1) - z(m+1,n+1,m_0,n_0,z_2)| \\ &= |F(m,n,z(m,n,m_0,n_0,z_1)) - F(m,n,z(m,n,m_0,n_0,z_2))| \\ &\le w(m,n,|z(m,n,m_0,n_0,z_1) - z(m,n,m_0,n_0,z_2)|) \\ &= w(m,n,p(m,n)). \end{aligned} \tag{5.6.7}$$

Now a suitable application of Theorem 5.2.1, part (a_2) to (5.6.7) and (5.6.5) we get (5.6.6), which shows the dependency of solutions of (5.6.1) on initial values.

Next, we apply the inequality given in Theorem 5.2.3, part (c_1) to obtain a bound on the solution of sum-difference equation of the form

$$z^2(m,n) = h(m,n) + \sum_{\sigma=0}^{m-1}\sum_{\tau=0}^{n-1} F(m,n,\sigma,\tau,z(\sigma,\tau)), \tag{5.6.8}$$

for $(m,n) \in N_0^2$, where $h : N_0^2 \to R,$, $F : E \times R \to R$, in which $E = \{(m,n,\sigma,\tau) \in N_0^4 : 0 \le \sigma \le m < \infty, 0 \le \tau \le n < \infty\}$.

Theorem 5.6.2. Suppose that the functions h, F in equation (5.6.8) satisfy the conditions

$$|h(m,n)| \le c, \tag{5.6.9}$$

$$|F(m,n,\sigma,\tau,z)| \le k(m,n,\sigma,\tau)|z|, \tag{5.6.10}$$

where c and $k(m,n,\sigma,\tau)$ are as in Theorem 5.2.3. Let $\Delta_1 k(m,n,\sigma,\tau)$, $\Delta_2 k(m, n,\sigma,\tau)$, $\Delta_2\Delta_1 k(m,n,\sigma,\tau)$ be as in Theorem 5.2.3 and $Q(m,n)$ is defined by (5.4.14). If $z(m,n)$ is a solution of equation (5.6.8) for $(m,n) \in N_0^2$, then

$$|z(m,n)| \le \sqrt{c} + \frac{1}{2}\sum_{s=0}^{m-1}\sum_{t=0}^{n-1} Q(s,t), \tag{5.6.11}$$

for $(m,n) \in N_0^2$.

Proof. Using the fact that $z(m,n)$ is a solution of equation (5.6.8), the conditions (5.6.9), (5.6.10) and making use of the inequality in Theorem 5.2.3, part (c_1) we get the required inequality in (5.6.11).

5.6.2 Volterra type sum-difference equation

In this section we present applications of the inequality in Theorem 5.5.2, part (b_2), which provide estimates on the solutions of sum-difference equation of the form

$$z(m,n) = h(m,n) + \sum_{\sigma=0}^{m-1}\sum_{\tau=0}^{n-1} F(m,n,\sigma,\tau,z(\sigma,\tau)), \tag{5.6.12}$$

for $(m,n) \in N_0^2$,where $h : N_0^2 \to R$, $F : E \times R \to R$, in which $E = \{(m,n,\sigma,\tau) \in N_0^4 : 0 \le \sigma \le m < \infty, 0 \le \tau \le n < \infty\}$.

The following theorem deals with the estimate on the solution of equation (5.6.12).

Theorem 5.6.3. Suppose that

$$|h(m,n)| \le a(m,n), \tag{5.6.13}$$

$$|F(m,n,\sigma,\tau,z)| \le k(m,n,\sigma,\tau)\, g(|z|), \tag{5.6.14}$$

where $a(m,n), k(m,n,\sigma,\tau), g(u)$ are as in Theorem 5.2.2, part (b_2). Let G, G^{-1}, $\Delta_1 k(m,n,\sigma,\tau), \Delta_2 k(m,n,\sigma,\tau), \Delta_2\Delta_1 k(m,n,\sigma,\tau)$ be as in Theorem 5.2.2, part (b_2) and $A(m,n)$ and $Q(m,n)$ are defined by (5.2.18) and (5.2.14) respectively. If $z(m,n)$ is any solution of (5.6.12) for $(m,n) \in N_0^2$, then for $0 \le m \le m_1, 0 \le n \le n_1$; $m, m_1, n, n_1 \in N_0$,

$$|z(m,n)| \le a(m,n) + G^{-1}\left[G(A(m,n)) + \sum_{\sigma=0}^{m-1}\sum_{\tau=0}^{n-1} Q(s,t,)\right], \tag{5.6.15}$$

and $m_1, n_1 \in N_0$ are chosen so that

$$G(A(m,n)) + \sum_{\sigma=0}^{m-1}\sum_{\tau=0}^{n-1} Q(s,t,) \in Dom\left(G^{-1}\right),$$

for all m, n lying in $0 \le m \le m_1, 0 \le n \le n_1$.

Proof. Let $z(m,n) \in D\left(N_0^2, R\right)$ be a solution of equation (5.6.12). Using the fact that $z(m,n)$ is a solution of (5.6.12) and the conditions (5.6.13), (5.6.14) we have

$$|z(m,n)| \le a(m,n) + \sum_{\sigma=0}^{m-1}\sum_{\tau=0}^{n-1} k(m,n,\sigma,\tau)\, g(|z(\sigma,\tau)|). \tag{5.6.16}$$

Now an application of the inequality in Theorem 5.2.2, part (b_2) to (5.6.16) yields the desired estimate in (5.6.15).

In the following theorem we obtain estimate on the solution of equation (5.6.12) by assuming that the function F satisfies the Lipschitz type condition.

Theorem 5.6.4. Suppose that

$$|F(m,n,\sigma,\tau,z) - F(m,n,\sigma,\tau,\bar{z})| \le k(m,n,\sigma,\tau)\, g(|z-\bar{z}|), \qquad (5.6.17)$$

where $k(m,n,\sigma,\tau)$ and $g(u)$ are as in Theorem 5.2.2, part (b_2). Let G, G^{-1}, $\Delta_1 k$ (m,n,σ,τ) , $\Delta_2 k(m,n,\sigma,\tau)$, $\Delta_2\Delta_1 k(m,n,\sigma,\tau)$ and $Q(m,n)$ be as in Theorem 5.2.2, part (b_2) and

$$e(m,n) = \sum_{\sigma=0}^{m-1}\sum_{\tau=0}^{n-1} |F(m,n,\sigma,\tau,h(\sigma,\tau))|, \qquad (5.6.18)$$

$$\bar{A}(m,n) = \sum_{\sigma=0}^{m-1}\sum_{\tau=0}^{n-1} k(m,n,\sigma,\tau)\, g(e(\sigma,\tau)). \qquad (5.6.19)$$

If z(m,n) is a solution of equation (5.6.12) for $(m,n) \in N_0^2$, then for $0 \le m \le m_2, 0 \le n \le n_2; m, m_2, n, n_2 \in N_0$,

$$|z(m,n) - h(m,n)| \le e(m,n) + G^{-1}$$

$$\times \left[G(\bar{A}(m,n)) + \sum_{s=0}^{m-1}\sum_{t=0}^{n-1} Q(s,t) \right], \qquad (5.6.20)$$

and $m_2, n_2 \in N_0$ are chosen so that

$$G(\bar{A}(m,n)) + \sum_{s=0}^{m-1}\sum_{t=0}^{n-1} Q(s,t) \in Dom\left(G^{-1}\right),$$

for all m, n lying in $0 \le m \le m_2, 0 \le n \le n_2$.

Proof. Let $z(m,n)$ be a solution of equation (5.6.12). Using the fact that $z(m,n)$ is a solution of (5.6.12) and (5.6.17) we observe that

$$|z(m,n) - h(m,n)| = \left| \sum_{\sigma=0}^{m-1}\sum_{\tau=0}^{n-1} \{F(m,n,\sigma,\tau,z(\sigma,\tau)) \right.$$

$$\left. -F(m,n,\sigma,\tau,h(\sigma,\tau)) + F(m,n,\sigma,\tau,h(\sigma,\tau))\} \right|$$

$$\le e(m,n) + \sum_{\sigma=0}^{m-1}\sum_{\tau=0}^{n-1} k(m,n,\sigma,\tau) g(|z(\sigma,\tau) - h(\sigma,\tau)|). \qquad (5.6.21)$$

Now a suitable application of the inequality in Theorem 5.2.2, part (b_2) to (5.6.21) yields (5.6.20).

5.6.3 Partial finite sum-difference equation

In this section we present applications of the inequality in Theorem 5.3.4, part(d_1) to study certain properties of solutions of partial finite sum-difference equation of the form

$$\Delta_2\Delta_1 z(m,n) = F\left(m,n,z(m,n),\sum_{\sigma=0}^{m-1}\sum_{\tau=0}^{n-1} P(m,n,\sigma,\tau,z(\sigma,\tau)),\right.$$

$$\left.\sum_{\sigma=0}^{m-1}\sum_{\tau=0}^{n-1}\left(\sum_{x=0}^{\sigma-1}\sum_{y=0}^{\tau-1} H(m,n,\sigma,\tau,x,y,z(x,y))\right)\right), \tag{5.6.22}$$

with the given initial conditions at $m=0, n=0$ as

$$z(m,0) = d(m), z(0,n) = e(n), z(0,0) = 0, \tag{5.6.23}$$

where $d,e : N_0 \to R, P : E_1 \times R \to R, H : E_2 \times R \to R, F : N_0^2 \times R^3 \to R$ in which $E_1 = \{(m,n,s,t) \in N_0^4 : 0 \le s \le m < \infty, 0 \le t \le n < \infty\}$, $E_2 = \{(m,n,s,t,\sigma,\tau) \in N_0^6 : 0 \le \sigma \le s \le m < \infty, 0 \le \tau \le t \le n < \infty\}$.

The following theorem deals with the uniqueness of solutions of the problem (5.6.22)-(5.6.23).

Theorem 5.6.5. Suppose that the functions F, P, H in (5.6.22) satisfy the conditions

$$|F(m,n,u,v,w) - F(m,n,\bar{u},\bar{v},\bar{w})|$$

$$\le b(m,n)|u-\bar{u}| + |v-\bar{v}| + |w-\bar{w}|, \tag{5.6.24}$$

$$|P(m,n,\sigma,\tau,u) - P(m,n,\sigma,\tau,\bar{u})| \le k(m,n,\sigma,\tau)|u-\bar{u}|, \tag{5.6.25}$$

$$|H(m,n,\sigma,\tau,x,y,u) - H(m,n,\sigma,\tau,x,y,\bar{u})|$$

$$\le h(m,n,\sigma,\tau,x,y)|u-\bar{u}|, \tag{5.6.26}$$

where $b(m,n)$, $k(m,n,\sigma,\tau)$, $h(m,n,\sigma,\tau,x,y)$ are as in Theorem 5.3.4, part (d_1). Then the problem (5.6.22)-(5.6.23) has at most one solution on N_0^2.

Proof. It is easy to observe that the problem (5.6.22)-(5.6.23) is equivalent to the following sum-difference equation

$$z(m,n) = d(m)+e(n)+\sum_{s=0}^{m-1}\sum_{t=0}^{n-1} F\left(s,t,z(s,t),\sum_{\sigma=0}^{s-1}\sum_{\tau=0}^{t-1} P(s,t,\sigma,\tau,z(\sigma,\tau)),\right.$$

$$\left.\sum_{\sigma=0}^{s-1}\sum_{\tau=0}^{t-1}\left(\sum_{x=0}^{\sigma-1}\sum_{y=0}^{\tau-1} H(s,t,\sigma,\tau,x,y,z(x,y))\right)\right). \tag{5.6.27}$$

Let $u(m,n)$ and $v(m,n)$ be two solutions of problem (5.6.22)-(5.6.23) for $(m,n) \in N_0^2$. Using the facts that $u(m,n)$ and $v(m,n)$ are the solutions of (5.6.27) and the conditions (5.6.24)-(5.6.26) we have

$$|u(m,n) - v(m,n)| \le \sum_{s=0}^{m-1}\sum_{t=0}^{n-1} b(s,t)\,|u(s,t) - v(s,t)|$$

$$+\sum_{s=0}^{m-1}\sum_{t=0}^{n-1}\left(\sum_{\sigma=0}^{s-1}\sum_{\tau=0}^{t-1} k(s,t,\sigma,\tau)\,|u(\sigma,\tau) - v(\sigma,\tau)|\right)$$

$$+\sum_{s=0}^{m-1}\sum_{t=0}^{n-1}\left(\sum_{\sigma=0}^{s-1}\sum_{\tau=0}^{t-1}\left(\sum_{x=0}^{\sigma-1}\sum_{y=0}^{\tau-1} h(s,t,\sigma,\tau,x,y)\,|u(x,y) - v(x,y)|\right)\right). \quad (5.6.28)$$

Now a suitable application of Theorem 5.3.4, part (d_1) (when $c = 0$) to (5.6.28) yields $u(m,n) = v(m,n)$, i.e., there is at most one solution to the problem (5.6.22)-(5.6.23) on N_0^2.

The next theorem shows the dependency of solutions of equation (5.6.22) on given initial values.

Theorem 5.6.6. Let $z_1(m,n)$ and $z_2(m,n)$ be the solutions of equation (5.6.22) with rhe given initial conditions at $m = 0, n = 0$ as

$$z_1(m,0) = d_1(m),\, z_1(0,n) = e_1(n),\, z_1(0,0) = 0, \quad (5.6.29)$$

and

$$z_2(m,0) = d_2(m),\, z_2(0,n) = e_2(n),\, z_2(0,0) = 0, \quad (5.6.30)$$

respectively, where $d_1, d_2, e_1, e_2 : N_0 \to R$ and

$$|d_1(m) + e_1(n) - d_2(m) - e_2(n)| \le c, \quad (5.6.31)$$

where $c \ge 0$ is a real constant.Suppose that the functions F, P, H in (5.6.22) satisfy the conditions (5.6.24), (5.6.25), (5.6.26).Then

$$|z_1(m,n) - z_2(m,n)| \le c \prod_{s=0}^{m-1}\left[1 + \sum_{t=0}^{n-1} \bar{Q}(s,t)\right], \quad (5.6.32)$$

for $(m,n) \in N_0^2$,where

$$\bar{Q}(m,n) = b(m,n) + \sum_{\sigma=0}^{m-1}\sum_{\tau=0}^{n-1} k(m,n,\sigma,\tau)$$

$$+\sum_{\sigma=0}^{m-1}\sum_{\tau=0}^{n-1}\left(\sum_{x=0}^{\sigma-1}\sum_{y=0}^{\tau-1} h(m,n,\sigma,\tau,x,y)\right). \quad (5.6.33)$$

Proof. From the hypotheses, it is easy to observe that

$$|z_1(m,n) - z_2(m,n)| \le |d_1(m) + e_1(n) - d_2(m) - e_2(n)|$$

$$+\sum_{s=0}^{m-1}\sum_{t=0}^{n-1}\left|F\left(s,t,z_1(s,t),\sum_{\sigma=0}^{m-1}\sum_{\tau=0}^{n-1}P(s,t,\sigma,\tau,z_1(\sigma,\tau)),\right.\right.$$

$$\left.\sum_{\sigma=0}^{s-1}\sum_{\tau=0}^{t-1}\left(\sum_{x=0}^{\sigma-1}\sum_{y=0}^{\tau-1}H(s,t,\sigma,\tau,x,y,z_1(x,y))\right)\right)$$

$$-F\left(s,t,z_2(s,t),\sum_{\sigma=0}^{m-1}\sum_{\tau=0}^{n-1}P(s,t,\sigma,\tau,z_2(\sigma,\tau)),\right.$$

$$\left.\left.\sum_{\sigma=0}^{s-1}\sum_{\tau=0}^{t-1}\left(\sum_{x=0}^{\sigma-1}\sum_{y=0}^{\tau-1}H(s,t,\sigma,\tau,x,y,z_2(x,y))\right)\right)\right|$$

$$\le c+\sum_{s=0}^{m-1}\sum_{t=0}^{n-1}b(s,t)\,|z_1(s,t)-z_2(s,t)|$$

$$+\sum_{s=0}^{m-1}\sum_{t=0}^{n-1}\left(\sum_{\sigma=0}^{s-1}\sum_{\tau=0}^{t-1}k(s,t,\sigma,\tau)\,|z_1(\sigma,\tau)-z_2(\sigma,\tau)|\right)$$

$$+\sum_{s=0}^{m-1}\sum_{t=0}^{n-1}\left(\sum_{\sigma=0}^{s-1}\sum_{\tau=0}^{t-1}\left(\sum_{x=0}^{\sigma-1}\sum_{y=0}^{\tau-1}h(s,t,\sigma,\tau,x,y)\,|z_1(x,y)-z_2(x,y)|\right)\right). \quad (5.6.34)$$

Now an application of Theorem 5.3.4, part (d_1) to (5.6.34) yields the estimate (5.6.32), which shows the dependency of solutions of (5.6.22) on given initial values.

5.6.4 Sum-difference equations of Volterra-Fredholm type

First we present an application of Theorem 5.5.2 to obtain a bound on the solution of sum-difference equation of the form

$$z(m,n) = f(m,n) + \sum_{s=0}^{m-1}\sum_{t=0}^{n-1}A(m,n,s,t,z(s,t))$$

$$+\sum_{s=0}^{\infty}\sum_{t=0}^{\infty}B(m,n,s,t,z(s,t)), \quad (5.6.35)$$

for $(m,n) \in N_0^2$, where $f : N_0^2 \to R$, $A, B : E \times R \to R$ are the given functions and $E = \{(m,n,s,t) \in N_0^4 : 0 \le s \le m < \infty, 0 \le t \le n < \infty\}$.

Theorem 5.6.7. Suppose that the functions f, A, B in equation (5.6.35) satisfy the conditions

$$|f(m,n)| \le c, \tag{5.6.36}$$

$$|A(m,n,s,t,z)| \le a(m,n,s,t)\,|z|, \tag{5.6.37}$$

$$|B(m,n,s,t,z)| \le b(m,n,s,t)\,|z|, \tag{5.6.38}$$

where $c, a(m,n,s,t), b(m,n,s,t)$ are as in Theorem 5.2.2. Let $r(m,n)$ be as in (5.5.33). If $z(m,n)$ is a solution of (5.6.35) for $(m,n) \in N_0^2$,then

$$|z(m,n)| \le \frac{c}{1-r(m,n)} \prod_{s=0}^{m-1}\left[1+\sum_{t=0}^{n-1} a(m,n,s,t)\right], \tag{5.6.39}$$

for $(m,n) \in N_0^2$.

Proof. Using the fact that $z(m,n)$ is a solution of (5.6.35) and the conditions (5.6.36)-(5.6.38) we have

$$|z(m,n)| \le c + \sum_{s=0}^{m-1}\sum_{t=0}^{n-1} a(m,n,s,t)\,|z(s,t)|$$

$$+\sum_{s=0}^{\infty}\sum_{t=0}^{\infty} b(m,n,s,t)\,|z(s,t)|. \tag{5.6.40}$$

Now an application of Theorem 5.5.2 to (5.6.40) yields the required estimate in (5.6.39).

We next consider the following sum-difference equations

$$\Delta_2\Delta_1 z(m,n) = F\left(m,n,z(m,n),\sum_{\sigma=0}^{\infty}\sum_{\tau=0}^{\infty} r(m,n,\sigma,\tau,z(\sigma,\tau)),\mu\right), \tag{5.6.41}$$

$$\Delta_2\Delta_1 z(m,n) = F(m,n,z(m,n),$$

$$\left.\sum_{\sigma=0}^{\infty}\sum_{\tau=0}^{\infty} r(m,n,\sigma,\tau,z(\sigma,\tau)),\mu_0\right), \tag{5.6.42}$$

with the given initial conditions at $m=0, n=0$ as

$$z(m,0) = \beta_1(m),\, z(0,n) = \beta_2(n),\, \beta_1(0) = \beta_2(0) = 0, \tag{5.6.43}$$

where $\beta_1, \beta_2 : N_0 \to R, r : E \times R \to R, F : N_0^2 \times R^3 \to R$ and μ, μ_0 are real parameters,in which $E = \{(m,n,\sigma,\tau) \in N_0^4 : 0 \le \sigma \le m < \infty, 0 \le \tau \le n < \infty\}$.

The following theorem shows the dependency of solutions of problems (5.6.41)-(5.6.43) and (5.6.42)-(5.6.43) on parameters.

Theorem 5.6.8. Suppose that

$$|r(m,n,\sigma,\tau,z) - r(m,n,\sigma,\tau,\bar{z})| \le e(m,n)\,h(\sigma,\tau)\,|z-\bar{z}|, \tag{5.6.44}$$

$$|F(m,n,z,w,\mu) - F(m,n,\bar{z},\bar{w},\mu)| \le f(m,n)\,(|z-\bar{z}| + |w-\bar{w}|), \tag{5.6.45}$$

$$|F(m,n,z,w,\mu) - F(m,n,z,w,\mu_0)| \le d(m,n)\,|\mu-\mu_0|, \tag{5.6.46}$$

where $f(m,n), h(m,n), e(m,n), d(m,n) \in D\left(N_0^2, R_+\right)$ and $e(m,n) \ge 1$,

$$\sum_{s=0}^{m-1}\sum_{t=0}^{n-1} d(s,t) \le M, \tag{5.6.47}$$

$M \ge 0$ is a real constant. Let

$$r_0 = \sum_{\sigma=0}^{\infty}\sum_{\tau=0}^{\infty} h(\sigma,\tau) \prod_{\xi=0}^{\sigma-1}\left[1 + \sum_{\eta=0}^{\tau-1} f(\xi,\eta)e(\xi,\eta)\right] < 1. \tag{5.6.48}$$

If $z_1(m,n)$ and $z_2(m,n)$ are the solutions of problems (5.6.41)-(5.6.43) and (5.6.42)-(5.6.43) , then

$$|z_1(m,n) - z_2(m,n)| \le \frac{k}{1-r_0} \prod_{s=0}^{m-1}\left[1 + \sum_{t=0}^{n-1} f(s,t)\,e(s,t)\right], \tag{5.6.49}$$

for $(m,n) \in N_0^2$, where $k = |\mu-\mu_0|\,M$.

Proof. Let $z(m,n) = z_1(m,n) - z_2(m,n)$ for $(m,n) \in N_0^2$. As in the proof of Theorem 5.6.5 we observe that

$$z(m,n) = \sum_{s=0}^{m-1}\sum_{t=0}^{n-1}\left\{F\left(s,t,z_1(s,t),\sum_{\sigma=0}^{\infty}\sum_{\tau=0}^{\infty} r(s,t,\sigma,\tau,z_1(\sigma,\tau)),\mu\right)\right.$$

$$-F\left(s,t,z_2(s,t),\sum_{\sigma=0}^{\infty}\sum_{\tau=0}^{\infty} r(s,t,\sigma,\tau,z_2(\sigma,\tau)),\mu\right)$$

$$+F\left(s,t,z_2(s,t),\sum_{\sigma=0}^{\infty}\sum_{\tau=0}^{\infty} r(s,t,\sigma,\tau,z_2(\sigma,\tau)),\mu\right)$$

$$\left.-F\left(s,t,z_2(s,t),\sum_{\sigma=0}^{\infty}\sum_{\tau=0}^{\infty} r(s,t,\sigma,\tau,z_2(\sigma,\tau)),\mu_0\right)\right\}. \tag{5.6.50}$$

Using (5.6.44)-(5.6.47) in (5.6.50) we observe that

$$|z(m,n)| \le \sum_{s=0}^{m-1}\sum_{t=0}^{n-1} f(s,t)\left(|z(s,t)| + \sum_{\sigma=0}^{\infty}\sum_{\tau=0}^{\infty} e(s,t)\,h(\sigma,\tau)\,|z(\sigma,\tau)|\right)$$

$$+\sum_{s=0}^{m-1}\sum_{t=0}^{n-1} d\left(s,t\right)\left|\mu-\mu_0\right|$$

$$\leq k+\sum_{s=0}^{m-1}\sum_{t=0}^{n-1} f\left(s,t\right)e\left(s,t\right)\left(\left|z\left(s,t\right)\right|+\sum_{\sigma=0}^{\infty}\sum_{\tau=0}^{\infty} h\left(\sigma,\tau\right)\left|z\left(\sigma,\tau\right)\right|\right). \quad (5.6.51)$$

Now a suitable application of Theorem 5.2.3 to (5.6.51) yields (5.6.49), which shows the dependency of solutions of problems (5.6.41)-(5.6.43) and (5.6.42)-(5.6.43) on parameters μ and μ_0

5.7 Notes

The study of partial finite difference equations has gained noticable importance during the past few years. Such equations arise frequently in combinatorics and in the approximation of solutions of partial differential equations by finite difference methods. In fact, we need new theory and methods for the study of various types of partial finite difference equations. The material in sections 5.2-5.5 contains a number of new finite difference inequalities involving functions of two independent variables recently developed by Pachpatte [36,40,41,45,48,49,53,55,56,62,66,68,71,76]. These inequalities can be used in the theory of partial finite difference equations in essentially the same capacity as the finite difference inequaliies with explicit estimates are used in the theory of ordinary finite difference equations. Section 5.6 is devoted to applications of some of the inequalities given in earlier sections.

References

1. Agarwal, R.P., Difference Equations and Inequalities, Marcel Dekker, Inc., New York, 1992.
2. Akhmedov, K.T., A.M.Yakubov and I.A.Veisov, Some integral inequalities, Izv.Akad.Nauk UzSSR 1(1972), 18-24 (in Russian).
3. Bainov, D. and P.Simeonov, Integral Inequalities and Applications, Kluwer Academic Publishers, Dordrecht, 1992.
4. Beckenbach, E.F. and R.Bellman, Inequalities, Springer-Verlag, Berlin, New York, 1965.
5. Beesack, P.R., Gronwall Inequalities, Carleton Mathematical Lecture Notes, No.11, 1975.
6. Bellman, R., Stability Theory of Differential Equations, McGraw-Hill, New York, 1953.
7. Bellman, R. and K.L.Cooke, Differential-Difference Equations, Academic Press, New York, 1963.
8. Bihari, I., A generalization of a lemma of Bellman and its application to uniqueness of differential equations, Acta Math.Acad.Sci.Hunger.7(1956),71-94.
9. Bykov, Ya.V. and Kh.M.Salpagarov, On the theory of integro-differential equations, in, Investigations in Integro-Differential Equations, Khirghizia 2, Izd.Akad.Nauk.Khirghizia SSR, 1962 (in Russian).
10. Dafermos, C.M., The second law of Thermodynamics and stability, Arch.Rat.Mech.Anal.70(1979), 167-179.
11. Dragomir, S.S., On discrete generalization of Pachpatte's inequality and applications, Bull.Math.Soc.Math.Roumanie 36(1992), 45-58.
12. Dragomir, S.S., Some Gronwall Type Inequalities and Applications, Nova Science Publishers, Inc., New York, 2003.
13. El'sgol'ts, L.F. and S.B.Norkin, Introduction to the theory and applications of differential equations with deviating arguments, Academic Press, New York, 1973.
14. Filatov, A.N. and L.V.Sharov, Integral inequalities and the theory of nonlinear oscillations, Moscow, Nauka, 1976 (in Russian).
15. Gamidov, Sh.G., Integral inequalities for boundary value problems for differential equations, Diff.Eqs. 5(1969), 463-475
16. Gronwall, T.H., Note on the derivatives with respect to a parameter of the solutions of a system of differential equations, Ann. of Math. 20(1919), 292-296.
17. Henry, D., Geometric Theory of Semilinear Parabolic Equations, Springer-Verlag, Berlin, New York, 1981.
18. Kung, Y., Delay differential equations with Applications in Population Dynamics, Academic Press, New York, 1993.
19. Lakshmikantham, V. and S.Leela, Differential and Integral Inequalities, Vols. I,II, Academic Press, New York, 1969.
20. LaSalle, J.P., Uniqueness theorems and successive approximations, Ann. of Math. 50(1949), 722-730.

21. Lipovan, O., A retarded Gronwall-like inequality and applications, J.Math.Anal.Appl. 252(2000), 389-401.
22. Lipovan, O., A retarded integral inequality and its applications, J.Math.Anal.Appl. 285(3003), 436-443.
23. Martynyuk, A.A. and R.Gutowski, Integral Inequalities and stability of motion, Naukov Dumka, Kiev, 1979 (in Russian).
24. Medveď, M., A new approach to an analysis of Henry type integral inequalities and their Bihari type versions, J. Math.Anal.Appl. 214(1997), 349-366.
25. Medveď, M., Singular integral inequalities and stability of semilinear parabolic equations, Archivum Mathematicum (Brno) 34(1998), 183-190.
26. Medveď, M., Nonlinear singular integral inequalities for functions of two and n independent variables, J.Inequal.Appl. 5(2000), 287-308.
27. Medveď, Nonlinear singular difference inequalities suitable for discretizations of Parabolic equations, Demonstratio Mathematica 33(2000), 517-525.
28. Medveď, M., Integral inequalities and global solutions of semilinear evolution equations, J.Math.Anal.Appl. 267(2002), 643-650.
29. Mikeladze, Sh.E., De la résolution numéique des équations intégrales, Bull.Acad.Sci.URSS VII (1935), 255-257 (in Russian).
30. Mitrinović, D.S., Analytic Inequalities, Springer-Verlag, Berlin, New York, 1970.
31. Mitrinović, D.S., J.E.Pečarić and A.M.Fink, Classical and new inequalities in analysis, Kluwer Academic Publishers,Dordrecht, 1993.
32. Mitrinović, D.S., J.E.Pečarić and A.M.Fink, Inequalities for functions and their integrals and derivatives, Kluwer Academic Publishers, Dordrecht, 1994.
33. Ou-Iang, L., The boundedness of solutions of linear differential equations $y'' + A(t)\, y = 0$, Shuxue Jinzhan 3(1957), 409-415.
34. Pachpatte, B.G., Inequalities for Differential and Integral Equations, Academic Press, New York, 1998.
35. Pachpatte, B.G., On some new inequalities related to a certain inequality arising in the theory of differential equations, J.Math.Anal.Appl. 251(2000), 736-751.
36. Pachpatte, B.G., On some finite difference inequalities in two independent variables, J.Math.Anal.Appl. 254(2001), 587-598.
37. Pachpatte, B.G., Inequalities for certain finite difference and sum-difference equations, Math.Ineq.Appl. 4(2001), 499-506.
38. Pachpatte, B.G., Inequalities applicable in the theory of partial finite difference equations, Tamkang J.Math. 32(2001), 155-166.
39. Pachpatte, B.G., On some fundamental finite difference inequalities, Tamkang J.Math. 32(2001), 217-223.
40. Pachpatte, B.G., Inequalities useful for certain partial differential and difference equations, Radovi Math. 10(2001), 5-15.
41. Pachpatte, B.G., On some fundamental integral inequalities and their discrete analogues, J.Inequal.Pure and Appl.Math. 2(2)Art.15, 2001.
42. Pachpatte, B.G., Inequalities for Finite Difference Equations, Marcel

Dekker, Inc., New York, 2002.
43. Pachpatte, B.G., Explicit bounds on certain integral inequalities, J.Math. Anal.Appl. 267(2002), 48-61.
44. Pachpatte, B.G., A note on certain integral and discrete inequalities, An.Sti.Univ.Al.I.Cuza Iasi XLVIII(2002), 221-228.
45. Pachpatte, B.G., Integral inequalities of Bihari type, Math.Inequal.Appl. 5(2002), 649-657.
46. Pachpatte, B.G., Inequalities useful in the theory of certain partial differential equations, Fasc.Math. Nr.33(2002), 49-56.
47. Pachpatte, B.G., On some retarded integral inequalities and applications, J.Inequal.Pure and Appl.Math. 3(2)Art.18, 2002.
48. Pachpatte, B.G., On certain integral inequalities and their discrete analogues, Demonstratio Mathematica 35(2002), 243-250.
49. Pachpatte, B.G., On nonlinear finite difference inequalities in two independent variables, Tamkang J.Math. 33(2002), 67-75.
50. Pachpatte, S.B. and B.G.Pachpatte, On some useful integral inequalities and their discrete analogues, Tamkang J.Math. 33(2002), 139-148.
51. Pachpatte, S.B. and B.G.Pachpatte, Inequalities for terminal value problems for differential equations, Tamkang J.Math.33(2002), 199-208.
52. Pachpatte, B.G., A note on certain integral inequality, Tamkang J.Math. 33(2002), 353-358.
53. Pachpatte, B.G., Bounds on certain integral inequalities, J.Inequal.Pure and Appl.Math. 3(3)Art.47, 2002.
54. Pachpatte, B.G., On an inequality to general Volterra-Fredholm integral equation, An.Sti.Univ.Al.I.Cuza Iasi XLIX(2003), 221-228.
55. Pachpatte, B.G., Explicit bounds on Volterra type integral inequalities, Tamsui Oxford J.Math.Sci. 19(2003), 13-25.
56. Pachpatte, B.G., Comparison inequalities for partial finite difference equations, Math.Inequal.Appl. 6(2003), 393-396.
57. Pachpatte, B.G., On some basic finite difference inequalities, Fasc. Math., Nr.34 (2004), 65-71.
58. Pachpatte, B.G., On a certain retarded integral inequality and its applications, J.Inequal.Pure and Appl.Math. 5(1)Art.19, 2004.
59. Pachpatte, B.G., Inequalities applicable to certain partial differential equations, J.Inequal.Pure and Appl.Math.5(2)Art.27, 2004.
60. Pachpatte, B.G., On some new nonlinear retarded integral inequalities, J.Inequal.Pure and Appl.Math. 5(3)Art.80, 2004.
61. Pachpatte, B.G., Inequalities applicable to retarded Volterra integral equations, Tamkang J.Math. 35(2004), 285-291.
62. Pachpatte, B.G., On a new integral inequality applicable to certain partial integrodifferential equations, in, Inequality theory and Applications, Nova Science Publishers, New York, Vol.4, 2004.
63. Pachpatte, B.G., On higher order differential equation with retarded argument, in, Differential Equations and Applications, Nova Science Publishers, New York, Vol.4, 2004.
64. Pachpatte, B.G., Explicit bound on a retarded integral inequality, Math

Inequal.Appl. 7(2004), 7-11.
65. Pachpatte, B.G., Mathematical Inequalities, (North-Holland Mathematical Library) Elsevier Science B.V. 2005.
66. Pachpatte, B.G., Inequalities applicable to certain partial finite sum-difference equations, Demonstratio Mathematica 38(2005), 291-302.
67. Pachpatte, B.G., Growth estimates on certain discrete inequalities involving iterated sums, An.Sti.Univ.Al.I.Cuza, Iasi , LI, s.I, Matematică (2005), 65-73.
68. Pachpatte, B.G., On generalizations of Bihari's inequality, Soochow J.Math. 31(2005), 261-271.
69. Pachpatte, B.G., A note on certain retarded inequality, Tamkang J.Math. 36(4)(2005), to appear.
70. Pachpatte, B.G., Explicit bounds on Gamidov type integral inequalities, Tamkang J.Math.37(1)(2006), to appear.
71. Pachpatte, B.G., Explicit bound on a certain finite difference inequality, Bul.Inst.Polit.Iasi L(LIV)(1-2)(2004), 61-66.
72. Pachpatte, B.G., Growth estimates on certain integral inequalities in two variables involving iterated integrals, Tamkang J.Math. , 36(3)(2005), 179-191.
73. Pachpatte, B.G., A note on some discrete inequalities involving iterated sums, Tamusi Oxford J.Math.Sci., to appear.
74. Pachpatte, B.G., Explicit bounds on some retarded integral inequalities, Fasc.Math. 35(2005), 85-98.
75. Pachpatte, B.G., Inequalities applicable to Volterra-Fredholm integrodifferential equations, submitted
76. Pachpatte, B.G., On Gamidov type inequalities in two variables, submitted.
77. Pachpatte, B.G., On retarded integral inequalities for kernels of (L)-type, submitted.
78. Pachpatte, B.G., On iterated integral inequalities involving kernels of (L)-type, submitted.
79. Pachpatte, B.G., On certain nonlinear integral inequalities involving iterated integrals, Tamkang J.Math. , to appear.
80. Peano, G.,Sull'integrabilitǎ delle equazioni differenziali del primo odine, Atti.R.Acad.Sc.Toriao,Vol.21(1885-86), 677-685.
81. Schröder, J., Operator Inequalities, Academic Press, New York, 1980.
82. Sugiyama, S., On some problems of functional-differential equations with advanced argument, Proc.U.S.Japan Sem.on Diff.Funct.Eqs., Benjamin, 1967, pp. 367-382.
83.Szarski, J., Differential Inequalities, PWN, Warsaw, 1965.
84. Walter, W., Differential and Integral Inequalities, Springer-Verlag, Berlin, New York, 1970.
85. Yang, E.H., Generalizations of Pachpatte's integral and discrete inequalities, Ann.Diff.Eqs. 13 *1997), 180-188.

Index

www.ingramcontent.com/pod-product-compliance
Ingram Content Group UK Ltd.
Pitfield, Milton Keynes, MK11 3LW, UK
UKHW020241290926
15470UKWH00002B/661

9 780444 527622